NEUROMETHODS

Series Editor
Wolfgang Walz
University of Saskatchewan
Saskatoon, Canada

For further volumes:
http://www.springer.com/series/7657

Neuromethods publishes cutting-edge methods and protocols in all areas of neuroscience as well as translational neurological and mental research. Each volume in the series offers tested laboratory protocols, step-by-step methods for reproducible lab experiments and addresses methodological controversies and pitfalls in order to aid neuroscientists in experimentation. *Neuromethods* focuses on traditional and emerging topics with wide-ranging implications to brain function, such as electrophysiology, neuroimaging, behavioral analysis, genomics, neurodegeneration, translational research and clinical trials. *Neuromethods* provides investigators and trainees with highly useful compendiums of key strategies and approaches for successful research in animal and human brain function including translational "bench to bedside" approaches to mental and neurological diseases.

Cognitive Behavioral Therapy with Youth

Tradition and Innovation

Second Edition

Edited by

Robert D. Friedberg

Independent Training Consultant in CBT, San Jose, CA, USA

Micaela A. Thordarson

Outpatient Acute Mental Health Services, Children's Hospital of Orange County, Orange, CA, USA

Humana Press

Editors
Robert D. Friedberg
Independent Training
Consultant in CBT
San Jose, CA, USA

Micaela A. Thordarson
Outpatient Acute Mental Health Services
Children's Hospital of Orange County
Orange, CA, USA

ISSN 0893-2336 ISSN 1940-6045 (electronic)
Neuromethods
ISBN 978-1-0716-5133-9 ISBN 978-1-0716-5134-6 (eBook)
https://doi.org/10.1007/978-1-0716-5134-6

This Humana imprint is published by the registered company Springer Science+Business Media, LLC, part of Springer Nature.
The registered company address is: 1 New York Plaza, New York, NY 10004, U.S.A.

Preface to the Series

Experimental life sciences have two basic foundations: concepts and tools. The Neuromethods series focuses on the tools and techniques unique to the investigation of the nervous system and excitable cells. It will not, however, shortchange the concept side of things as care has been taken to integrate these tools within the context of the concepts and questions under investigation. In this way, the series is unique in that it not only collects protocols but also includes theoretical background information and critiques which led to the methods and their development. Thus it gives the reader a better understanding of the origin of the techniques and their potential future development. The Neuromethods publishing program strikes a balance between recent and exciting developments like those concerning new animal models of disease, imaging, in vivo methods, and more established techniques, including, for example, immunocytochemistry and electrophysiological technologies. New trainees in neurosciences still need a sound footing in these older methods in order to apply a critical approach to their results.

Under the guidance of its founders, Alan Boulton and Glen Baker, the Neuromethods series has been a success since its first volume published through Humana Press in 1985. The series continues to flourish through many changes over the years. It is now published under the umbrella of Springer Protocols. While methods involving brain research have changed a lot since the series started, the publishing environment and technology have changed even more radically. Neuromethods has the distinct layout and style of the Springer Protocols program, designed specifically for readability and ease of reference in a laboratory setting.

The careful application of methods is potentially the most important step in the process of scientific inquiry. In the past, new methodologies led the way in developing new disciplines in the biological and medical sciences. For example, Physiology emerged out of Anatomy in the nineteenth century by harnessing new methods based on the newly discovered phenomenon of electricity. Nowadays, the relationships between disciplines and methods are more complex. Methods are now widely shared between disciplines and research areas. New developments in electronic publishing make it possible for scientists that encounter new methods to quickly find sources of information electronically. The design of individual volumes and chapters in this series takes this new access technology into account. Springer Protocols makes it possible to download single protocols separately. In addition, Springer makes its print-on-demand technology available globally. A print copy can therefore be acquired quickly and for a competitive price anywhere in the world.

Saskatoon, Canada *Wolfgang Walz*

Preface: First, Be a Painter!

> You can't learn techniques, and then try to become a painter. Techniques are a result.
>
> –Jackson Pollock

Like the first edition of *Cognitive Behavioral Therapy with Youth: Tradition and Innovation*, this second edition begins with a quote from a legendary artist. Indeed, this is apropos because all the chapter authors are artists in their own right. In each chapter, the individual authors root their innovative approaches in well-founded theory and solid science. The clever modifications of traditional approaches are governed by a thorough understanding of cognitive behavioral therapy.

Relying on science and sound conceptual design seems to be a lost practice and art. In this fast-paced age with dizzying technological advances, it is quite tempting to search for quick fixes and bags of tricks. However, like the first edition, this new volume eschews the stop-gap procedures. Rather, the chapters provide time-efficient and effective approaches that are fully embedded within robust theoretical backgrounds and based on data. Accordingly, they reflect authentic cognitive behavioral therapy and paint full pictures of the approach.

It is our sincere hope that readers of this second edition adopt the stance presented by the chapter authors and embrace the notion of becoming cognitive behavioral artists.

San Jose, CA USA — *Robert D. Friedberg*
Orange County, CA USA — *Micaela A. Thordarson*

Acknowledgments

First and always, I am indebted to the brilliant Barbara A Friedberg for her, loving care, guidance, and endless patience. I would never have accomplished anything without her. She is both my north and shining star. Second, special thanks to my gifted co-editor and treasured friend, Dr. Micaela Thordarson, whose keen intelligence, boundless energy, thoughtfulness, and good humor made this project terrifically enjoyable. I am hugely grateful to the many colleagues and friends who contributed to this text. Their expertise and wisdom are genuinely humbling. Finally, thanks to Wolfgang Walz, Patrick Marton, Anna Rakovsky, and the entire Springer Protocols team for inviting this second edition and shepherding the work through the process. RDF

It would frankly be criminal not to begin with the MOST enthusiastic and heartfelt acknowledgment of and boundless gratitude for my co-editor, mentor, and dear friend Bob Friedberg. It is only through Bob's guidance, collaboration, instruction, and low-key nagging that I (a) managed to wrap up this project and (b) find myself in such a position in my career to have the opportunity to work on incredible projects such as this!! Speaking of…it is profoundly humbling to receive submissions from people I have been citing for years and get to see their continued brilliance and clinical leadership firsthand. Thank you to each of you for producing such wonderful contributions to this text and sharing your wisdom with readers. Finally, a huge thank you to my family for their forever support of me, especially my mom who is always mad at me for not sharing my accomplishments. I hope you don't get to read this, mama (or maybe this time I will share). MAT

Contents

Contributors

JONATHAN S. ABRAMOWITZ • *Department of Psychology and Neuroscience, University of North Carolina at Chapel Hill, Chapel Hill, NC, USA*

EMILY BADIN • *Department of Child and Adolescent Psychiatry, NYU Langone Health, New York, NY, USA*

DOROTHY H. BALSER • *Department of Psychology and Center for Children and Families, Florida International University, Miami, FL, USA*

KELLY N. BANNEYER • *Department of Pediatrics/Psychology Section, Baylor College of Medicine, Houston, TX, USA*

DEEPIKA BOSE • *Department of Psychiatry and Behavioral Sciences, University of California, San Francisco, San Francisco, CA, USA*

MELISSA A. BROTMAN • *Neuroscience and Novel Therapeutics Unit, Emotion and Development Branch, National Institutes of Mental Health, National Institutes of Health, Bethesda, MD, USA*

BRIAN C. CHU • *Department of Clinical Psychology, Graduate School of Applied and Professional Psychology, Rutgers University, Piscataway, NJ, USA*

STEPHANIE CRUTHIRDS • *CARES Institute, Rowan University School of Osteopathic Medicine, Stratford, NJ, USA*

LISA CULLINS • *Neuroscience and Novel Therapeutics Unit, Emotion and Development Branch, National Institutes of Mental Health, National Institutes of Health, Bethesda, MD, USA*

ESTHER DEBLINGER • *CARES Institute, Rowan University School of Osteopathic Medicine, Stratford, NJ, USA*

JILL EHRENREICH-MAY • *Department of Psychology, University of Miami, Coral Gables, FL, USA*

JAMAL H. ESSAYLI • *Department of Pediatrics and Psychiatry and Behavioral Health, Penn State College of Medicine, Hershey, PA, USA*

RACHEL FEIN • *Legacy Community Health, Houston, TX, USA*

LISA FELBER • *Neuroscience and Novel Therapeutics Unit, Emotion and Development Branch, National Institutes of Mental Health, National Institutes of Health, Bethesda, MD, USA*

KRISTER W. FJERMESTAD • *Department of Psychology, University of Oslo, Oslo, Norway*

ROBERT D. FRIEDBERG • *Independent Training Consultant in CBT, San Jose, CA, USA*

RAMARIS E. GERMAN • *Neuroscience and Novel Therapeutics Unit, Emotion and Development Branch, National Institutes of Mental Health, National Institutes of Health, Bethesda, MD, USA*

ELIZABETH R. HALLIDAY • *Department of Psychology, University of Miami, Coral Gables, FL, USA*

BENTE STORM MOWATT HAUGLAND • *Department of Clinical Psychology, University of Bergen, Bergen, Norway; Regional Center for Child and Youth Mental Health and Child Welfare, NORCE, Bergen, Norway*

EMILY K. JUEL • *Department of Psychology and Neuroscience, University of North Carolina at Chapel Hill, Chapel Hill, NC, USA*

FRANCESCA KASSING • *California School of Professional Psychology, Alliant International University, Fresno, CA, USA*

RACHEL E. KIM • *The Baker Center for Children and Families and Harvard Medical School, Boston, MA, USA*

CIERA KORTE • *Oregon Health & Sciences University, Portland, OR, USA*

JULIETTE M. LIBER • *Department of Developmental Psychology, University of Amsterdam, Amsterdam, The Netherlands*

JOHN E. LOCHMAN • *Department of Psychology, Center for Youth Development, The University of Alabama, Tuscaloosa, AL, USA*

LAILA A. MADNI • *Department of Psychology, Dana and David Dornsife College of Letters, Arts and Sciences, University of Southern California, Los Angeles, CA, USA*

YAHAIRA I. MÁRQUEZ • *CARES Institute, Rowan University School of Osteopathic Medicine, Stratford, NJ, USA*

BRYCE D. MCLEOD • *Department of Psychology, Virginia Commonwealth University, Richmond, VA, USA*

MELISSA PADRON • *Department of Psychology and Center for Children and Families, Florida International University, Miami, FL, USA*

ALAYNA L. PARK • *Department of Psychology, University of Oregon, Eugene, OR, USA*

JEREMY W. PETTIT • *Department of Psychology and Center for Children and Families, Florida International University, Miami, FL, USA*

SANDRA S. PIMENTEL • *Department of Psychiatry and Behavioral Sciences, Albert Einstein College of Medicine, Bronx, NY, USA*

EILI N. RIISE • *Department of Child and Adolescent Psychiatry, District General Hospital of Førde, Førde, Norway*

DEVON E. ROMERO • *Department of Counseling, The University of Texas at San Antonio, San Antonio, TX, USA*

RANDYE J. SEMPLE • *Department of Psychiatry and the Behavioral Sciences, Keck School of Medicine, University of Southern California, Los Angeles, CA, USA*

AIMEE SIROIS-GEDDIE • *CARES Institute, Rowan University School of Osteopathic Medicine, Stratford, NJ, USA*

SARA L. STROMEYER • *The Birmingham Psychology Group, Birmingham, AL, USA*

MICAELA A. THORDARSON • *Outpatient Acute Mental Health Services, Children's Hospital of Orange County, Orange, CA, USA*

CLARISSA V. VELEZ • *Department of Psychology, University of Oregon, Eugene, OR, USA*

STEPHANIE VIOLANTE • *Department of Child and Adolescent Psychiatry, University of California, Los Angeles, Los Angeles, CA, USA*

GRO JANNE H. WERGELAND • *Department of Child and Adolescent Psychiatry, Haukeland University Hospital, Bergen, Norway; Department of Clinical Medicine, University of Bergen, Bergen, Norway*

JAMELL D. WHITE • *Neuroscience and Novel Therapeutics Unit, Emotion and Development Branch, National Institutes of Mental Health, National Institutes of Health, Bethesda, MD, USA*

Chapter 1

Overview of CBT Spectrum Approaches

Stephanie Violante, Bryce D. McLeod, Krister W. Fjermestad, and Juliette M. Liber

Abstract

Cognitive-behavioral therapy (CBT) is a psychosocial treatment with strong scientific evidence supporting its use with youth for a variety of emotional and behavioral problems. This chapter provides a broad overview of CBT with youth and is divided into three sections. The first section describes the behavioral and cognitive theories that underlie the CBT approach. We also discuss how the two theories have been integrated into theories used to guide CBT. The second section describes specific techniques commonly found in CBT and reviews the typical modalities employed to deliver CBT. Finally, the third section focuses on the delivery of CBT.

Key words CBT, Youth, Behavior, Cognitive, Evidence-based treatment

1 Introduction

Cognitive-behavioral therapy (CBT) is a specific psychosocial treatment that has garnered strong scientific evidence supporting its use with children and adolescents (hereafter called youth) for a variety of emotional and behavioral problems. The CBT model is based on the premise that cognitions, emotions, and behaviors interact to produce and maintain emotional and behavioral problems in youth. CBT has received empirical support for a variety of problems experienced by youth, including depression [1, 2], anxiety [3], post-traumatic stress disorder [4], eating disorders [5], externalizing disorders including attention deficit hyperactivity disorder [6], functional and chronic pain [7, 8], and substance abuse [9]. Notably, the evidence also includes effectiveness research, i.e., when CBT has been delivered as part of regular clinical healthcare systems [5, 10]. Furthermore, there is emerging evidence for CBT's effectiveness with various vulnerable groups, such as autistic children

Robert D. Friedberg and Micaela A. Thordarson (eds.), *Cognitive Behavioral Therapy with Youth: Tradition and Innovation*, Neuromethods, vol. 231, https://doi.org/10.1007/978-1-0716-5134-6_1,

[11], child refugees [12], LGBTQ+ youth [13], and youth with mild intellectual disabilities or borderline intellectual functioning [14]. CBT can thus be used to treat a wide variety of problems.

This chapter aims to provide a broad overview of CBT, divided into three sections. The first section focuses on the theories underlying the CBT approach. CBT represents the integration of behavioral and cognitive theories. Since the important behavioral (operant, classical conditioning, social learning) and cognitive (Beck's cognitive theory) theories emerged separately before the CBT model's development, we describe these theories and how they are used in CBT. We then conclude the section with a brief description of how these theories were eventually integrated into a theory designed to guide CBT.

The second section presents information about how CBT works to promote symptom reduction in youth. It describes the behavioral and cognitive techniques commonly found in CBT programs and reviews the common modalities (e.g., individual, family-focused) used to deliver CBT to youth and their families.

The final section provides specific information about factors to consider when delivering CBT to youth with emotional and behavioral problems. The section is intended to describe the factors that promote the effectiveness of CBT by helping understand what therapist behaviors may optimize youth clinical outcomes.

2 Theoretical Background

In this section, we detail the theoretical underpinnings of CBT, discuss the main behavioral and cognitive theories used to develop it, and describe how the cognitive and behavioral theories were integrated to inform the CBT model.

2.1 Behavioral Theory

Behavioral theory is characterized by three types of learning processes: classical conditioning, operant conditioning, and social learning. Each of these theories has been used to develop the therapeutic techniques detailed in the next section.

2.1.1 Classical Conditioning

The theory of classical conditioning, based on the work of Pavlov [15], posits that learning occurs through repeated pairings between a neutral stimulus and a stimulus that naturally produces a behavior. An unconditioned stimulus (UCS; something that triggers a naturally occurring response) leads to an unconditioned response (UCR; naturally occurring response following the UCS). When the UCS is closely preceded by a conditioned stimulus (CS; an initially neutral stimulus), the CS will eventually produce the same response, termed the conditioned response (CR; the acquired response to the formerly neutral stimulus) [16].

Classical conditioning is used to explain the development and maintenance of certain psychological problems, such as anxiety disorders. If a youth is bitten by a dog (UCS), they will likely experience a fear response (UCR). The subsequent experience of seeing a dog (a previously neutral or positive stimulus) after being bitten might lead to an association between seeing a dog (CS) and a feeling of fear (CR), potentially leading to the development of a specific phobia of dogs. CBT techniques that target these emotions are based on the principles of classical conditioning. For example, exposure, a therapeutic technique, is based on the principles of classical conditioning theory. To reduce a youth's anxiety about dogs, treatment would involve repeated exposure to the CS (being around dogs) in the absence of the UCS (being bitten by a dog). After extended periods of exposure to the CS in the absence of the UCS, then the CS will no longer produce the CR (fear; i.e., extinction).

2.1.2 Operant Conditioning

The theory of operant conditioning, generally credited to Skinner [17], emphasizes the role of events in the environment on behaviors. Specifically, it suggests that the likelihood that a behavior will be repeated is a function of events that immediately precede or follow the behavior. In operant theory, behaviors are influenced through positive reinforcement, defined as the addition of a stimulus that leads to an increase in behavior (e.g., giving a youth a toy to reward desired behavior), negative reinforcement defined as the removal of a stimulus that leads to an increase in behavior (e.g., removing a chore to reward desired behavior), positive punishment defined as the addition of a stimulus that leads to a decrease in behavior (e.g., adding a chore to discipline undesired behavior), and negative punishment defined as the removal of a stimulus that leads to a decrease in behavior (e.g., removing a toy to discipline undesired behavior) [18]. The contingencies of reinforcement are a main focus of operant conditioning: antecedents defined as a cue for youth to engage in a behavior (e.g., caregiver instructions to do homework), behaviors defined as the behavior that is to be altered (e.g., doing homework), and consequences defined as what follows the behavior that makes it more or less likely to happen again (e.g., caregiver verbal praise). Understanding how these three components operate in unison to explain behavior is an important step in understanding the factors that serve to cause (antecedent) and maintain (consequence) a target behavior. This understanding, in turn, is used in CBT to inform the development of techniques that target antecedents and/or consequences intended to alter behavior. Operant conditioning thus provides the foundation for many assessment and technique procedures used in CBT (e.g., functional analysis) designed to increase or decrease the frequency of behaviors.

2.1.3 Social Learning

Social learning theory, historically developed through the work of Bandura [19], focuses on the influence of social contexts of behavior on learning. Its central premise is that behaviors are learned from the environment through observational learning. Specifically, new behaviors, skills, or information are acquired, and existing behaviors are altered through observing others. While reinforcement and punishment are not necessary aspects of social learning, social learning theory and operant theory can interact to explain behavior change. Rotter [20] proposed that an individual's behavior is determined not only by observing another but by their expectation of receiving a reward or punishment. For example, a youth may be more likely to repeat an observed behavior if that behavior was rewarded and less likely to repeat the behavior if it was punished because they will expect the associated reward or punishment in the future. Social learning theory has been applied to youth-focused CBT through therapist modeling techniques to teach new skills, educating caregivers about the role modeling can play in influencing youth behaviors, and teaching caregivers to model desired behaviors for their youth [21]. Such techniques will be described in more detail in the next section.

2.2 Cognitive Theory

Traditional behavior theory only focuses on observable behavior and eschews a focus on internal processes. However, some scientific practitioners were not convinced that behavior could be explained solely by observable stimuli and turned to the role of internal cognitions on behavior and emotions. The resulting "cognitive revolution" was a paradigm shift in psychology away from a sole reliance on observable behavior to an incorporation of unobservable cognitive processes [22, 23]. The new emphasis on the role of cognitions in psychology prompted the development of Beck's cognitive theory of depression [24], later extended to other psychological disorders. His theory assumes that cognitive processes become habitual and automatic over time, producing cognitive schemas (i.e., organizational frameworks based on patterns of internal experience), which shape our interpretations of events. These schemas may lead to inaccurate interpretations of innocuous experiences [21]. For example, if a youth holds a schema that they are "weird" or "awkward," they might interpret a group of youth laughing as being directed toward them, ignoring alternative explanations (e.g., somebody told a joke). Beck described these inaccurate interpretations as "illogical thinking processes" [25], and maintained that they underlie emotional disorders.

Beck's cognitive theory suggests that different psychological disorders can be distinguished by their cognitive content. For example, the theory assumes that depression develops and is maintained through negative core beliefs about the self, the world, and the future (i.e., the cognitive triad), which interact with the environment to generate situation-specific negative cognitions

[25]. Distinctly, anxiety is more likely to be characterized by cognitions about physical and psychological threats or dangers [21]. Although initially conceptualized as relevant to anxiety and depression, Beck's cognitive theory was later applied to a wide range of disorders, such as eating disorders [26] and substance use disorders [27]. In the 1970s, Beck applied his cognitive theory to the development of cognitive therapy, eventually resulting in CBT's emergence.

3 Integration: Cognitive-Behavioral Theory

CBT integrates theories that focus on internal (cognitive) and observable (behavioral) processes to explain the development and maintenance of emotional and behavioral problems [28]. Specifically, CBT assumes that our cognitions, behaviors, and emotions are reciprocally linked and that altering one will result in changes in the others [29]. These reciprocal relationships serve as the central foundation for CBT. CBT integrates strategies that emphasize the influence of cognitive factors and behavioral contingencies, to alter an individual's cognitions and behaviors to reduce negative emotions and encourage positive behaviors. In the last decade, so-called "third wave" CBTs have been developed [30]. These CBT variants are typically characterized by a cross-diagnostic approach and the inclusion of elements like metacognition, mindfulness, acceptance, decentering, self-compassion, values-focused behavior, and perspective taking [31, 32].

4 CBT Techniques

Practicing CBT involves a mix of specific behavioral and cognitive techniques delivered within a collaborative relationship with the youth and their family. This section reviews specific behavioral and cognitive techniques typically found in CBT programs for youth emotional and behavioral problems. Our coverage is designed to help highlight some of the most common CBT techniques and does not represent a complete list. The section also provides a brief description of the different modalities that have been used to deliver CBT, such as individual and group.

4.1 Behavioral Techniques

Behavioral techniques are primarily designed to increase the frequency of desired behaviors and decrease the frequency of undesired behaviors. The behavioral techniques are guided by operant, classical conditioning, and social learning theories (see previous section). Some of the most prominent behavioral techniques are described below.

4.1.1 Relaxation Training

This involves teaching a youth a method for self-calming. Relaxation training is found in many CBT programs and can range in type and focus. The simplest form of relaxation is diaphragmatic breathing, which involves training youths to breathe more deeply from the diaphragm. This form of relaxation training is brief and easy to use, making it a practical tool for youth in many situations. A somewhat more complicated relaxation strategy is progressive muscle relaxation, which involves teaching youth to tense and relax different muscle groups. This can be paired with a personalized cue word (e.g., "calm") in cue-controlled relaxation. These more involved strategies help youths recognize tension in their body, so they can identify tension and engage in relaxation strategies.

4.1.2 Activity Planning

Activity planning is a common technique used to reduce depression; however, it is also used with youth who may experience chaotic life situations or perfectionism issues. Activity planning can be done by setting up a regular schedule with a youth. The schedule can include specific activities planned for each day and a discussion about ways to increase the likelihood that the youth will be able to complete the activity. More specific planning increases the likelihood that an activity will be done. For example, instead of planning "hanging with friends," specifying which friend(s), what to do, where, and how/when to contact the friend increases the probability that the activity will be done.

4.1.3 Exposure

Exposure is a behavioral technique mainly used with youth who experience problems with anxiety. Exposure aims to teach youth to approach and cope with a feared stimulus. Exposure typically comprises four phases [33]. The first phase is preparation and psychoeducation, in which the youth is provided a rationale for exposure. Hierarchy development is the second phase, in which a list of feared stimuli is created. The third phase involves repeated gradual exposure to feared stimuli with post-processing following each exposure task. Generalization and maintenance is the final phase, during which exposures are conducted in various contexts to promote generalization. The negative emotions associated with that stimulus are eventually extinguished through repeated practice facing a feared stimulus. Repeated practice allows for habituation toward the feared stimulus. Ratings of anxiety during the exposure tasks help the youth learn that anxiety reduces over time. Exposure can be conducted imaginally (i.e., the youth visualizes the feared stimulus), in role-play (i.e., an analog situation that is intended to approximate the feared situation), or in vivo (i.e., exposure to an actual feared stimulus). Currently, new insights in exposure research conducted with adult populations show that the mechanism behind exposure linked to anxiety reduction may not be through habituation to the anxiety-provoking stimulus but through the acquisition of a new association with the originally anxiety-provoking stimulus (i.e., inhibitory learning) [34, 35]. The inhibitory learning approach to exposure is based on the premise that new learning requires a mismatch between expectancy and outcome [36]. An important clinical implication is that exposure is no longer hierarchal. Instead, individuals are encouraged to engage in exposures that maximally violate their expectation that a feared outcome will occur. In addition, there is a change from focusing on anxiety levels pre- and post-gradual exposure to consolidating learning by emphasizing the disconfirmation of the expected feared outcome following the exposure.

4.2 Contingent Reinforcement Plans

These techniques are used to increase or decrease the frequency of behavior and often are used with youth who engage in unwanted behavior. Contingent reinforcement plans (also called reward plans) can be set up to directly shape a youth's behavior using positive reinforcement (e.g., rewards). The therapist can use reward plans in session; however, it is more common for the therapist to teach the caregiver or the youth to administer the rewards. A reward plan consists of two components: (a) a list of clearly defined target behaviors that will be rewarded and (b) a menu of rewards and a schedule for administration. Based on the operant conditioning principles, reward charts can involve positive reinforcement (i.e., providing a desired consequence, such as a token or money, when the desired behavior is performed) or negative reinforcement (i.e., removing or reducing something undesirable, such as a chore or punishment, when the desired behavior is performed). The reward plan can sometimes include negative punishment (also called response cost) when a desired consequence or situation is removed for an undesired behavior. The simplest example is when a youth has a privilege removed following behavior, such as having screen time reduced or removed for a day for a non-desirable behavior.

4.2.1 Cognitive Techniques

Cognitive techniques are designed to address thinking patterns that influence the behavior and emotions of youth. An essential notion guiding cognitive techniques is that cognitions may negatively or positively influence our emotions or behaviors. This is particularly so in situations that involve intense or difficult emotions. Cognitive techniques fall into two general categories. First, cognitive techniques can help build cognitive processes designed to help youth cope with difficult situations or emotions (e.g., problem-solving) by providing a new way of processing information about the world (e.g., psychoeducation). Second, cognitive techniques can help change unhelpful ways of thinking and counteract cognitive errors (e.g., cognitive restructuring). When using cognitive techniques with youth, it is essential to consider their level of cognitive development. Metacognition (i.e., thinking about thinking) generally develops around age eight and becomes more sophisticated throughout adolescence [37]. Thus, cognitive techniques may need to be simplified or may only sometimes be appropriate for younger children.

4.2.2 Problem-Solving

Teaching problem-solving skills is a central technique that has cognitive and behavioral components. With problem-solving skills, the therapist teaches a youth to cope with difficult situations or emotions using specific problem-solving steps. Problem-solving typically involves identifying the problem, listing multiple solutions to the problem, considering the pros and cons of each solution, selecting and using a solution, and then reviewing the results. Youths are often encouraged to reward themselves for performing productive problem-solving.

4.2.3 Psychoeducation

Psychoeducation is a commonly employed technique with cognitive components. Psychoeducation provides new information about symptoms, diagnoses, and the treatment process. Information about these components helps the youth (or their caregivers) better understand what they are experiencing, normalize their experience, and instill hope regarding the treatment.

4.2.4 Cognitive Restructuring

This is a core cognitive technique used in CBT programs for various presenting problems. This entails attempting to alter dysfunctional cognitions into more functional cognitions. This process can include three stages: (a) identification of current cognitions and the negative feelings and maladaptive behavior that emerge from them along with underlying assumptions and schemas, (b) cognitive challenging of the dysfunctional cognitions, and (c) generating alternative cognitions that do not lead to dysfunctional feelings and behaviors, but help the youth cope with certain situations better.

Determining which cognitions lead to improved or impaired feelings is an essential first step. It is critical to teach youth that focusing on some cognitions may help control and manage a difficult situation, whereas focusing on others may escalate and complicate feelings and subsequent coping in the situations that elicit the cognitions. The second and third steps involve challenging dysfunctional cognitions and generating alternate cognitions. In practice, these two steps are often included in the same exercises. These steps include asking for evidence, evaluating how much the youth believes in the cognition(s), and checking for alternative cognitions that modify the initial cognition into a more helpful self-instruction. Besides asking for evidence, lists of questions are often provided, which youth can pick from to challenge their cognitions. Examples of challenging cognitions for anxiety or depression are: "Is it really?" "What proof do I have?" "Am I forgetting the positive?" Examples of challenging cognitions for aggression are: "What would ... (a cool role model) think or do in this situation?" "What would I think about this 5 years from now?" By teaching youth to challenge their cognitions, they automatically move forward to the third step, which is to help them generate alternative adaptive cognitions. Asking the youth to rate the credibility of their "original cognition" (i.e., anxious, depressed, or aggressive) and the "new cognition," one can explore the power of the "new cognition." This also invites the youth to consider and evaluate the usefulness of the new cognition carefully.

4.2.5 Acceptance

In contrast to cognitive restructuring, acceptance-based approaches discourage thought challenging and modification. Instead, acceptance-based approaches encourage individuals to nonjudgmentally notice their thoughts and feelings and accept them without trying to change them [38]. These approaches are common in "third wave" CBTs such as acceptance and commitment therapy [39] and dialectical behavior therapy [40] and aim to promote psychological flexibility (i.e., the ability to respond to uncomfortable thoughts or feelings in a functional and value-based manner) [38].

4.2.6 Self-Monitoring

Monitoring can help increase awareness of feelings, behaviors, and cognitions. Monitoring such topics may increase the youth's (and the therapist's) insight into the topic's quality, frequency, and intensity. Self-monitoring can be administered in many variations. Two common approaches are using a diary or brief measures that permit data collection to be used to build graphs. Monitoring is often a starting point for treatment, and it helps establish the baseline and offers room to discuss realistic goal setting. Most CBT programs involve self-monitoring activities. For example, CBT programs for depression include mood monitoring and CBT for disruptive behavior problems includes monitoring of positive and negative behaviors [41].

4.3 Techniques Targeting Emotion Regulation

Emotions, including awareness and regulation, are a CBT target that involves cognitive and behavioral techniques. A common first step is evaluating emotional competence to determine the youth's basic understanding and awareness of emotions. Emotive education involves building the prerequisites for emotion regulation, recognition, or understanding. For young children, emotive education will likely include information about the difference between cognitions, feelings, and behaviors and recognition of nonverbal behavior and facial expressions. Education about emotions can also include helping youth understand their emotions' physiological basis. This may take the form of psychoeducation about the nervous system and body mapping tasks to identify where physiological signals derive from in the body.

The second component involves grading emotions, typically done using various forms of scaling (e.g., 0–8 to rate the level of anxiety). Teaching youth to grade their emotions is essential for improved regulation. That is, increased awareness of the intensity of emotion is critical to helping youth learn how to regulate their emotions. Developing a system to grade emotions in terms of intensity is a valuable tool in subsequent evaluations of CBT tasks (e.g., Did an exposure reduce the youth's fear?) and is also helpful in addressing youth expectations. For example, if it turns out that a youth always expects to feel no sadness (a "0") following CBT for depression, psychoeducation about realistic outcome expectations may be required. Finally, developing a benchmarking system for youth's emotional experiences helps prepare them for behavioral tasks. For example, suppose a youth with agoraphobia rates their anxiety for taking a 1-hour bus ride during rush hour with the same intensity as approaching a bus station outside rush hour. In that case, more work may be needed to develop a more nuanced fear hierarchy. Learning how to grade emotions is thus an important target as a CBT technique with youth, and it is an essential means for more successful delivery of subsequent CBT tasks.

Imagery techniques (mental imagery or imagery rescripting) are techniques often included for adolescents and young adults with maladaptive coping strategies for negative emotions, such as self-harm, depression, or trauma [42, 43]. These techniques involve the change or replacement of a mental image causing harm or suffering with a different mental image. Building a helpful mental image typically includes the integration of multiple senses (e.g., sight, sound), making it emotionally more evocative than pure verbal interactions.

5 Modalities of Cognitive-Behavioral Therapy

CBT can be delivered in many different modalities that involve youth, caregivers, and other family members to varying degrees. Depending on the presenting problems, some modalities may better fit than others. Other factors, such as cost or number of available therapists, can determine the best modality.

5.1 Individual CBT

In individual CBT, the primary focus of treatment is the youth. Thus, this modality is a one-on-one collaborative and interactive process between the therapist and the youth [21]. With this modality, there is a variable level of caregiver involvement. The level is likely higher for younger children, as children below age 12 cannot be relied upon for accurate information and depend on their caregivers to get to the treatment location. Moreover, this offers a natural opportunity to involve caregivers in treatment (e.g., by explaining homework assignments to youth and their caregivers).

5.2 Group CBT

In group CBT, treatment is delivered to multiple individuals in a group setting. Several factors make group CBT desirable: cost-effectiveness, opportunities for socialization, normalization of psychopathology, and positive peer influence on the group process [21]. Group CBT can be either youth-, caregiver-, or family-focused: that is, the groups can consist of youth only, caregivers and youth together, or caregivers and youth separately. In youth with antisocial behavior, participation in group treatment may be contraindicated due to the "peer-contagion effect" or "deviancy training" [44].

5.3 CBT with Families

CBT with families involves regular and intensive caregiver involvement. Sometimes, other family members, such as siblings, may be involved. There are several variants of this approach. In some versions of family CBT, the caregiver and the youth meet with a therapist. In other versions, the caregiver and the youth meet individually with the therapist. In CBT, caregivers can be regarded as co-clients or co-therapists. Family CBT often includes concepts and techniques that can be traced back to general systems theory [45]. This often involves prompting families to interact in sessions and show behaviors that may be regarded as problematic. Patterson explains that by using operant procedures, desirable behaviors can be positively reinforced, and undesirable behaviors can be extinguished. Family members may be invited to provide feedback, ideas on new behaviors, and reflections on interactions, thus adding to the reinforcement of positive interactions and the weakening of negative or maladaptive interactions.

5.4 CBT and New Technologies

Blended CBT refers to using Internet- or computer-delivered CBT with face-to-face CBT. Ebert et al. [46] explain that using the computer or the Internet to provide CBT may overcome some limitations of traditional treatment services. Advantages of computer- and Internet-based CBT include (a) availability, (b) anonymity, (c) accessibility at any time and place, (d) flexibility in self-direction and self-pacing, (e) reduced travel time and costs for both clients and therapists, and (f) the appeal of interactivity and visual attractiveness of Internet-based programs [46].

Tele-CBT refers to CBT using a telephone or other communication mode that is not face-to-face. Tele-CBT has been evaluated primarily with adults for various problems, but evaluation of the approach with youth has begun. Tele-CBT is particularly well suited for rural and underserved areas and thus shares some advantages with blended CBT. *Serious games* are alternative education tools that go beyond entertainment to enable learning digitally and interactively (i.e., a game) [47]. A serious game tends to be brief and implemented for a specific objective and can be used to enhance desired outcomes. *Virtual reality therapy* includes using a three-dimensional computer-generated environment in which a person can move around and interact as if they were in that world. Due to the world's reality, techniques such as desensitization and exposure are facilitated. Virtual reality meta-analytic findings have supported exposure therapy [48]. For youth, serious games and virtual reality therapy are not restricted to the treatment of anxiety. For example, virtual reality therapy can be implemented to reduce pain and anxiety in pediatric patients undergoing medical procedures [49], social skills training for youth with autism [50], or for the treatment of violence in youth with schizophrenia [51]. Serious games can be implemented in educational settings, increasing awareness, or with a therapeutic aim. The findings regarding the use of serious games for health promotion are significant but limited [52]. Also, what therapeutic ingredients are included in serious games is typically not described.

6 CBT Delivery

Our final section focuses on general factors that need to be considered when using CBT to address emotional and behavioral problems experienced by youth. The CBT techniques outlined above can be delivered with varying levels of therapist competence, which can influence the effectiveness of CBT for youth with emotional and behavioral problems. Therapist competence is the extent to which the therapist balances technical and relational skills and uses them to facilitate therapeutic change. This broad concept encompasses skillfulness and responsiveness in delivering specific CBT techniques [53]. The past decade has witnessed an increased focus on accountability and the importance of establishing therapist competency in the education and practice of psychology [54] and other areas of mental health [55]. This section, therefore, focuses on aspects of CBT delivery that can help foster the development of therapist competence.

6.1 Assessment

Assessment plays a vital role in the delivery of CBT and is defined as the process used to collect, interpret, and use clinical information to describe a youth [56]. Most CBT models emphasize the importance of continuous data collection throughout treatment. Typically, these data help shape and guide the treatment process from intake to termination. When grounded in science and theory, assessment can help make CBT more efficient and effective [57]. For example, symptom and diagnostic measures that help a therapist arrive at the correct diagnosis can inform the development of a case conceptualization and facilitate the subsequent delivery of CBT. Assessment can thus play an essential role in guiding CBT throughout treatment.

The principles of the evidence-based assessment movement have informed the use of outcome monitoring and evaluation in CBT. This method is defined as an approach to clinical evaluation that utilizes science and theory to guide the assessment process [57]. A critical aim of the movement is developing and promoting guidelines to direct research, structure training, and inform clinical practice [58]. The principles represented in the movement include a scientific approach to assessment, a strong emphasis on the score reliability and validity of tools, and a data-informed approach to clinical practice. While it is beyond the scope of the present chapter to provide a thorough review of evidence-based assessment, readers interested in learning how the principles of this movement can be used to guide the treatment selection and planning process can see Youngstrom et al. [57].

6.2 Case Conceptualization

Case conceptualization represents an important tool used to guide the delivery of CBT and is defined as a set of hypotheses about the causes, antecedents, and maintaining factors of a youth's presenting problems [59, 60]. See also Chapter 2 of this text. Considered an essential component of evidence-based practice by the American Psychological Association, the association states that, "The purpose of evidence-based practice in psychology is to promote effective psychological practice and enhance public health by applying empirically supported principles of psychological assessment, case conceptualization, therapeutic relationship, and intervention" (p. 273) [61]. A good case conceptualization identifies the factors that cause and maintain a youth's target problem(s) and helps the therapist understand how best to translate findings from the empirical literature for use with a particular youth. Case conceptualization following a systematic approach can be helpful when CBT needs to be tailored to fit heterogeneous populations [62]. Assessment tools used throughout treatment are critical to generating data to inform the development and refinement of a case conceptualization [63]. Thus, the two skills—assessment and case conceptualization—are ideally used in CBT.

6.3 Cultural Considerations

CBT delivery should be informed by knowledge of how culture and diversity can influence the experience and expression of youth's emotional and behavioral problems (*see* also Chapter 14 of this text). Culture is "an integrated pattern of human behavior that includes thought, language, action, and artifacts and depends on man's capacity for learning and transmitting knowledge to succeeding generations" (p. 5) [64]. The failure to take a person's nationality, ethnicity, acculturation level, socioeconomic status, and gender identity into consideration can lead to poor clinical outcomes. For example, behavior may be labeled pathological when cultural factors account for the behavior. Therapists, thus, need to consider culture when determining how to deliver CBT. First, therapists must understand how the expression of symptoms or distress may be influenced by culture. The available evidence suggests that the expression of psychological distress may vary across cultures [65, 66]. This variation may be due to value systems that find different symptoms more or less acceptable. Second, therapists must be aware that cultural factors may influence reporting practices. The acceptability of specific symptoms may influence what symptoms are reported as problematic. Understanding how symptom expression and reporting practices may be influenced by culture is an essential component of conducting a culturally sensitive assessment. Therapists can use the Diagnostic and Statistical Manual of Mental Disorders Cultural Formulation Interview [67] with youth and their families to understand how cultural and diversity issues may influence the treatment process. Finally, it is essential to consider cultural factors in treatment planning and delivery. Some treatment components may need to be modified or replaced to be relevant and appropriate for the individual. For example, many cultures do not approach thoughts and feelings as firmly distinct concepts, so general psychoeducation about the relation between thoughts, feelings, and behaviors may need to be adjusted.

6.4 Developmental Considerations

One of the distinguishing characteristics of youth is rapid change, which directly impacts the delivery of CBT. Differentiating between developmentally appropriate behavior and disorder-specific behavior is essential. At specific age periods, typically developing youth may show anxiety for ghosts, rigidity about rules, oppose limit-setting, cross social boundaries, get absorbed in computer games, or have difficulty planning homework. Youth with emotional and behavioral problems may show such behaviors with an inappropriate intensity or frequency or at an inappropriate age. When behaviors exceed what is typical for normal development and lead to clinical impairment, CBT may help, either directed at the youth, their caregivers, or their family. Understanding family dynamics and triggers that precede inappropriate behavior is crucial to targeting the right "problem" in treatment. Likewise, many factors are essential to take into account to determine whether CBT is appropriate or for whom CBT is appropriate.

How CBT is delivered should be influenced by knowledge of developmental norms. Developmental factors determine what delivery approaches (e.g., play-based, talk) will provide the most accurate and valid information and increase client involvement and motivation. Language development and comprehension directly influence a youth's ability to understand questions, report and reflect upon their experience, and be comfortable in different clinical situations (e.g., play versus interviews). Therapists should, therefore, take a developmentally informed approach to delivery and select an approach from the perspective of developmental norms.

In general, CBT is offered to youth aged eight or above, though some have questioned the use of CBT with younger youth. Grave and Blissett [68] argued that the cognitive capacities of young children are still developing and that a rational approach, as with CBT, may not be suitable. Treatment programs for younger children tend to include adaptations to fit the participants' developmental levels. Finding a fit for the developmental level of youth and their family may require adaptations to language, materials, activities, or timing of treatment [69]. If and how to involve caregivers may change throughout a youngster's life from toddler to young adult. For instance, The Incredible Years program stresses the importance of spending positive time together (e.g., participating in youth's play) [70]. It would be a misjudgment to ask caregivers to involve an adolescent in building a tower. Translating "shared play" into "shared activities" may fit younger adolescents, whereas translating into "shared time" is most fitting for older adolescents.

6.5 Relational Skills

The delivery of specific CBT techniques does not occur in a vacuum but needs to be adapted to the youth's developmental stage, mood, temperament, and personality. All specific CBT techniques are intertwined with the therapists' relational skills and abilities to create a facilitative therapeutic process (*see* also Chapter 3 of this text). The alliance is commonly defined as the quality of the youth-therapist affective bond and the degree of collaboration between the youth and therapist on therapeutic activities [71]. The alliance has consistently been shown to predict treatment outcomes with youth [72, 73]. A stronger alliance is also associated with consistent youth attendance [74] and higher treatment satisfaction ratings [75, 76].

Given the link between a strong alliance and positive clinical outcomes, an essential issue for therapists is how to enhance the alliance. Broad therapist competencies associated with forming a strong therapeutic relationship in CBT have been identified [69, 77]. These include instilling hope and optimism for change and engaging youth with developmentally appropriate activities. The suggestions also include emphasizing alliance building early in treatment, reflective statements, collaborative language (e.g., *we*, *us*, and *let's*), and tracking and validating feelings. Working with youth and caregivers is essential to achieve mutually agreed-upon treatment goals and activities. This is because youth rarely refer themselves to treatment, so they may not initially agree with their caregivers about the goals for treatment, especially in youth with externalizing disorders. It is also recommended that therapists provide a treatment rationale linking the purpose of the treatment activities to the overall treatment goal each time a new treatment activity is introduced [77]. In terms of instilling hope and optimism, therapists can assess for unrealistic treatment expectancies, provide psychoeducation about the expected length and outcome of treatment, explain the treatment rationale, and set up therapeutic activities that help build youths' self-efficacy. These approaches to strengthen the alliance can help maximize youth involvement in treatment, which is defined as the youths' active participation in therapeutic activities, including homework compliance.

7 Conclusions

CBT is a potent treatment approach that integrates two separate theoretical traditions: behavioral theory and cognitive theory. CBT has been tested in several individual, family, and group formats. Evidence supporting CBT has been amassed for several problems common in youth, including anxiety disorders, depression, disruptive behavior disorders, and eating disorders.

Conflicts of Interest

The authors have no conflicts of interest to declare that are relevant to the content of this chapter.

References

1. Liang JH, Li J, Wu RK et al (2021) Effectiveness comparisons of various psychosocial therapies for children and adolescents with depression: a Bayesian network meta-analysis. Eur Child Adolesc Psychiatry 30(5):685–697. https://doi.org/10.1007/s00787-020-01492-w
2. Ma L, Zhang Y, Huang C et al (2020) Resilience-oriented cognitive behavioral interventions for depressive symptoms in children and adolescents: a meta-analytic review. J Affect Disord 270:150–164. https://doi.org/10.1016/j.jad.2020.03.051
3. James AC, Reardon T, Soler A et al (2020) Cognitive behavioural therapy for anxiety disorders in children and adolescents. Cochrane Database Syst Rev 11(11):CD013162. https://doi.org/10.1002/14651858.CD013162.pub2
4. Danzi BA, La Greca AM (2020) Treating children and adolescents with posttraumatic stress disorder: moderators of treatment response. J Clin Child Adolesc Psychol 50(4):510–516. https://doi-org.ezproxy.uio.no/10.1080/15374416.2020.1823849
5. Wergeland GJ, Ghaderi A, Fjermestad K et al (2024) Family therapy and cognitive behavior therapy for eating disorders in children and adolescents in routine clinical care: a systematic review and meta-analysis. Eur Child Adolesc Psychiatry. https://doi.org/10.1007/s00787-024-02544-1
6. Lambez B, Hardwood-Gross A, Golumbic EZ et al (2020) Non-pharmacological interventions for cognitive difficulties in ADHD: a systematic review and meta-analysis. J Psychiatr Res 120:40–55. https://doi.org/10.1016/j.jpsychires.2019.10.007
7. Chen JY, Chen SN, Lee CH et al (2023) A systematic review and meta-analysis of randomized control trials: efficacy of cognitive behavioral therapies for the management of functional and recurrent abdominal pain disorders in children and adolescents. Cogn Behav Ther 52(5):438–459. https://doi.org/10.1080/16506073.2023.2200562
8. Fisher E, Villanueva G, Henschke N et al (2022) Efficacy and safety of pharmacological, physical, and psychological interventions for the management of chronic pain in children: a WHO systematic review and meta-analysis. Pain 163(1):e1–e19. https://doi.org/10.1097/j.pain.0000000000002297

9. Hogue A, Henderson CE, Ozechowski TJ et al (2014) Evidence base on outpatient behavioral treatments for adolescent substance use: updates and recommendations 2007–2013. J Clin Child Adolesc Psychol 43:695–720. https://doi.org/10.1080/15374416.2014.915550
10. Riise EN, Wergeland GJH, Njardvik U et al (2021) Cognitive behavior therapy for externalizing disorders in children and adolescents in routine clinical care: a systematic review and meta-analysis. Clin Psychol Rev 83:101954. https://doi.org/10.1016/j.cpr.2020.101954
11. Wang X, Zhao J, Huang S et al (2021) Cognitive behavioral therapy for autism spectrum disorders: a systematic review. Pediatrics 147(5):e2020049880. https://doi.org/10.1542/peds.2020-049880
12. Lawton K, Spencer A (2021) A full systematic review on the effects of cognitive behavioural therapy for mental health symptoms in child refugees. J Immigr Minor Health 23(3): 624–639. https://doi.org/10.1007/s10903-021-01151-5
13. Craig SL, Austin A, Alessi E (2013) Gay affirmative cognitive behavioral therapy for sexual minority youth: a clinical adaptation. Clin Soc Work J 41:258–266. https://doi.org/10.1007/s10615-012-0427-9
14. Wright F, Hronis A, Roberts R et al (2024) Assessing the abilities of children and adolescents with intellectual disabilities to engage in cognitive behaviour therapy: a pilot study. J Educ Dev Psychol 42(1):13–21. https://doi.org/10.1080/20590776.2024.2389076
15. Pavlov IP (1927) Conditioned reflexes: an investigation of the physiological activity of the cerebral cortex. Oxford University Press, Oxford
16. Benjamin CL, Puleo CM, Settipani CA et al (2011) History of cognitive-behavioral therapy in youth. Child Adolesc Psychiatr Clin North Am 20(2):179–189. https://doi.org/10.1016/j.chc.2011.01.011
17. Skinner BF (1928) The behavior of organisms: an experimental analysis. Appleton-Century, New York
18. Skinner BF (1969) Contingencies of reinforcement: a theoretical analysis. Meredith, New York
19. Bandura A (1969) Social learning theory of identificatory processes. In: Goslin DA (ed) Handbook of socialization theory and research. McNally & Company, Chicago
20. Rotter JB (1954) Social learning and clinical psychology. Prentice-Hall, Inc., New York
21. Southam-Gerow MA, McLeod BD, Brown RC et al (2011) Cognitive-behavioral therapy for adolescents. In: Brown BB, Prinstein MJ (eds) Encyclopedia of adolescence, vol 3. Academic Press. https://doi.org/10.1016/B978-0-12-373951-3.00106-X
22. Miller GA (2003) The cognitive revolution: a historical perspective. Trends Cogn Sci 7(3): 141–144. https://doi.org/10.1016/S1364-6613(03)00029-9
23. Neisser U (1967) Cognitive psychology. Appleton-Century-Crofts, New York
24. Beck AT, Rush AJ, Shaw BF et al (1979) Cognitive therapy of depression. Guilford Press, New York
25. Beck AT (1967) Depression: clinical, experimental, and theoretical aspects. Guilford Press, New York
26. Williamson DA, Muller SL, Reas DL et al (1999) Cognitive bias in eating disorders: implications for theory and treatment. Behav Modif 23(4):556–577. https://doi.org/10.1177/0145445599234003
27. Wright FD, Beck AT, Newman CF et al (1993) Cognitive therapy of substance abuse: theoretical rationale. NIDA Res Monogr 137:123–146
28. Lochman JE, Pardini DA (2008) Cognitive-behavioral therapies. In: Rutter M, Bishop D, Pine D et al (eds) Rutter's child and adolescent psychiatry. Wiley-Blackwell, New York, pp 1026–1045
29. Gresham FM, Lochman JE (2008) Methodological issues in research using cognitive-behavioral interventions. In: Mayer MJ, Van Acker R, Lochman JE et al (eds) Cognitive-behavioral interventions for emotional and behavioral disorders: school-based practice. Guilford, New York
30. Perkins AM, Meiser-Stedman R, Spaul SW et al (2023) The effectiveness of third wave cognitive behavioural therapies for children and adolescents: a systematic review and meta-analysis. Br J Clin Psychol 62(1):209–227. https://doi.org/10.1111/bjc.12404
31. Brown LA, Gaudiano BA, Miller IW (2011) Investigating the similarities and differences between practitioners of second- and third-wave cognitive behavioural therapies. Behav Modif 35(2):187–200. https://doi.org/10.1177/0145445510393730
32. Neff K, Tirch D (2013) Self-compassion and ACT. In: Kashdan T, Ciarrochi V (eds) Mindfulness, acceptance, and positive psychology: the seven foundations of well-being. New Harbinger Publications, Oakland

33. Ollendick TH, Hovey LD (2009) Competencies for treating phobic and anxiety disorders in children and adolescents. In: Thomas J, Hersen M (eds) Handbook of clinical psychology competencies. Springer, New York, pp 1219–1244
34. Craske MG, Treanor M, Conway CC et al (2014) Maximizing exposure therapy: an inhibitory learning approach. Behav Res Ther 58:10–23. https://doi.org/10.1016/j.brat.2014.04.006
35. McGuire JF, Storch EA (2019) An inhibitory learning approach to cognitive-behavioral therapy for children and adolescents. Cogn Behav Pract 26(1):214–224. https://doi.org/10.1016/j.cbpra.2017.12.003
36. Rescorla RA, Wagner AR (1972) A theory of Pavlovian conditioning: variations in the effectiveness of reinforcement and non-reinforcement. In: Prokasy AH (ed) Classical conditioning II: current research and theory. Appleton-Century-Croft, New York
37. Garber J, Frankel SA, Herrington CG (2016) Developmental demands of cognitive behavioral therapy for depression in children and adolescents: cognitive, social, and emotional processes. Annu Rev Clin Psychol 12(1):181–216. https://doi.org/10.1146/annurev-clinpsy-032814-112836
38. Hayes SC, Luoma JB, Bond FW et al (2006) Acceptance and commitment therapy: model, processes and outcomes. Behav Res Ther 44(1):1–25. https://doi.org/10.1016/j.brat.2005.06.006
39. Hayes SC, Strosahl KD, Wilson KG (2012) Acceptance and commitment therapy: the process and practice of mindful change, 2nd edn. Guilford Press, New York
40. Linehan MM, Wilks CR (2015) The course and evolution of dialectical behavior therapy. Am J Psychother 69(2):97–110. https://doi.org/10.1176/appi.psychotherapy.2015.69.2.97
41. Maalouf FT, Brent DA (2012) Depression and suicidal behavior. In: Szigethy E, Weisz JR (eds) Cognitive-behavioral therapy for youth and adolescents. American Psychiatric Association, Washington, DC, pp 163–184
42. Schwarz S, Grasmann D, Schreiber F et al (2020) Mental imagery and its relevance for psychopathology and psychological treatment in children and adolescents: a systematic review. Int J Cogn Ther 13(4):303–327. https://doi.org/10.1007/s41811-020-00092-5
43. Schmied E, Hack L, Connemann B et al (2024) You're not alone: imagery rescripting for adolescents who self-harm. Front Psychol 15:1395603. https://doi.org/10.3389/fpsyg.2024.1395603
44. Gifford-Smith M, Dodge KA, Dishion TJ et al (2005) Peer influence in children and adolescents: crossing the bridge from developmental to intervention science. J Abnorm Child Psychol 33:255–265. https://doi.org/10.1007/s10802-005-3563-7
45. Patterson T (2014) A cognitive behavioral systems approach to family therapy. J Fam Psychother 25(2):132–144. https://doi.org/10.1080/08975353.2014.910023
46. Ebert DD, Zarski AC, Christensen H et al (2015) Internet and computer based cognitive-behavioral therapy for anxiety and depression in youth: a meta-analysis of randomized controlled outcome trials. PLoS One 10(3):e0119895. https://doi.org/10.1371/journal.pone.0119895
47. Gros B (2017) Game dimensions and pedagogical dimension in serious games. In: Zheng R, Gardner MK (eds) Handbook of research on serious games for educational applications. IGI Global, pp 402–417. https://doi.org/10.4018/978-1-5225-0513-6.ch019
48. Morina N, Ijntema H, Meyerbroker K et al (2015) Can virtual reality exposure therapy gains be generalized to real-life? A meta-analysis of studies applying behavioral assessments. Behav Res Ther 74:18–24. https://doi.org/10.1016/j.brat.2015.08.010
49. Eijlers R, Utens EM, Staals LM et al (2019) Systematic review and meta-analysis of virtual reality in pediatrics: effects on pain and anxiety. Anesth Analg 129(5):1344–1353. https://doi.org/10.1213/ANE.0000000000004165
50. Ke F, Moon J, Sokolikj Z (2022) Virtual reality-based social skills training for children with autism spectrum disorder. J Spec Educ Technol 37(1):49–62. https://doi.org/10.1177/0162643420945603
51. Dellazizzo L, Potvin S, Bahig S et al (2019) Comprehensive review on virtual reality for the treatment of violence: implications for youth with schizophrenia. NPJ Schizophr 5(1):11. https://doi.org/10.1038/s41537-019-0079-7
52. David OA, Costescu C, Cardos R et al (2020) How effective are serious games for promoting mental health and health behavioral change in children and adolescents? A systematic review and meta-analysis. Child Youth Care Forum 49:817–838. https://doi.org/10.1007/s10566-020-09566-1
53. Barber JP, Sharpless B, Klostermann S et al (2007) Assessing intervention competence

and its relation to therapy outcome: a selected review derived from the outcome literature. Prof Psychol Res Pr 38:493–500. https://doi.org/10.1037/0735-7028.38.5.493

54. Southward MW, Kushner ML, Terrill DR et al (2024) A review of transdiagnostic mechanisms in cognitive behavior therapy. Psychiatr Clin North Am 47(2):343–354. https://doi.org/10.1016/j.psc.2024.02.003
55. Unutzer J, Schoenbaum M, Druss BG et al (2006) Transforming mental health care at the interface with general medicine: report for the president's commission. Psychiatr Serv 57(1):37–47. https://doi.org/10.1176/appi.ps.57.1.37
56. Hunsley J (2002) Psychological testing and psychological assessment: a closer examination. Am Psychol 57:139–140. https://doi.org/10.1037/0003-066X.57.2.139
57. Youngstrom EA, Prinstein MJ, Mash EJ et al (2022) Assessment of disorders in childhood and adolescence. The Guilford Press, New York
58. Hunsley J, Mash EJ (2007) Evidence-based assessment. Annu Rev Clin Psychol 3:29–51. https://doi.org/10.1146/annurev.clinpsy.3.022806.091419
59. Eells TD (1997) Psychotherapy case formulation: history and current status. In: Handbook of psychotherapy case formulation. Guilford Press, New York, pp 1–25
60. Nezu AM, Nezu CM, Peacock MA et al (2004) Case formulation in cognitive-behavioral therapy. In: Haynes SN, Heiby EM, Hersen M (eds) Comprehensive handbook of psychological assessment, vol 3: behavioral assessment. Wiley, New York, pp 402–426
61. American Psychological Association (2006) Evidence-based practice in psychology. Am Psychol 61(4):271–285
62. Sauter FM, van den Bogaard M, van Vilet C et al (2023) An AAIDD-informed framework for cognitive behavioral case formulation and cognitive behavior therapy for young people with mild intellectual disabilities or borderline intellectual functioning. Clin Psychol 30(3): 299–311. https://doi.org/10.1037/cps0000151
63. Christon LM, McLeod BD, Jensen-Doss A (2015) Evidence-based assessment meets evidence-based treatment: an approach to science-informed case conceptualization. Cogn Behav Pract 22(1):36–48. https://doi.org/10.1016/j.cbpra.2013.12.004
64. Frisby CL, Reynolds CR (2005) Comprehensive handbook of multicultural school psychology. Wiley, New York
65. Lansford JE, Godwin J, Bornstein MH et al (2018) Parenting, culture, and the development of externalizing behaviors from age 7 to 14 in nine countries. Dev Psychopathol 30(5):1937–1958. https://doi.org/10.1017/S0954579418000925
66. Patriarca GC, Rey Y, Marin CE et al (2022) Parent involvement enhances CBT for anxiety disorders in Hispanic/Latino youth: acculturation as a moderator. J Consult Clin Psychol 90(10):827–836. https://doi.org/10.1037/ccp0000770
67. Lewis-Fernández R, Aggarwal NK, Hinton L et al (2016) DSM-5® handbook on the cultural formulation interview. American Psychiatric Publishing Inc, Washington, DC
68. Grave J, Blissett J (2004) Is cognitive behavior therapy developmentally appropriate for young children? A critical review of the evidence. Clin Psychol Rev 24:399–420. https://doi.org/10.1016/j.cpr.2004.03.002
69. Sblurlati ES, Schniering CA, Lyneham HJ et al (2011) A model of therapist competencies for empirically supported cognitive behavioral treatment of child and adolescent anxiety and depressive disorders. Clin Child Fam Psychol Rev 14(1):89–109. https://doi.org/10.1007/s10567-011-0083-6
70. Webster-Stratton C, Reid MJ (2003) The incredible years parents, teachers, and children training series: a multifaceted treatment approach for young children with conduct problems. In: Kazdin AE, Weisz JR (eds) Evidence-based psychotherapies for children and adolescents. Guilford Press, New York, pp 224–240
71. Elvins R, Green J (2008) The conceptualization and measurement of therapeutic alliance: an empirical review. Clin Psychol Rev 28(7): 1167–1187. https://doi.org/10.1016/j.cpr.2008.04.002
72. Karver MS, De Nadai AS, Monahan M et al (2018) Meta-analysis of the prospective relation between alliance and outcome in child and adolescent psychotherapy. Psychotherapy 55:341–355
73. Roest JJ, Welmers-Van de Poll MJ, Peer Van der Helm GH et al (2023) A three-level meta-analysis on the alliance-outcome association in child and adolescent psychotherapy. Res Child Adolesc Psychopathol 51(3):275–293. https://doi.org/10.1007/s10802-022-00986-2
74. Shirk SR, Karver MS, Brown R (2011) The alliance in child and adolescent psychotherapy. Psychotherapy 48(1):17–24. https://doi.org/10.1037/a0022181

75. Fjermestad KW, Lerner MD, McLeod BD et al (2017) Motivation and treatment credibility predict alliance in cognitive behavioral treatment for youth with anxiety disorders in community clinics. J Clin Psychol 74(6):793–805. https://doi.org/10.1002/jclp.22551

76. Ormhaug SM, Shirk SR, Wentzel-Larsen T (2015) Therapist and client perspectives on the alliance in the treatment of traumatized adolescents. Eur J Psychotraumatol 6(1): 27705

77. Brown RC, Southam-Gerow MA, McLeod BD et al (2017) The global therapist competence scale for youth psychosocial treatment: development and initial validation. J Clin Psychol 8: 1–16. https://doi.org/10.1002/jclp.22537

Chapter 2

Case Conceptualization in CBT: Clinical Maps Helping Practitioners and Patients Find Their Way

Ciera Korte and Robert D. Friedberg

Abstract

Cognitive behavioral therapy (CBT) is often misunderstood as a treatment approach characterized by a mash-up of techniques. However, CBT is fundamentally guided by case conceptualization. In this chapter, case conceptualization is viewed as a clinical Global Positioning System piloting the selection, sequencing, and timing of interventions. The unique and transtheoretical components of case formulations are presented. The process of drafting a case conceptualization is outlined. Finally, a detailed treatment plan directed by the case conceptualization is illustrated.

Key words Cognitive behavioral therapy, Children, Adolescents, Case conceptualization, Treatment planning

1 Introduction

I wisely started with a map. (J.R.R. Tolkien)

Maps represent the basic stuff of navigation. Without them, paths follow uncertain directions and travelers become easily lost. Advances in technology have created modern maps called Global Positioning System (GPS) devices that provide immediate feedback. These GPS devices tell users when they take wrong turns and need to reroute their course. Wouldn't it be great if working clinicians owned a GPS device to alert them to proper directions and missed calculations? Well, they do! Case conceptualizations are clinical GPS devices.

Competent CBT practice is guided by case formulation [1–3] and represents the essence of evidence-based practice [4]. Kendall and colleagues [5] agreed that case conceptualization enlivens CBT. Perhaps, this is the reason case formulation is seen as the nucleus of CBT [1].

Robert D. Friedberg and Micaela A. Thordarson (eds.), *Cognitive Behavioral Therapy with Youth: Tradition and Innovation*, Neuromethods, vol. 231, https://doi.org/10.1007/978-1-0716-5134-6_2,

This chapter emphasizes case conceptualization as the keystone for effective CBT practice. The chapter begins with a discussion of the importance of case conceptualization. The unique CBT and transtheoretical components of case conceptualization are delineated. Rudimentary principles for drafting the case formulation are explained. The chapter concludes with a case example illustrating the case conceptualization and subsequent treatment planning process.

2 Why Is Conceptualization So Important?

> If you do not know where you want to go, it doesn't matter which path you take. (Lewis Carroll)

First and foremost, case conceptualization directs CBT [4]. Case formulation catalyzes fundamental CBT practice including choice of techniques as well as their timing [1, 6]. CBT also fosters relevance matching. Relevance matching refers to the "analytic framework for approaching questions about the applicability of the treatment or sets of treatments concurrently to a given service population" [7, p. 424]. Moreover, case formulations enable clinicians to predict and repair therapeutic alliance ruptures, manage nonadherence, monitor treatment progress, as well as provide better supervision to trainees [6].

Delivering techniques that are divorced from case formulation yields an emaciated approach. Simply, as the Lewis Carroll quote implies, this misguided strategy is far too mechanical [8]. Additionally, this clinical decision robs CBT of its "conceptual guts" [9, p. 153]. Shirk [10] concluded that applying CBT without a case conceptualization is analogous to cooking without a recipe. Similar to developing a case conceptualization requires mixing various separate ingredients together to create a coherent whole product. In the following section, the unique and common components in CBT conceptualization are outlined.

3 Unique and Transtheoretical Components of Case Formulation

> A theory is the more impressive the greater simplicity of its premises, the more different are the kinds of things it relates to and the more evidence of the range of its application. (Albert Einstein)

Case conceptualization is an exercise in pattern recognition [4, 11]. Freeman et al. [11, p. 254] explained, "Effective conceptualization starts with the premise that behavior has a reason, meaning, or purpose. Behavior is not random but guided by what has been learned or what has been acquired through conditioning.

Behavior is affected by biological, social, and psychological systems and serves a function within these systems." The pattern often emerges as a complex mosaic consisting of individuals' learning history, genetic, biological, and contextual circumstances, as well as their thoughts, feelings, and behaviors. Accordingly, learning and information processing theories guide the formulation process.

4 Unique CBT Elements of Conceptualization

4.1 Learning Theory

CBT leans heavily on learning theory. In particular, classical [12], operant [13], and social learning paradigms [14] inform case conceptualization. Classical conditioning is especially helpful in understanding anxiety and phobias. Operant conditioning is used when constructing functional analyses. Antecedents as well as discriminative stimuli are identified and then consequences (positive reinforcement, negative reinforcement, response cost, and punishment) which make responses such as approach and avoidance behaviors more or less likely are determined. Social learning theory offers perspective on vicarious learning experiences and the development of self-efficacy.

4.2 Information Processing

The hierarchical organizational model (HOM) and the content specificity hypothesis (CSH) explain the way individuals process information [15–17]. This system is hierarchically organized with schemata or core beliefs placed squarely in the center of three concentric circles and underlying assumptions and automatic thoughts (ATs) lying at the periphery. The layers vary from each other on two pivotal dimensions: degree of explicit awareness and level of emotional intensity.

Schemata command the center of individuals' personality. They represent core beliefs or meaning-making systems [18, 19]. Accordingly, schemata are considered cognitive structures which direct cognitive processes and products. Legendary developmental psychologist, Jerome Kagan [20, p. 121], explained that a schema is "the cognitive unit that will store experience in a form so faithful a person can recognize a past event." Schema content is generally absolutistic and devoid of gradients. Not surprisingly, schematic material packs a powerful emotional punch and lives at the most implicit level of information processing. Both these characteristics combine to make schema change quite challenging.

These cognitive structures develop over time through maturation and environmental influences. Hammen and Zupan [21] noted that these core meaning frameworks do not crystallize until late adolescence and early adulthood. Consequently, nascent schemata contribute to but do not determine young patients' behavior until their mid-twenties. Simply, they may be considered diatheses or vulnerability factors.

Table 1
Content specificity hypothesis

Feeling state	Sample thought	Content specificity domain
Depression	I'm a loser The world sucks Things will never change	Negative view of self Negative view of others and one's experiences Negative view of the future
Anxiety	When things are uncertain, something awful *will* happen If I don't get a date to the dance, my life is completely ruined There's nothing I can do to cope with this! If bad things happen, no one can help me	Overestimation of the probability of the danger Overestimation of the magnitude of the danger Ignoring coping capacity Neglect of rescue factors
Panic	This dizzy feeling I get sometimes means I have a brain tumor	Catastrophic misinterpretation of normal bodily sensations
Social anxiety	When people see my art project up on the bulletin board, they will be so very salty to me	Fear of negative evaluation
Anger		Hostile attributional bias (confusing accidental with deliberate) Labeling others Conflating fair with what is desired Violation of personal rules

ATs are streams of conscious thoughts, explanations, judgments, images, and appraisals that live at the level of explicit awareness. Although they can be more readily accessed than schemata and underlying assumptions, ATs often are not easy to identify. The CSH [21–25] is a crucial guide for case conceptualization. The CSH posits that specific clusters of cognition are associated with different emotional states. McClure and colleagues [26] refer to these groupings as unique mental parcels (UMPs). In this way, the CSH highlights the path toward identifying meaningful interpretations, attributions, explanations, conclusions, and images. Table 1 provides examples of the specific content that accompanies depression, anger, anxiety, panic, and social anxiety.

4.3 Transtheoretical Elements in CBT Case Formulation

Ethnocultural Cultural Context Cognitive behavioral approaches inherently incorporate contextual components and embrace systemic thinking [2]. Nonetheless, appreciating these elements is not unique to CBT and most, if not all, psychotherapeutic approaches do so as well. Fortunately, due to CBT's robust conceptual paradigm, these potential moderators are seamlessly integrated into the model [2, 27–29]. For instance, recent work [30] asserts beliefs about one's racial group will shape beliefs about oneself, others, and the world. Internalized racism presents as beliefs exhibiting themes of inferiority, inadequacy, and powerlessness, which are often associated with avoidance or other problematic coping strategies.

Developmental History Similar to ethnocultural variables, young patients' developmental history puts presenting problems into a proper backdrop. Developmental milestones and achievements are viewed from a biological and learning theory perspective. They yield vital information regarding patients' experiences with emotional/behavioral regulation (eating, sleeping, etc.) and responsiveness to environmental demands (school, home, peers, etc.). Medical and physical health conditions as well as any medications can also be included in the patient's developmental histories. Moreover, they provide information about patients' strengths, capacities, and vulnerabilities.

Family Functioning Children live in systems so including familial factors into the conceptualization is fundamental. Clearly, clinicians are well-advised to collect information on parental psychiatric and general medical conditions. Indeed, this will generate data on possible genetic loading as well as providing a context for patients' development. Determining discipline habits and parent-child contingencies is also essential. Identifying the parameters surrounding sibling relationships and interactions is also key.

School/Social Functioning Young patients functioning at school yield additional valuable data points. As Friedberg and McClure [2] noted, patients' behavior at school informs clinicians regarding how they navigate demands, transitions between tasks, and peer relationships. Additionally, obtaining information about school behavior provides insight into how patients separate from caregivers [3]. Finally, identifying the ways children and adolescents adjust to various situations on buses, in the cafeteria, and during gym class fosters a 360-degree view.

4.4 Drafting the Case Formulation

Receive with simplicity everything that happens to you. (Rashi)

Integrating the unique and transtheoretical elements into case conceptualization is a complex information processing task. Yet, parsimony is valued and embraced in this process. Kuyken and colleagues [4] recommended that "conceptualization should be as simple as possible without losing essential meaning (p. 316)." Case conceptualization is guided by a hypothesis testing approach [30]. Reviewing and revising clinical assumptions throughout the treatment process is common practice. Therefore, drafts of case formulation are written in sand rather than stone [6]. Manassis [3] added that since they are tentative suppositions, case formulations are open to both confirmation and disconfirmation via data collection. Patient and caregiver input is valued and hypotheses are adjusted via feedback. In this way, the case conceptualization process precisely aligns with collaborative empiricism. Consequently, flexibility, open-mindedness, and critical thinking are necessary.

When formulating the case, patients' psychological portraits are sketched in a compelling and individualized manner. All the unique and transtheoretical components described above should be funneled into a short paragraph centered around three domains (view of self, view of others, and view of the world) [17, 31, 32]. The various elements considered in the CBT case formulation intersect bidirectionally to form a prismatic picture [2, 6].

The following section provides a fictionalized case conceptualization of an 11-year-old patient and the resultant treatment plan.

4.5 Case Example

4.5.1 Presenting Information

Jayden is an 11-year-old biracial cisgender male who lives in two separate homes located in urban areas. Jayden presents with symptoms of depression accompanying his existing diagnosis of functional abdominal pain (FAP). Per parental and self-report, Jayden's depression reflects five symptom domains of major depression, including the core symptoms of (1) low mood nearly every day and (2) loss of interest in activities in addition to (3) insomnia, (4) excessive guilt, and (5) difficulty concentrating. These five symptoms have persisted for two or more weeks. Over the past few weeks, Jayden has become increasingly withdrawn and disinterested in activities he enjoyed at home, such as playing with his dog or building Lego models. Jayden shares that he has been waking early in the morning and struggles to fall asleep at night.

4.5.2 Initial Metrics

Jayden and his mother both completed the Children's Depression Inventory second Edition (CDI-2) [33], Screen for Child Anxiety Related Disorders (SCARED) [34], Functional Disability Inventory (FDI) [35], and Abdominal Pain Index (API) [36] prior to their intake.

SCARED Jayden's total anxiety score showed mild but nonsignificant clinical elevation (Total Score = 25), indicating possible presence of an anxiety disorder with low specificity. Examination of subscales revealed clinically nonsignificant somaticizing/panic symptoms (Panic = 2), separation-related worries (Separation Anxiety = 3), and social anxiety (Social Anxiety = 5). The general worries scale revealed mild clinical elevation (Generalized Anxiety = 10) and included concerns about the future, worry related to performance, worry about being "as good as other kids," worry about not being liked by others, and being "somewhat" nervous. School avoidance was also elevated above clinical cutoff (School Avoidance = 5), with noted stomachaches at school and worries and fear related to school attendance. Jayden's mother reported similar levels of overall anxiety (Total Score = 26), with no clinical elevation in somaticizing/panic symptoms (Panic = 3) and separation-related worries (Separation Anxiety =3). She endorsed minimal social anxiety within normal limits, noting some avoidance of social events (Social Anxiety = 7) and did not endorse concerns related to general worries (Generalized Anxiety = 7). Like her son, Jayden's mother endorsed significant school avoidance (School Avoidance = 6). In sum, results indicate worry-related barriers to school engagement with concerns related to performance and evaluation.

CDI-2 Jayden's total self-reported depression score is "Very Elevated" ($T = 85$) for boys his age, indicating clinically significant depression. Among the subscales, Emotional Problems ($T = 84$), Negative Mood/Physical Symptoms ($T = 82$), Negative Self-Esteem ($T = 79$), Functional Problems ($T = 85$), and Ineffectiveness ($T = 86$) fall within the "Very Elevated" range indicating significant distress, impairment, and disturbance in self-cognitions consistent with depression. Notably, Jayden endorses somewhat lower yet "Elevated" Interpersonal Problems ($T = 67$). Despite irritability Jayden reports getting along with friends during interactions and feeling important to family. Jayden's mother's CDI score reflects clinically significant depressive symptoms punctuated by higher Interpersonal and Negative Mood/Physical Symptom factor scores.

FDI Jayden endorsed a total score of 20 on the FDI (Range = 0–60), indicating that GI symptoms contribute to mild-moderate interference with engagement in chores, school, physical activity, appetite, and sleep. His mother endorsed a total score of 19, noting similar areas of impairment related to GI distress. Notably, she comments during the intake interview that sleep and appetite have worsened since the onset of depressive symptoms, indicating that mood disturbance may be aggravating observed impact of GI symptoms.

API Jayden's Abdominal Pain Index indicates a moderate total pain score of 2.3 (Range = 0–4), with pain occurring 2–3 times for 1–2 hours at a time most days. Jayden rates his pain at a 6/10 on average, with 10 being the most pain possible. Jayden's mother endorses slightly lower levels of pain, reporting a total pain score of 1.9, suggesting that Jayden may not always share his pain symptoms with his mother and/or his mother may not be fully attuned to his physical discomfort.

4.5.3 Ethnocultural and Sociocultural Background

Jayden is the biracial child of a White father and Black mother. He attends a predominately White, non-Hispanic school. Following the pandemic, Jayden's parents wished for him to attend school close to his father's home which is in a generally White school district that is well-funded and produces higher school academic outcomes. Not surprisingly, the lack of racial and ethnic diversity creates various obstacles, which Jayden and his parents are challenged to navigate. For instance, Jayden disclosed that he experiences a sense of marginalization, due to being in class with nearly all White students and teachers, reading the history books which emphasize majority culture historical figures and provide less focus on the contributions of Black and other diverse individuals. On more than one occasion, Jayden reports being the target of racist slurs following school reentry after the pandemic. Consequently, Jayden holds beliefs "Black people don't count at this school," "I don't think the school is very welcoming to me," and "there is not a safe place for Black kids like me here."

Per both mother's and Jayden's note, school adjustment has been a struggle since he is one of the few students of color despite his parents' diligent efforts to promote a positive racial identity. Additionally, stress related to marginalization at school heightened since the start of the pandemic and media coverage of police violence over the last several years. Jayden's mother tries to shield him from the news. His father takes a different approach in his home and often talks about world events such as wars abroad and politics. Thus, Jayden is becoming increasingly aware of tension in the world and expresses both curiosity and concern about global events ("The world is F---d up. Things are out of control; there's nothing we can do").

Jayden's parents both work full-time and live separately. Their divorce is accompanied by some financial strain, which has disproportionately impacted Jayden's mother as a primary caregiver living in an urban area. She navigates her own health concerns related to diabetes, which aggravates financial strain particularly when needing to take time from work for health appointments. She denies food insecurity but worries periodically about housing security as rent prices for her studio apartment increased during the pandemic. Jayden's father has also felt the financial impact of divorce and the pandemic, endorsing working weekends to cover rent and help support Jayden.

4.5.4 Developmental Data

Jayden was born full-term and his mother indicated the prenatal, perinatal, and postnatal periods were unremarkable. Additionally, all developmental milestones were met within normal limits (WNL) per maternal report. No problems adhering to environmental demands as a young child were noted.

4.5.5 Family History and Functioning

Jayden's mother and father divorced 6 years ago and share custody. He lives predominantly with his mother, spending occasional weekends and holidays with his father. Communication between his parental figures are quite conflictual and limited. They interact predominantly at drop-off and pickups at school.

Jayden's family psychiatric history is significant for maternal posttraumatic stress disorder (PTSD), which amplifies his mother's concerns about Jayden's safety as a person of color. His father's psychiatric history is remarkable for episodic depression which is now partially managed by medication. However, during episodes of depression, his father is somewhat withdrawn and has less energy to engage with Jayden.

4.5.6 School History and Functioning

Jayden attends public school where he is enrolled in the sixth grade. Jayden's concentration is suspect especially when he "spaces" in class. His school day begins by waking early (e.g., 5:30 AM) so that his mother can drive to work while he has breakfast at school. Jayden dreads going to school and feels lonely at breakfast since his friends arrive later. He picks at his food and wishes that he could stay at home later like his friends do. By the time his best friend joins him to walk to their first class, Jayden is quiet and reserved. His friend notices and looks to other peers to joke with, as Jayden doesn't appear to be in the mood. Jayden's mother is becoming increasingly concerned about Jayden's mood, noting he has been more irritable at home after school. Sometimes, his mother will let him stay home from school when his pain is severe and she questions whether this is the right thing to do since his repeated absences are associated with falling grades. Previously, Jayden excelled in his classes and showed off his report card to his parents. However, recently he has been reluctant to share his grades since the start of the new academic year with school. Jayden's mother initially thought he just needed time to "readjust" but is now worried that he is not getting the support he needs to succeed at school. When Jayden's mother or his father tries to help him with homework, Jayden frequently states that he is "bad at everything." When queried further about what this means to Jayden, he looks down and states that he is "a horrible kid." According to his teachers, Jayden first began to struggle academically after being bullied in fourth grade but note that his grades have further declined since the beginning of the school year. His math teacher expresses frustration that Jayden is "not putting in the effort" as he used to and reports that he frequently must remind him to complete his work.

Jayden experiences periodic FAP flare-ups at school. He stated that recently these episodes seem to "get in the way of everything." For example, when Jayden is feeling down about his pain, he opts to sit out of activities during recess and physical education and stops joking around with his friends. Additionally, Jayden believes he is the subject of his teacher's increased scolding and negative attention. He disclosed that he sometimes thinks he is "dumb" and is starting to wonder, "What is the point of trying?" After all, his teachers must know he is "no good." Nonetheless, Jayden expresses a desire to "be able to enjoy things [he] used to enjoy" and "do better in school."

Despite his growing disengagement, Jayden still demonstrates moments of curiosity and motivation when given encouragement and individualized attention. For instance, his science teacher noted that he perked up during a recent hands-on experiment and asked insightful questions, underscoring that Jayden's self-doubt and faltering performance is more reflective of his emotional state than his actual abilities. While these moments have become further and farther over the past months, they are a powerful reminder of Jayden's potential. Such moments are highlighted by experiences that challenge distressing thought content.

4.5.7 Medical History

His medical history is significant for FAP diagnosed at 9 years old. His pain is recurrent and episodic, waxing and waning. On days that his symptoms are especially active, pain compromises sleep and attention and may contribute to insomnia and difficulty concentrating. Jayden takes cyproheptadine for pain which he reports is "not very helpful" recently, despite having worked well in the past. Jayden has a lactose sensitivity and tries to avoid dairy in his diet, especially rich or creamy foods that he associates with increased abdominal pain and diarrhea.

4.5.8 Emotional Functioning and Automatic Thoughts

Jayden's emotional life is characterized by multiple distressing feelings including sadness, hopelessness, and frustration. Jayden endorsed over the last week "wishing [he] was not here anymore," indicating some passive suicidal ideation (SI). These thoughts have occurred on two separate occasions, once when Jayden was teased for getting the answer wrong when called on in class; antecedents of suicidal ideation are comprised of situations in which Jayden's feelings that he does not belong and others are judgmental or uncaring are exacerbated. Suicidal ideation causes Jayden to socially withdraw, consequently, furthering a sense of isolation. He believes these thoughts are "bad" and "scary" and denies current or previous active SI as well as any current or previous acts of non-suicidal self-injury. Jayden admits that he is spending less time with his few friends who seem to have "moved on" and "don't care about me anymore because I'm no fun." When asked about a time he felt differently, Jayden shares that "things were different before the pandemic" and that the world is "full of bad things now," referencing the pandemic and wars abroad. Jayden concludes that things will never be the same.

Jayden's negative automatic thoughts appear in the classroom ("I'm dumb"), at home ("I'm bad at everything"), when among his peers ("[they] don't care about me anymore" and "I'm not good enough"), and in relationship to his symptoms, which "get in the way of everything." Jayden holds absolutistic beliefs about success (*there is no point in trying if it does not lead to success*). These thoughts extend into the larger world and future, as he believes the world, permanently damaged, "will never be the same" and concludes "what's the point of trying."

4.5.9 Psychological Assets

Despite his distressing psychological, systemic, and contextual experience, Jayden owns numerous psychological assets. He is creative, enjoys building things, possesses a good sense of humor, is an eager learner, and formed friendships despite social marginalization. Additionally, his parents provide affection and academic support. Both of his parents are active in his medical care and management.

4.6 Case Conceptualization

Jayden is presenting with significant distress associated with causally interactive symptoms in physiological, emotional, cognitive, behavioral, and interpersonal domains. His history indicates several complicating and contributing factors to his psychological difficulties including family members carrying a PTSD diagnosis, ongoing parental conflict, financial difficulties, reaction to in-person schooling, increased academic pressures, and experiences of racially based discrimination at school. Jayden's emotional life is punctuated by significant depression, hopelessness, and secondary irritability which serves as compensatory strategy for perceived powerlessness.

Jayden's depression and hopelessness are accompanied by a surplus of depressogenic cognitions. He suffers from harsh self-recriminations (*I'm bad, I'm dumb, I'm a horrible kid, I'm no good*), negative view of others/experiences (*people don't like me, others are unsafe and unforgiving, I'm always failing at school*), and a pessimistic view of the future (*things won't change, I am helpless*). Moreover, he has worries about loss of control (*I'm not in charge of anything—not even my pain*). He dreads disappointments and suffers from contingent self-worth (*if I disappoint others, I am not worthwhile and unlikeable*). Jayden is also prone to isolation and alienation-mediated perceptions (*I don't belong, I don't fit in, others dislike and bully me*). Based on all the above data, Jayden's case formulation takes the form of "***I am unwanted, helpless, unlovable, and out of place in a world that is unpredictable and uncaring where people are unforgiving, racist, mean, and see difference as bad.***"

Jayden's default behavioral strategy is avoidance. Jayden tends to take a passive approach to coping with his stressors which perpetuate his perceived helplessness. In the following paragraphs, we delineate three representative scenarios reflecting Jayden's safety behaviors. The first scenario involves classroom behavior.

Jayden's reaction in class is to "give up" when he believes he is not doing well. Historical antecedents to this behavior pattern are likely past academic struggles following bullying in fourth grade. Recent precipitants include unwanted negative attention in class and regular reminders to stay on task. Jayden interprets these instructions as his teacher believing he is "dumb," which is coupled with sadness and frustration. His behavioral response is to reduce effort and engage in painful self-recrimination. Consequently, Jayden's academic performance continues to decline. Thus, he is left with a sense he does not measure up and is resigned to a state of failure.

The second example surrounds Jayden's functional abdominal pain. Jayden's historical diagnosis of FAP reflects somewhat unpredictable pain without an organic etiology which heightens a sense of uncertainty about his symptoms and may contribute to a sense of helplessness. Current precipitants include increased frequency and severity of pain at recess and in physical education class. Jayden interprets his pain signals as alarms indicating that he will not be able to engage in activities without the pain worsening. Thus, Jayden's behavioral response is to withdraw from physical and interpersonal activities. While Jayden may be negatively reinforced by temporary alleviation of physical pain, this temporary relief is emotionally costly. Unfortunately, withdrawing from activities positively reinforces Jayden's belief that his symptoms "get in the way of everything" and limit opportunities for social engagement necessary for maintaining friendships.

Jayden's painful physical and psychological symptoms interact with and maintain depressogenic thought content not only at school but also at home. When Jayden experiences challenges with homework or painful symptoms at home (precipitants), his behavioral reaction is to withdraw from the situation. This response is maintained by negative reinforcement in the short term, as Jayden's comments about not being able to complete homework or go to school are periodically met with family accommodation and decreased demands. However, his behavior perpetuates the avoidance of challenges, which undermines self-efficacy and task approach skills. Consequently, Jayden feels increasingly hopeless, experiences more chronic negative affect, and becomes more convinced of his depressogenic beliefs.

4.7 Treatment Planning

During the initial phase of treatment, psychoeducation is recommended to explain the diagnosis of depression and orient Jayden and his parents to the cognitive-behavioral model of care. This model will be utilized to collaboratively identify and prioritize treatment goals. At present, passive suicidal ideation and hopelessness are the biggest threats to Jayden's immediate safety and most distressing to him and his family. Providing the family with crisis line information and developing a safety plan will help Jayden and his parents identify individual, familial, school, and community support outside of sessions. The cognitive behavioral conceptualization can help inspire hope and support Jayden and his family in understanding how suicidal ideation can arise as ATs. Such ATs serve the function of trying to problem-solve an "escape" from painful feelings, both emotional and physical. In the first visit, the cognitive behavioral triangle might be introduced in a manner that encompasses both feelings of depression and physical sensations of pain. Learning how Jayden's depressive symptoms and abdominal pain intersect in his daily life can help him and his family recognize that while his symptoms may currently impact his thoughts and feelings, they have agency.

Hopelessness and suicidal ideation can be targeted with behavioral activation and monitored using brief age-appropriate measures such as the Children's Depression Inventory (CDI-2). Implementing behavioral activation to increase opportunities for mood uplifts and more hopeful thinking is likely a good initial strategy. Working with Jayden and his family to select realistic activities that help facilitate positive affect and a sense of mastery may chip away at a sense of hopelessness and helplessness. For example, Jayden might try petting his dog, playing a game, and completing one homework assignment. The activities selected should be consistent with Jayden's personal goals of feeling better and doing well in school. The concept of activity pacing is key in the context of Jayden's FAP: selecting smaller activities for immediate goals and working up to larger activities, such as participating in physical activities. Once the suicidal ideation and hopelessness are quieted down, helping Jayden build a reasons-to-live inventory and leverage his personal values/strengths to temper his distress is a central clinical objective.

During this next phase of treatment, the case formulation guides the clinician when deploying cognitive restructuring procedures. An initial goal might be supporting Jayden in identifying and labeling his feelings as well as behavioral reactions. Completing a thought diary can help him begin to become aware of depressive thought processes as well as the accompanying emotions. Encouraging Jayden to identify alternative ways of questioning and responding his self-limiting thoughts is a logical next step. For example, Jayden might join the clinician as a "thought detective" to collaboratively and curiously explore how his thoughts, feelings, and behaviors are connected. The clinician and he could work together to begin interrogating his cognitive biases through Socratic dialogues; re-engineering his "all-or-nothing thinking" and other unhelpful thought patterns can facilitate broader perspective-taking and cognitive flexibility. Additionally, added sense of control and better self-efficacy are other benevolent potential effects. Perhaps, the Private I technique [2] or Fact Hunter [6] could be employed.

Jayden's conditional assumptions and beliefs can also be questioned and challenged. For example, the conditional assumption *if I do not do as well as people expect, then I am no good* is an example of contingent self-worth. Applying the Cut the Knot [37] procedure which works to loosen the chains of contingent self-worth seems like a good strategy for Jayden.

Jayden's self-deprecating thoughts, such as "I am bad," are likely supported by systemic racism and coded messages in his community. Throughout the process, it will be vital for the clinician to respond effectively to issues related to minority discrimination as well as clarify distinctions between externally experienced racism and internalized racism. As previously noted, CBT interventions are well equipped to address these complexities due to the flexible nature of the approach [2, 38]. Using the decision tree outlined by Park and colleagues [38], the clinician deliberately considered why, when, and how to modify the conventional approach to appreciate Jayden's unique circumstances. Accordingly, based on Jayden's past and current lived experiences, the clinician and Jayden could engage in a collaborative guided discovery process marked by genuine empathy, Socratic dialogues, and productive problem-solving. Socratic dialogues aimed at reducing internalized racism and clinical sequelae of the racial bullying are likely reasonable approaches. Therefore, implementing reattribution designed to mitigate self-disparaging biases such as "I'm less than 'cause I'm Black" and "Black people don't count" seems clinically indicated. Additionally, delivering a child-friendly reattribution procedure such as the Responsibility Pizza [2] is a reasonable option. Finally, objective stressors such as the presence of racial bullying and social marginalization require problem-solving. Accordingly, spaces within neighborhoods, at school, and in the broader community were identified via problem-solving. Mining social support with like-minded friends, teachers, and administrators is yet another goal for treatment.

After successful cognitive restructuring and rational analysis, behavioral experiments may be introduced. For example, if medically indicated, a doctor's note and school accommodation allow Jayden to act as an "assistant coach" or scorekeeper during this time rather than having him participate in overly strenuous exercise or sit out. This flexibility helps restructure Jayden's environment to create roles in which he can engage and reduces risk of increased isolation and withdrawal from activities.

4.8 Conclusion

> Truth is ever to be found in simplicity, and not in the multiplicity and confusion of things. (Isaac Newton)

Case conceptualization is indispensable in CBT. Unfortunately, there is no gene for case conceptualization [39]. Therefore, clinicians require good training and supervision in order to develop competent formulation skills. One of the favorite things I (RDF) heard from Dr. Aaron T. Beck early in my training as a cognitive therapist was him saying, "there is an infinite amount of information to learn about a patient but a finite amount of time to intervene." As clinicians we need to discern the most relevant information for treating patients (e.g., signals) from interesting but peripheral data (noise). Differentiating signal from noise is fundamental to the pattern recognition process explained earlier in the chapter. Nonetheless, simplicity and efficiency should not be confused with reductionism which attenuates full understanding. Therefore, theoretical and empirically based training which alerts practitioners to the most salient information to glean from patients and caregivers is pivotal. Finally, a direct link from conceptualization to intervention is essential. Case conceptualizations that lie inertly divorced from clinical delivery of CBT procedures are merely intellectual exercises. The gold standard for good case formulation is that it leads to improved clinical outcomes and young patients living emotionally healthier and productive lives. Indeed, they are maps that ideally propel the routes of progress.

Conflicts of Interest

Ciera Korte has no conflicts of interest to declare that are relevant to the content of this chapter.

Robert D. Friedberg *has received speaker and consultant honoraria from* the Beck Institute for Cognitive and Behavioral Therapy and Research, Kinark Family and Youth Services, and the Rockland County Department of Mental Health. He receives royalties from Springer, Guilford, John Wiley, Routledge, and Professional Resource Press.

References

1. Friedberg RD (2015) Where's the beef: concrete elements in supervision with CBT with youth. J Am Acad Child Adolesc Psychiatr 54: 527–531
2. Friedberg RD, McClure JM (2015) Clinical practice of cognitive therapy with children and adolescents: the nuts and bolts. Guilford Press, New York
3. Manassis K (2014) Case formulation with children and adolescents. Guilford Press, New York
4. Kuyken W, Padesky CA, Dudley R (2009) Collaborative case conceptualization: working effectively with clients in cognitive behavioral therapy. Guilford Press, New York
5. Kendall PC, Chu B, Gifford A et al (1998) Breathing life into a manual. Cogn Behav Pract 5:89–104
6. Friedberg RD, Gorman AA, Wilt LH et al (2011) Cognitive behavior therapy for busy child psychiatrists and other mental health professionals. Routledge Publications, New York
7. Chorpita BF, Bernstein A, Daleiden EL (2011) Empirically guided coordination of multiple evidence-based treatments: an illustration of relevance mapping in children's mental health services. J Consult Clin Psychol 79:470–480
8. Friedberg RD, Fidaleo RA (1992) Training hospital staff in cognitive therapy. J Cogn Psychother 6:105–112
9. Friedberg RD, Taylor LA (1994) Perspectives on cognitive therapy supervision. J Rat-Emot Cogn Beh Ther 12:147–161
10. Shirk SR (1999) Integrated child psychotherapy: treatment ingredients in search of a recipe. In: Russ SW, Ollendick TH (eds) Handbook of psychotherapies with children and families. Springer, New York, pp 369–384
11. Freeman A, Felgoise SH, Davis D (2008) Clinical psychology: integrating science and practice. Wiley, New York
12. Watson JB (1924) Behaviorism. People's Institute, New York
13. Skinner BF (1966) Behavior of organisms. Prentice-Hall, New York
14. Bandura A (1977) Social learning theory. Prentice-Hall, New York
15. Beck AT, Clark D (1988) Anxiety and depression: an information processing perspective. Anxiety Res 1:23–33
16. Ingram R, Kendall PC (1986) Cognitive clinical psychology: implications of an information processing perspective. In: Ingram RE (ed) Information processing approaches to clinical psychology. Academic Press, Orlando, pp 3–21
17. Padesky CA (1994) Schema change processes in cognitive behavior therapy. Clin Psychol Psychother 31:267–278

18. Guidano VF, Liotti G (1983) Cognitive processes and emotional disorders: a structural approach to psychotherapy. Guilford Publications, New York
19. Guidano VF, Liotti G (1985) A constructionist foundation for cognitive therapy. In: Mahoney MJ, Freeman A (eds) Cognition and psychotherapy. Plenum, New York, pp 101–142
20. Kagan J (1986) Rates of change in psychological processes. J Appl Dev Psychol 7:125–130
21. Hammen C, Zupan BA (1984) Self-schemas, depression, and the processing of personal information in children. J Exp Chld Psychol 37:598–608
22. Alford BA, Beck AT (1997) The integrative power of cognitive therapy. Guilford Press: New York
23. Beck AT (1976) Cognitive therapy and the emotional disorders. International Universities Press, New York
24. Jolly J (1993) A multi-method test of the cognitive content-specificity hypothesis in young adolescents. J Anxiety Dis 7:223–233
25. Laurent J, Stark KD (1993) Testing the cognitive content-specificity hypothesis with anxious and depressed youngsters. J Abnorm Psychol 102:226–237
26. McClure JM, Friedberg RD, Thordarson MA et al (2019) CBT express. Guilford Publications, New York
27. Huey SJ, Park AL, Galan CA et al (2023) Culturally-responsive cognitive-behavioral therapy for ethnically diverse populations. Ann Rev Clin Psychol 19:51–78
28. Iwamasa GY, Hays PA (2019) Culturally responsive cognitive behavior therapy: practice and supervision, 2nd edn. American Psychological Association, Washington, DC
29. Park AL, Rith-Najarian LR, Saifan D et al (2023) Strategies for incorporating culture into psychological interventions for youth of color. Evid Based Pract Child Adolesc Ment Health 8:181–193
30. Steele JM, Moore BA (2024) Racism and African American mental health: using cognitive behavior therapy to empower healing. Routledge, New York
31. Beck JS (2021) Cognitive behavioral therapy: the basics and beyond, 3rd edn. Guilford Publications, New York
32. Persons JB (2012) Case-formulation approach to cognitive-behavior therapy. Guilford Publications, New York
33. Kovacs M (2012) Children's depression inventory, 2nd edn. APS, New York. TM [Dataset]
34. Birmaher B, Khetarpal S, Brent D et al (1997) The screen for child anxiety related emotional disorders (SCARED): scale construction and psychometric characteristics. J Am Acad Child Adolesc Psychiatry 36:545–553
35. Walker LS, Greene JW (1991) The functional disability inventory: measuring a neglected dimension of child health status. J Pediatr Psychol 16:39–58
36. Walker LS, Smith CA, Garber J et al (1997) Development and validation of the abdominal pain index using a revised scoring method. Psychol Assess 9:392–405
37. Friedberg RD, McClure JM, Hillwig-Garcia J (2009) Cognitive therapy techniques for children and adolescents: tools for enhancing practice. Guilford Publications, New York
38. Park AL, Velez CV, Kim RE (in press) Cognitive behavioral therapy with diverse youth. In: Friedberg RD, Thordarson MA (eds) Cognitive behavioral therapy with youth: tradition and innovation, 2nd edn
39. Padesky CA (2007) The next frontier: building positive qualities with cognitive behavior therapy. Invited address presented at the 5th World congress of behavioral and cognitive therapies, Barcelona, Spain, July

Chapter 3

Therapeutic Alliance in Cognitive Behavioral Therapy for Youth

Deepika Bose and Jeremy W. Pettit

Abstract

Therapeutic alliance plays a significant role in youth cognitive behavioral therapy (CBT) by enhancing therapy participation and engagement. Alliance research has demonstrated a predictive effect of alliance on youth therapy outcomes across a range of diagnoses, with the strength of association varying by problem type and treatment setting. The tripartite model of alliance, emphasizing the emotional bond, collaborative goal setting, and task agreement, serves as a useful framework for understanding and fostering a strong alliance in youth CBT. In this chapter, we highlight key strategies for enhancing each of the three alliance-building components. Additionally, we discuss strategies for fostering a strong parent-provider alliance, given that parents play a key role in youth CBT. Employing the strategies discussed in this chapter fosters alliance-building, which in turn can enhance engagement, improve outcomes, and strengthen the overall effectiveness of youth CBT interventions.

Key words Therapeutic alliance, Child/adolescent, Cognitive behavioral therapy, Parent alliance, Youth alliance, Alliance-building strategies

1 Introduction and Overview

Due to its foundation in behaviorism and cognitive science, as well as the development of manualized protocols focusing on specific techniques, cognitive behavioral therapy (CBT) has at times faced criticism for its perceived emphasis on technical aspects of treatment at the expense of the provider-patient relationship, a phenomenon Mahoney [1] termed "technolatry." In recent decades, theoretical advancements and empirical research have resoundingly responded to such criticism. Therapeutic alliance is now recognized as an integral component for change in CBT for a range of psychological disorders [2, 3]. In the youth literature, therapeutic alliance is commonly defined using Bordin's [4] tripartite model of alliance, which proposes that alliance is comprised of (a) the development of an emotional *bond* between provider and patient, (b) collaborative agreement on intervention *goals*, and

Robert D. Friedberg and Micaela A. Thordarson (eds.), *Cognitive Behavioral Therapy with Youth: Tradition and Innovation*, Neuromethods, vol. 231, https://doi.org/10.1007/978-1-0716-5134-6_3,

(c) collaborative agreement on therapeutic *tasks* or activities. A strong alliance is considered to be one in which patients perceive a trusting and supportive relationship with their provider, mutual agreement on intervention goals, and mutual agreement on therapeutic activities that will be conducted to achieve the agreed-upon goals.

Therapeutic alliance is believed to enhance youth CBT outcomes in part by increasing engagement and participation in CBT activities [5]. More specifically, alliance is theorized to increase treatment expectancy (the belief that therapy will lead to desired outcomes), motivation for change, session attendance, and active participation in and completion of assigned therapy tasks [5–7]. A strong alliance is particularly important for youth therapy, given that youth typically do not self-refer to treatment. Instead, caregivers or other adult figures decide that children require treatment and subsequently bring them to therapy [8]. As a result, providers may be confronted by youths who do not identify a problem or need for help, and may subsequently experience lower therapy engagement (e.g., decreased motivation to change and/or participate in therapy tasks). Further, a strong alliance may be especially vital in the context of CBTs. This is because behavioral assignments in CBTs often involve putting youth in situations that they actively avoid (e.g., exposures for anxiety), have low motivation to complete (e.g., pleasant activity scheduling for depression), and/or lead to temporary distress (e.g., response prevention for obsessive-compulsive disorder, increasing food consumption while refraining from compensatory behaviors in eating disorders). A strong, positive alliance may promote engagement and participation in these assigned and challenging tasks, which in turn improves outcomes [9, 12].

In the proceeding sections of this chapter, we summarize research on therapeutic alliance and youth CBT outcomes and discuss the current state and future directions of youth alliance research. Then, we describe specific alliance-building strategies for providers to consider in the context of youth CBTs. The specific alliance-building strategies described are organized by the three components of alliance described above (bond, goals, tasks) and include considerations for the parent alliance in youth CBT.

2 Research Summary

Therapeutic alliance has been shown to predict positive CBT outcomes across a range of psychological disorders in youth including anxiety, obsessive-compulsive disorder (OCD), depression, eating disorders, and post-traumatic stress disorder. Alliance is typically measured using youth, parent, provider, and/or observer ratings.

Common youth alliance measures used in research on youth CBT include the Therapeutic Alliance Scales for Children [11], Working Alliance Inventory [12], and Therapy Process Observation Coding System-Alliance scale [13].

Systematic reviews and meta-analyses have documented that the overall effect of alliance on youth psychotherapy outcomes across research studies is small and positive [5, 14], though individual studies demonstrate mixed findings. These mixed findings can be explained in part by differences between studies in sample composition (especially youth problem type), measurement approach, and intervention characteristics and settings. For example, a meta-analysis of alliance-outcome associations in youth psychotherapy found that the predictive effect of alliance on youth outcomes was stronger for internalizing disorders vs. substance use and eating disorders, outpatient vs. inpatient treatment, and behavioral vs. mixed behavioral-nonbehavioral treatments [5]. Another meta-analysis examining alliance-outcome associations in psychotherapy for youth anxiety, depression, and OCD found that the strength of the alliance-outcome effect was stronger in treatment for youth OCD, when alliance was measured between sessions 4 and 6, and in studies based in the United States [10].

Given that studies have found differences in the strength of alliance-outcome associations across studies, researchers have started to move away from asking "Does alliance contribute to outcomes?" to instead asking "*For whom* and *under what conditions* does alliance contribute most strongly to outcomes?" Answers to the latter question align with calls for precision treatment [15] and provide insight into how to tailor alliance-building strategies based on child characteristics, intervention characteristics and settings, and other factors that may influence outcomes.

In addition to determining *for whom* and *under what conditions* alliance contributes to outcomes, researchers have also begun to explore questions related to *how* alliance contributes to outcomes. Although there is a strong theoretical basis for how alliance promotes positive outcomes in youth CBT (i.e., by increasing youth engagement and participation in therapy), there is little empirical research to support these theories. Most studies that have examined the mechanisms by which alliance contributes to outcomes have failed to find support for theorized mechanisms (e.g., [16–19]), likely due to small sample sizes and suboptimal alliance or outcome measurements [5]. One recent study that examined therapeutic alliance, emotion regulation, and depression outcomes in internet-based therapy for adolescent depression found that the relationship between youth alliance and depression reductions may be partially explained by changes in emotion regulation [20]. In other words, higher alliance predicted increased emotion regulation in adolescents, which in turn predicted lower depression. This promising finding suggests that a strong alliance may help teens feel more regulated and comfortable within the therapeutic environment, which can then allow them to engage in therapeutic tasks and activities required for positive change [20].

Researchers are also increasingly recognizing therapeutic alliance in youth psychotherapy as a dynamic construct that may be characterized by shifts, including ruptures and repairs [21]. A therapeutic "rupture" occurs when there is tension in the alliance that may be due to strains in the emotional bond, disagreement on goals, and/or perceived lack of collaboration on tasks [22]. In contrast to the adult literature, there is scant empirical research on how to repair ruptures in youth psychotherapy. Nof and colleagues [23] outline a comprehensive approach for identifying and repairing ruptures in youth therapy, based on the work by Safran and Muran [24] in adults. They propose first recognizing when a rupture occurs, which serves as a prompt both to engage in self-regulation and reflect on the child's underlying communication message (e.g., Is the child trying to communicate that they are not ready to speak about sensitive topics?). Next, they advocate for openly addressing the rupture with the child using descriptive, objective language devoid of causal inferences (e.g., "I noticed you stopped responding to me and turned your body away"). Further, they stress the significance of taking accountability for the providers' role in the rupture, whether it involves acknowledging unhelpful behavior or misunderstanding a situation, while expressing appreciation for the child's efforts to communicate their discomfort. Finally, they recommend resolving the rupture via change strategies such as taking breaks to play games if the rupture stems from excessive cognitive or emotional demands during a session, or developing a narrative story with fictional characters to illustrate the process of rupture and subsequent repair.

Another topic that has garnered attention in youth alliance research is the examination of youth characteristics and provider behaviors that promote a strong alliance. With respect to youth characteristics, studies have shown that initial symptom severity, perceived security in caregiver relationships, socioemotional functioning, and perceived social support predict alliance [25–27]. Provider behaviors including support, collaboration, exploration of youth's subjective experiences, and responsiveness positively predict alliance, whereas behaviors such as pushing the youth to talk, being overly formal, failing to acknowledge youth's emotions, criticizing, and excessive recall of information negatively predict alliance [18, 28, 29]. In the sections below, we provide detailed and practical alliance-building strategies for providers to consider in youth CBT. The strategies described below are based on empirical findings and recommendations by experts in the field.

3 Alliance-Building Strategies

3.1 Developing an Emotional Bond

The emotional or affective bond refers to feelings of trust, comfort, support, and dependability within the therapeutic relationship. When the therapeutic bond is strong, children and adolescents are more likely to experience positive feelings toward their provider and may demonstrate increased willingness to disclose personal and vulnerable information. A strong therapeutic bond also fosters feelings of safety and security within the therapeutic environment. Below are specific therapeutic strategies that support the development and maintenance of a strong therapeutic bond in youth CBT.

3.1.1 Establish Confidentiality

For children and adolescents to develop a trusting relationship with their provider, it is vital that they understand what will be kept confidential in sessions. In fact, research has demonstrated that establishing trust and confidentiality may be one of the most important factors in promoting adolescent help-seeking behaviors [30]. Establishing confidentiality involves providing a clear explanation of what will and will not be shared with parents and/or outside professionals. In addition, it is also helpful to also explain the purpose of notetaking during sessions (if applicable), who may access the child's therapy records, and the conditions under which therapy records may be accessed by others.

3.1.2 Engage in Appropriate Levels of Empathy, Warmth, Concern, and Genuineness at the Outset of Therapy

While interacting with children and families, it is important for providers to convey genuine empathy and warmth while expressing appropriate concern for the issues that are brought up in the session [31]. Providers should consider how they are attending to each of the aforementioned variables starting from their first interaction with the child or family and throughout the course of therapy. For example, during the intake or initial evaluation process, is the provider seemingly reading off a diagnostic checklist with little concern, empathy, or interest in the child's or family's plight? Or does the provider actively listen to the child's and parent's report of symptoms and provide appropriate statements and validation to convey understanding and appropriate concern?

Providers should keep in mind that children and adolescents meeting a new provider for the first time may feel hesitant to answer specific questions or share personal information outright. In such scenarios, it is important for providers to avoid pushing the child to talk, as doing so can have a negative impact on alliance formation [28]. Instead, providers should allow children and adolescents the opportunity to skip questions or come back to them later if they prefer. For younger children who may hesitate to speak due to shyness/anxiety, providers may allow different options to respond such as nonverbal signals (head nods, pointing to "yes" or "no") or allowing their parents to respond to questions at first. As children and adolescents increase their comfortability with providers over time, they may be encouraged to gradually respond more frequently and independently.

3.1.3 Assess and Address Cultural and Diversity Factors

Another important consideration for developing a strong therapeutic bond is the understanding and assessment of how cultural and diversity factors may impact therapy processes or outcomes [32, 33]. During the initial evaluation phase and as therapy progresses, providers are encouraged to assess and acknowledge families' values and cultural beliefs to better understand the family's preferences and expectations. For example, families from varying backgrounds may have different expectations about the degree of formality in the therapeutic relationship or expectations around treatment outcomes [32]. Examples of cultural-based assessment questions are provided in the DSM-5 Cultural Formulation Interview (CFI; [34, 35]), which assesses how cultural factors influence patients' perspectives on their mental health experiences (e.g., causes, symptoms, supports, past and current help-seeking). The CFI is available online at www.psychiatry.org/dsm5.

Altogether, assessing and addressing therapy expectations and their cultural underpinnings can help families experience increased comfortability and security within the therapeutic relationship. Indeed, research has demonstrated that assessment of cultural values and preferences leads patients to feel better understood by their providers [33], thereby enhancing the patient-provider affective bond.

3.1.4 Actively Listen and Remember

Throughout therapy, providers should continuously engage in active listening and remembering. Active listening encompasses both nonverbal and verbal signals of attention. Nonverbal signals of attention include making appropriate eye contact, warm or neutral facial expressions (e.g., relaxed face, smile, or half smile), and head nods to convey agreement or understanding. Verbal signals of active listening include reflective statements (repetition or paraphrasing of what the child or family has said to clarify or convey understanding) and asking relevant and appropriate follow-up questions that convey curiosity and interest in the child's or family's subjective experiences. It is also important to remember information that children and families share during CBT sessions. For example, if a child communicates an interest in a particular hobby or TV show, remembering that information and then incorporating it into later examples during CBT activities can serve to increase the therapeutic bond. In this scenario, the child learns that the provider is listening and perceives that the provider cares and is knowledgeable about their hobbies and interests. Contrarily, providers forgetting facts and details that children and families have shared can negatively impact the therapeutic bond. Taking notes during or immediately after sessions can help busy providers remember details about their patients' hobbies, interests, and other information shared during sessions.

3.1.5 Provide Validation

Providing validating statements to children and families can positively impact the therapeutic bond [36]. Validation refers to the communication that another person's emotions, thoughts, or actions make sense in a particular context [37]. While providing validation, it is important to acknowledge the youth's emotions (e.g., "It makes sense that you feel…") while avoiding critical statements [5].

When encouraging children and adolescents to engage in changing thoughts and behaviors, CBT providers may inadvertently invalidate their patients' experiences. For example, a teen with social anxiety expresses a worry such as "I will make a fool of myself if I talk." If the provider immediately tells the teen that they are engaging in a thinking "error" (i.e., communicating that the teen's thought is flawed or unlikely to be untrue) *without* conveying understanding for the initial thought or providing appropriate context as to *why* the thought is considered an "error," the teen may be left feeling invalidated and as if their automatic thinking patterns are "wrong" or "bad." The following example illustrates a situation in which a teen feels invalidated by his provider:

Teen: *I didn't end up volunteering for the presentation because I kept thinking I would mess up and embarrass myself.*

Provider: *But last week we talked about how that's not true. That's "fortune telling," one of the thinking errors we talked about, remember? You can't predict with certainty that you'll mess up or embarrass yourself.*

Teen: *Yeah, I know you said that, but last time I did a presentation, it did go horribly. I told you—I froze and other kids were laughing. How can I just tell myself that won't happen again? I believe I will mess up; I can't change that. Maybe I'll just always have "errors" in my thinking. I can't do anything right.*

The next example illustrates how a provider may be validating when faced with the same situation:

Teen: *I didn't end up volunteering for the presentation because I kept thinking I would mess up and embarrass myself.*

Provider: *Yeah, I understand why you were thinking that. It makes sense to have those thoughts, especially because I remember what you told me about what happened the last time you presented. This makes me think about what we talked about last week—how the thoughts we tell ourselves can impact how we feel and what we do. Remember the thinking traps?*

Teen: *Yeah, I do. I realized that after. I know I was "fortune telling." I tried to tell myself that just because I messed up last time, that doesn't mean I'll mess up again. But at that point, I was too anxious already.*

Provider: *That's great you were thinking about the thinking traps. And it sounds like you made a real effort to change your thinking in the moment. But like you pointed out, your anxiety was pretty high by that point. Makes me wonder, the next opportunity you get to present, how can you remind yourself of the thinking traps* before *your anxiety gets too high?*

In the scenario above, the provider and teen can then collaboratively problem-solve how to set the teen up for increased chances of success next time, without making the teen feel like he was a failure.

3.1.6 Engage in Developmentally Appropriate Tasks and/or Play Activities

When engaging children and adolescents in CBT, it is vital to tailor CBT activities to the child's developmental level [38]. Providers should consider factors such as language, games/activities, materials, and pace of treatment [39]. Appropriate developmental tailoring of CBT activities can facilitate the therapeutic bond by making the therapeutic process more fun and enjoyable. In fact, research has demonstrated that behaving in an overly formal manner with children and adolescents negatively impacts therapeutic alliance [28]. Additionally, developmental tailoring increases engagement with CBT material, thereby increasing CBT knowledge and skill acquisition [38].

When working with younger children, providers should consider incorporating interactive games and other play-based activities to facilitate bonding and engagement. Play-based activities may be in the form of art (e.g., drawing and coloring), games (e.g., board games, card games), toys (e.g., blocks, Legos, dolls), and/or sports [40]. Providers can incorporate dedicated time (e.g., 5–10 min) for play-based activities at the beginning of sessions to increase a child's comfort with the therapeutic environment, and/or at the end of sessions as a reward for therapy participation and to end the session on a positive note. In addition, play activities may also be used to teach and consolidate learning of CBT-specific concepts (40 for additional information and examples). Children can also benefit from consistent labeled praise and rewards (e.g., stickers, candy, small toys) for therapy participation, particularly in the context of challenging behavioral tasks (e.g., exposures) in CBT [32].

When providing services to younger children, it is also helpful to use clear, simplified language and to provide concrete examples using basic terms to communicate ideas and information. Visually appealing handouts and the use of mnemonics, such as the FEAR acronym (Feeling frightened? Expecting bad things to happen? Actions and Attitudes that can help? Results and Rewards) used in the *Coping Cat* manual to describe anxiety recognition and subsequent skill use [41], can also be helpful for engaging younger children.

In contrast to younger children, mature or older adolescents may find simplified, concrete language to be "childish" or condescending [38]. Adolescents may benefit from more detailed explanations regarding the treatment rationale, with the support of metaphors or examples that are relevant to their own interests. When working with adolescents, it may be helpful to assess the types of activities they like to engage in and tailor CBT activities accordingly. Further, instead of dedicated play time before and after sessions, adolescents may prefer engaging in conversational topics that align with their values and interests. When providers notice they have shared interests or common ground with their adolescent patients, they should be careful not to overemphasize the similarities in a manner that comes off as inauthentic, as this negatively predicts alliance [5, 28].

3.2 Agreement on Goals

Goal agreement refers to both short- and long-term goals in therapy. Youth often present with multiple problems and/or diagnoses [42], so it is important to ensure that youth, parents, and providers are all in agreement on the specific goals of therapy.

For example, a 10-year-old girl named Layla presents with her mother, Susie, to a mental health clinic. Layla is diagnosed with co-occurring separation anxiety and social anxiety. Layla indicates a preference to address her social anxiety, given that she is missing out on social opportunities and feels motivated to make new friends. By contrast, Susie prefers that Layla's separation anxiety be addressed initially, given that Susie frequently engages in accommodating behaviors (e.g., picking up Layla early from school, excessive reassurance) that are causing significant disruptions in Susie's life. In such a scenario, it will be important for the provider to meet with Layla and Susie together and make a collaborative decision about the optimal course of treatment. The collaborative discussion may include a review of Layla's symptoms and their associated impairment. If Layla's social anxiety is causing more functional impairment than her separation anxiety (e.g., social anxiety is significantly impairing her across multiple aspects of life while separation anxiety is not), then perhaps Susie will agree to start by addressing Layla's social anxiety. During this discussion, it will also be helpful for the provider to explain how CBT skills may apply transdiagnostically. For example, cognitive restructuring and relaxation skills may be used and applied for both diagnoses simultaneously. And with respect to exposure hierarchies, Layla may agree to completing and implementing a social anxiety hierarchy first, and then move on to a separation anxiety hierarchy. Regardless of what the ultimate agreed-upon decision is, it is important that the provider communicates each goal *clearly* and *effectively* so all parties understand and agree with the goals.

The goal-setting stage of CBT is also an opportunity for providers to instill hope and optimism for positive change, an important component of alliance-building [31, 32]. Children and families may initiate CBT services with worries, concerns, or negative and unrealistic expectations about CBT [43]. Neglecting to assess and address these concerns could negatively impact therapeutic alliance and engagement. Therefore, providers should take time to understand any negative or unrealistic treatment expectancies that families may hold, and provide realistic expectations about therapeutic goals and their proposed timeline to completion. Examples of negative or unrealistic expectations include beliefs that CBT tasks may be unhelpful or harmful, concerns about the time-limited nature of CBT, and expectations that CBT will permanently and completely eliminate symptoms. In such scenarios, it will be helpful to provide detailed psychoeducation on the theory, rationale, and evidence in support of CBT while setting realistic expectations including explanations about how behavioral tasks in CBT (delivered correctly) are not harmful, explaining why CBT is time-limited and how the specific number of weeks a child is enrolled in therapy may vary from case to case, and reframing treatment goals to tolerating or managing symptoms while decreasing impairment (vs. complete symptom elimination).

3.3 Task Collaboration

Task collaboration refers to mutual and collaborative agreement between patients and providers on the specific roles and activities expected of both the provider and youth/family. In the context of youth CBT, discussion of role assignments may involve an explanation that the provider will function as a "coach" that will teach and model CBT skills for the child, while the child is expected to actively participate in the therapeutic activities and complete weekly homework assignments. For older children, it may also be helpful to frame CBT work as a collaborative partnership in which the provider is the expert in CBT while the child is the expert on themselves, and the plan is to work collaboratively to develop and personalize their treatment plan.

Task collaboration is particularly important in CBTs, given that behavioral assignments involve active, consistent participation and may at times be challenging for children and adolescents. For example, a young child with a phobia of dogs may initially balk at the idea of approaching and petting a Great Dane as part of his therapy assignments. Similarly, a teen with an eating disorder may initially refuse to increase their food consumption without engaging in compensatory behaviors. In these situations, it is vital that providers provide a detailed rationale that clearly links the purpose of the therapeutic tasks to the youth's broader intervention goals [32]. Using the examples above, a provider may explain the rationale for exposures and state that the child's ability to tolerate petting dogs can allow him to go to friends' houses who have dogs and public spaces he is missing out on because of the prospect of running into a dog. For the teen with the eating disorder, the provider can explain how increased food consumption without compensatory behaviors will allow the teen to safely engage in activities she enjoys (e.g., playing sports) and will prevent medical complications and hospitalizations.

While providing psychoeducation and tying tasks to goals, it may also be helpful to build in motivational enhancement activities (e.g., decisional balance exercises) and/or other activities to build self-efficacy. Children and adolescents also benefit from repetition, so providers should not hesitate to repeat their explanations about the theory and rationale for specific therapeutic tasks as often as needed.

While adopting a collaborative stance in CBT, it can be helpful to allow children and adolescents reasonable control over their therapy assignments. For example, the child with the phobia of dogs can help set the pace of treatment by providing suggestions about the intermediate exposure tasks that she will engage in as she works up to petting a Great Dane. To prevent alliance ruptures, providers should actively solicit feedback from children and adolescents to ensure that they understand and are willing to engage in the assigned therapeutic activities. At times, providers will need to be responsive and open to other suggestions about therapeutic goals and CBT tasks or plans.

Some providers may feel hesitant to challenge children and adolescents to difficult exposures out of fear of rupturing the alliance [44, 45]. However, research on alliance in exposure-based CBT in youth has demonstrated that exposures do not rupture the alliance, and that child, provider, and parent perspectives of alliance remain high regardless of whether youth practice in-session exposures or not [46]. Given that exposures are considered an "active" ingredient of CBT, we strongly recommend providers continue to deliver exposures when treating youth anxiety and OCD, while keeping the alliance-building strategies outlined above in mind.

4 Parent Alliance

Given that parents typically have some level of involvement in youth CBT, it is also important to consider the provider-parent alliance. A strong provider-parent alliance has been shown to positively influence youth outcomes [13, 47] and has been associated with fewer cancellations, no shows, and greater provider-parent consensus on therapy termination [48].

The role of parents in youth CBT varies by a number of factors, such as the child's developmental age, availability and/or willingness of parents to be involved in treatment, and the treatment type or modality being implemented (e.g., family CBT, individual CBT). The degree of parental involvement can range from not very involved to very involved in youth CBT. For example, some parents may simply consent to treatment (e.g., in school-based interventions) or limit their role to dropping off and picking up their children from appointments. Higher levels of involvement include providing updates and monitoring progress, tracking and overseeing homework completion, facilitating out-of-session behavioral assignments, and joining sessions to learn CBT skills alongside children. In some cases, parents may attend parent-only sessions to learn how to implement skills to manage their child's symptoms or manage how they respond to their child's symptoms [38].

Several alliance-building strategies are recommended when working with parents, largely centered on providing clarity about the process of therapy and establishing agreed-upon boundaries. At the outset of therapy, it is important to clarify parents' interest, willingness, and availability to participate actively in the youth's therapy. Early clarification ensures that parents and providers are aligned and that potential challenges with participation are identified and addressed early on. In addition to establishing parents' ability to participate, it is essential to ensure that parents' potential concerns about their child's treatment are assessed and understood. Providers can facilitate open dialogue with parents using active listening and validation of parents' concerns, along with psychoeducation about therapy, to identify and address potential concerns before therapy begins. After establishing parents' ability to participate and addressing concerns, providers should clearly communicate the expected role of parents in therapy, including the time and commitment that parental participation will require. Role expectations may need to be flexible and adjusted based on parents' willingness and/or ability to participate.

To maximize the alliance with both youth and parents, clear discussion and agreement on boundaries is essential. Providers must engage parents in a discussion about patient confidentiality, specifying what will and will not be shared with parents. This also includes clearly establishing boundaries regarding the frequency and modality of out-of-session contact. It can be helpful to specify the best method to contact the provider (phone, email, etc.), expected response times, instances in which a provider may not respond, and what to do or who to call in the case of emergencies. Finally, as with youth patients, it is important to ensure that parents agree with (and, as appropriate, have a say in) treatment goals, tasks, and timeline for therapy completion.

5 Conclusion

In conclusion, the therapeutic alliance plays a crucial role in youth CBT outcomes, enhancing engagement and participation in treatment activities. A robust alliance in CBT predicts positive outcomes across various psychological disorders, although the strength of this association varies depending on factors such as problem type and treatment setting. Understanding the mechanisms through which alliance contributes to outcomes is an ongoing area of scientific inquiry, with promising findings suggesting a link to improved emotion regulation and symptom reduction.

The tripartite model of alliance, emphasizing emotional bond, collaborative goal setting, and agreement on therapeutic tasks, serves as a framework for understanding and fostering a strong alliance in CBT for youth. Key strategies for enhancing the emotional bond include establishing confidentiality, conveying warmth and empathy, assessing cultural factors, active listening and remembering, providing validation, and developmental tailoring of CBT tasks and activities. Collaborative goal setting ensures alignment among providers, youth, and parents and provides opportunities to address negative treatment expectancies while instilling hope and optimism for change. Task collaboration, characterized by agreement on roles and activities, promotes active engagement in therapeutic tasks that may be challenging in CBT.

Additionally, fostering a strong parent-provider alliance in youth CBT can also facilitate youth engagement and positive outcomes in CBT. Alliance-building strategies with parents involve early clarification of parents' participation, addressing concerns, goal agreement, and task collaboration, and the establishment of role expectations, boundaries, and confidentiality.

By implementing alliance-building strategies, providers can enhance engagement, improve therapeutic outcomes, and promote positive growth in youth undergoing CBT interventions. Moving forward, continued research and implementation of alliance-building strategies hold promise for enhancing the effectiveness of CBT for youth and promoting positive therapeutic outcomes.

Conflicts of Interest

The authors have no conflicts of interest to declare that are relevant to the content of this chapter.

References

1. Mahoney MJ (1991) Human change processes: the scientific foundations of psychotherapy. Basic Books, New York
2. Castonguay LG, Constantino MJ, McAleavey AA et al (2010) The therapeutic alliance in cognitive-behavioral therapy. In: Muran JC, Barber JP (eds) The therapeutic alliance: an evidence-based guide to practice. Guilford, New York
3. Kazantzis N, Dobson KS (2022) Therapeutic relationships in cognitive behavioral therapy: theory and recent research. Psychother Res 32(8):969–971
4. Bordin ES (1979) The generalizability of the psychoanalytic concept of the working alliance. Psychother Theory Res Pract 16(3):252–260
5. Karver MS, DeNadai AS, Monahan M et al (2018) Meta-analysis of the prospective relation between alliance and outcome in child and adolescent psychotherapy. Psychotherapy 55(4):341–355
6. Hawley KM, Garland AF (2008) Working alliance in adolescent outpatient therapy: youth, parent and therapist reports and associations with therapy outcomes. Child Youth Care Forum 37(2):59–74
7. Karver MS, Handelsman JB, Fields S et al (2005) A theoretical model of common process factors in youth and family therapy. Ment Health Serv Res 7(1):35–51
8. Barmish AJ, Kendall PC (2005) Should parents be co-clients in cognitive-behavioral therapy for anxious youth? J Clin Child Adolesc Psychol 34(3):569–581
9. Chu BC, Kendall PC (2004) Positive association of child involvement and treatment outcome within a manual-based cognitive-behavioral treatment for children with anxiety. J Consult Clin Psychol 72(5):821–829
10. Bose D, Proenza DA, Costales G et al (2022) Therapeutic alliance in psychosocial interventions for youth internalizing disorders: a systematic review and preliminary meta-analysis. Clin Psychol Sci Pract 29(2):124–136
11. Shirk SR, Saiz CC (1992) Clinical, empirical, and developmental perspectives on the therapeutic relationship in child psychotherapy. Dev Psychopathol 4(4):713–728
12. Horvath AO, Greenberg LS (1989) Development and validation of the working alliance inventory. J Couns Psychol 36(2):223–233
13. McLeod BD, Weisz JR (2005) The therapy process observational coding system-alliance scale: measure characteristics and prediction of outcome in usual clinical practice. J Consult Clin Psychol 73(2):323–333
14. McLeod BD (2011) Relation of the alliance with outcomes in youth psychotherapy: a meta-analysis. Clin Psychol Rev 31(4):603–616
15. Lebowitz ER, Zilca-Mano S (2022) Not so common anymore? Beyond the common factor understanding of the role of alliance in youth psychotherapy. Clin Psychol Sci Pract 29(2):140–142
16. Anderson RE, Spence SH, Donovan CL et al (2012) Working alliance in online cognitive behavior therapy for anxiety disorders in youth: comparison with clinic delivery and its role in predicting outcome. J Med Internet Res 14(3):e88
17. Bose D, Pettit JW, Silk JS et al (2023) Therapeutic alliance, attendance, and outcomes in youths receiving CBT or client-centered therapy for anxiety. J Clin Child Adolesc Psychol 7:1–11
18. Karver M, Shirk S, Handelsman JB et al (2008) Relationship processes in youth psychotherapy: measuring alliance, alliance-building behaviors, and client involvement. J Emot Behav Disord 16(1):15–28
19. Shirk SR, Gudmundsen G, Kaplinski HC et al (2008) Alliance and outcome in cognitive-behavioral therapy for adolescent depression. J Clin Child Adolesc Psychol 37(3):631–639
20. Lindqvist K, Mechler J, Falkenström F et al (2023) Therapeutic alliance is calming and curing—the interplay between alliance and emotion regulation as predictors of outcome in internet-based treatments for adolescent depression. J Consult Clin Psychol 91(7):426–437
21. Cirasola A, Midgley N, Fonagy P et al (2022) The therapeutic alliance in psychotherapy for adolescent depression: differences between treatment types and change over time. J Psychother Integr 32(3):326–341
22. Eubanks CF, Muran JC, Safran JD (2018) Alliance rupture repair: a meta-analysis. Psychotherapy 55(4):508–519
23. Nof A, Dolev T, Leibovich L (2019) If you believe that breaking is possible, believe also that fixing is possible: a framework for ruptures and repairs in child psychotherapy. Res Psychother 22(1):364
24. Safran JD, Muran JC (2000) Resolving therapeutic alliance ruptures: diversity and integration. J Clin Psychol 56(2):233–243
25. Chu BC, Skriner LC, Zandberg LJ (2014) Trajectory and predictors of alliance in cognitive behavioral therapy for youth anxiety. J Clin Child Adolesc Psychol 43(5):721–734

26. Levin L, Henderson HA, Ehrenreich-May J (2012) Interpersonal predictors of early therapeutic alliance in a transdiagnostic cognitive-behavioral treatment for adolescents with anxiety and depression. Psychotherapy 49(2): 218–230
27. Whitehead M, Jones A, Bilms J et al (2018) Child social and emotion functioning as predictors of therapeutic alliance in cognitive-behavioral therapy for anxiety. J Clin Psychol 75(1):7–20
28. Creed TA, Kendall PC (2005) Therapist alliance-building behavior within a cognitive-behavioral treatment for anxiety in youth. J Consult Clin Psychol 73:498–505
29. Russell R, Shirk S, Jungbluth N (2008) First-session pathways to the working alliance in cognitive-behavioral therapy for adolescent depression. Psychother Res 18(1):15–27
30. Lynch L, Moorhead A, Long M et al (2021) What type of helping relationship do young people need? Engaging and maintaining young people in mental health care—a narrative review. Youth Soc 53(8):1376–1399
31. Sburlati ES, Schniering CA, Lyneham HJ et al (2011) A model of therapist competencies for the empirically supported cognitive behavioral treatment of child and adolescent anxiety and depressive disorders. Clin Child Fam Psychol Rev 14:89–109
32. Brown RC, Parker KM, McLeod BD et al (2014) Building a positive therapeutic relationship with the child or adolescent and parent. In: Sburlati ES, Lyneham HJ, Schniering CA, Rapee RM (eds) Evidence-based CBT for anxiety and depression in children and adolescents: a competencies-based approach. Wiley Blackwell, New York, pp 63–78
33. Sanchez AL, Jent J, Aggarwal NK et al (2022) Person-centered cultural assessment can improve child mental health service engagement and outcomes. J Clin Child Adolesc Psychol 51(1):1–22
34. American Psychiatric Association (2013) Diagnostic and statistical manual of mental disorders, 5th edn. American Psychiatric Publishing, Arlington
35. Lewis-Fernández R, Aggarwal NK, Kirmayer LJ (2020) The cultural formulation interview: progress to date and future directions. Transcult Psychol 57(4):487–496
36. Wasson Simpson KS, Gallagher A, Ronis ST et al (2022) Youths' perceived impact of invalidation and validation on their mental health treatment journeys. Adm Pol Ment Health Serv Res 49:1–14
37. Linehan M (1997) Validation and psychotherapy. In: Bohart A, Greenberg L (eds) Empathy reconsidered: new directions in psychotherapy. American Psychological Association, Washington, DC, pp 353–392
38. Sauter FM, Heyne D, Michiel Westenberg P (2009) Cognitive behavior therapy for anxious adolescents: developmental influences on treatment design and delivery. Clin Child Fam Psychol Rev 12:310–335
39. Friedberg RD, Gorman AA (2007) Integrating psychotherapeutic processes with cognitive behavioral procedures. J Contemp Psychother 37:185–193
40. Park AL, Kim RE (2022) Incorporating play into cognitive behavioral therapy for youth. In: Friedberg RD, Rozmid E (eds) Creative CBT with youth: clinical applications using humor, play, superheroes, and improvisation. Springer, Cham, pp 83–100
41. Kendall PC, Hedtke K (2006) Cognitive-behavioral therapy for anxious children: therapist manual, 3rd edn. Workbook, Ardmore
42. Bearman SK, Weisz JR (2015) Comprehensive treatments for youth comorbidity–evidence-guided approaches to a complicated problem. Child Adolesc Ment Health 20(3):131–141
43. Rabner J, Norris LA, Crane ME et al (2023) Impact of youth and caregiver worries about treatment on therapeutic process factors and posttreatment outcomes. J Anxiety Dis 10: 27–41
44. Deacon BJ, Farrell NR, Kemp JJ et al (2013) Assessing therapist reservations about exposure therapy for anxiety disorders: the therapist beliefs about exposure scale. J Anxiety Disord 27(8):772–780
45. Whiteside SP, Deacon BJ, Benito K et al (2016) Factors associated with practitioners' use of exposure therapy for childhood anxiety disorders. J Anxiety Disord 40:29–36
46. Kendall PC, Comer JS, Marker CD et al (2009) In-session exposure tasks and therapeutic alliance across the treatment of childhood anxiety disorders. J Consult Clin Psychol 77(3): 517–525
47. Keeley ML, Geffken GR, Ricketts E et al (2011) The therapeutic alliance in the cognitive behavioral treatment of pediatric obsessive–compulsive disorder. J Anxiety Disord 25(7):855–863
48. Hawley KM, Weisz JR (2005) Youth versus parent working alliance in usual clinical care: distinctive associations with retention, satisfaction, and treatment outcome. J Clin Child Adolesc Psychol 34(1):117–128

Chapter 4

Cognitive Behavioral Therapy for Depression in Youth

Elli N. Riise, Bente Storm Mowatt Haugland, and Gro Janne H. Wergeland

Abstract

Depression is a prevalent mental health disorder among youth, with rising rates and significant comorbidities. It severely impacts social life, school attendance, and family relations. Furthermore, depression is associated with peer difficulties, suicidality, increased risk of substance abuse, and a reduced quality of life. Cognitive behavioral therapy (CBT) is a well-established and effective treatment for depression in youth. Core principles in CBT for depression is to address thoughts and behaviors that perpetuate depressive symptoms. The current chapter presents key principles of CBT for depression in youth using examples, strategies for overcoming challenges, illustrations, and sample dialogues between youth and therapist. It offers valuable knowledge and resources for practitioners working with depressed youth, providing useful tools for clinical practice.

Key words Child and adolescent depression, Cognitive behavioral therapy, Youth, Depression, Cognitive restructuring, Behavioral activation, Psychotherapy

1 Introduction

Among US adolescents, depressive disorders are the third most common mental health condition, with a median age of onset at 13 years [1]. Globally, the prevalence of depressive disorder has been estimated to be 8% among children and adolescents (hereafter called youth) [2]. The prevalence of mood disorders is lower in childhood years [3] but increases substantially in adolescence [1, 4]. Furthermore, the percentage of youth with subthreshold depression is found to be significantly higher, also associated with considerable impairment and need for help [5]. Finally, time trends indicate that the prevalence of depressive disorders (and anxiety) has increased in the twenty-first century, particularly among teenage girls [6].

Depression is a complex disorder, involving multiple aspects of life as well as the personal functioning of children and adolescents. The youth's feelings (e.g., sadness), perceptions (e.g., focus on negative aspects of occurrences), cognition (e.g., self-critical),

Robert D. Friedberg and Micaela A. Thordarson (eds.), *Cognitive Behavioral Therapy with Youth: Tradition and Innovation*, Neuromethods, vol. 231, https://doi.org/10.1007/978-1-0716-5134-6_4,

behavior (e.g., withdrawal), and interpersonal functioning (e.g., avoiding social contact) may be heavily influenced by depression. Depression is associated with serious negative consequences, such as school absenteeism, peer problems, suicidality, risk for future substance abuse, and low quality of life [7, 8].

Multiple factors, such as genetic (e.g., temperament), individual (e.g., emotion regulation), and environmental/contextual (e.g., parental psychopathology, parenting style, poverty), have been identified as risk factors for the development of depression in youth. Depression in youth is very likely to co-occur with other mental health disorders, with as much as 40–90% comorbidity. The most common comorbid disorders are anxiety, behavioral problems, ADHD, eating disorders, and substance abuse [9–11].

The negative psychosocial impacts of depression and the multiple risk factors have consequences for how to apply CBT with these youth. In CBT, collaboration and active involvement by the youth are crucial for change to occur. This includes engaging the young person in exercises during and between session and exploring feelings, behavior, and cognitions. Due to the complexity of depressive disorders, treatment should be multifactorial, including different techniques, focused on cognitive, relational, emotional, and behavioral aspects. Furthermore, contextual factors need to be assessed and dealt with, such as ongoing or past bullying by peers, recent loss of persons or functions, failures at school, different types of negative life events such as parental divorce, or experiences of abuse and other types of potential trauma. Also, feelings of shame, guilt, hopelessness, and helplessness may have to be explicitly addressed during the treatment.

2 Theoretical Framework

The theoretical foundation of CBT for depression in youth is based on Aaron Beck's cognitive triad model of depression [12]. This model posits that negative thought patterns about oneself, the world and the future—such as "I will always be a failure", "Human beings do not care about anything but themselves" and "I won't be able to get a job or find someone who love me"—contribute to the onset and maintenance of depressive symptoms. In CBT, depression is understood as being maintained by a combination of maladaptive cognitions (e.g., self-critical or hopeless thoughts) and behaviors such as avoidance, withdrawal, and inactivity. These cognitions and behaviors diminish opportunities for experiences of joy, engagement, and success. CBT aims to change negative cycles by addressing both the cognitive and behavioral components of depression. Thus, interventions focus on supporting individuals in recognizing and challenging distorted thinking, building more realistic and adaptive thought patterns, and increasing engagement in pleasurable and meaningful activities.

3 Treatment Phases

CBT with depressed youth usually has an initial phase, a middle phase, and a final phase. In the initial phase the therapist conducts assessments, establishes therapeutic alliance, creates a case conceptualization, specifies treatment goals, and provides psychoeducation. The middle phase focuses on implementing cognitive and behavioral strategies. In the final phase, the therapist prepares the youth and the families for termination of therapy, emphasizing the importance of relapse prevention.

Several treatment manuals for depression in children have been developed (e.g., [13–15]). They provide structured guidance and useful tools, offering clinicians a practical framework to deliver effective and consistent care throughout the treatment process.

The following sections will introduce the initial stage of CBT for depression, followed by the key behavioral and cognitive interventions as well as skills training, before covering topics to address in the final sessions to support relapse prevention.

4 The Initial Sessions

4.1 Assessment

In the initial phase of CBT for depression, the therapist conducts assessments to identify primary areas of difficulty. An assessment of the symptoms, their severity, duration, level of distress, and interference with daily functioning is necessary to do a diagnostic evaluation and develop a case formulation and a treatment plan. Differences in symptom presentations can vary depending on age, developmental level, gender, cultural differences, and expectations. For example, adolescents have a more developed emotional vocabulary and are better able to verbalize their symptoms than young children. For younger children, depressive symptoms are sometimes expressed more as irritability or cranky moods rather than sadness. There are also other important differences in symptom presentation between school-age children and adolescents, including comorbid psychopathology. In younger children, depression is commonly accompanied by oppositional defiant disorder (ODD) and anxiety. In adolescence, anxiety remains a common comorbidity, while substance use becomes more prevalent. Additionally, adolescents face an increased risk of suicide behavior and school dropout compared to children [16–18].

Suicidal ideation and suicidal intent should be carefully considered, including how they manage mood and suicidal urges (*see* Chapter 5). Besides information from the young person, information and concerns from caregivers and teachers are important to provide a more comprehensive understanding of the situation.

A consultation with physicians is recommended to rule out that physical conditions are causing the symptoms or to provide medical treatment [19].

Assessment should include information regarding the familial, educational, and social functioning, including the quality of interpersonal relationships and support networks. Risk factors such as bullying, maltreatment, life events, trauma, lifestyle factors, and whether the young person belongs to a high-risk group (e.g., special education needs, chronic illness, gender minority group) should be considered, as well as identifying strengths and protective factors. Also, it is important to assess whether other difficulties are present, as comorbidity can complicate the clinical picture and may require adjustments to the treatment. Assessing the child or adolescent using a developmental perspective and within the cultural context is critical [16].

A structured diagnostic interview can help assess the presence, nature, and severity of a depressive disorder and identify comorbid conditions that should be considered when developing the treatment plan. Other psychological conditions that may present similarly to major depressive disorder include persistent depressive disorder, disruptive mood dysregulation disorder or post-traumatic stress disorder, anxiety disorders, conduct disorders, substance use disorder, oppositional defiant disorder, and attention-deficit-hyperactivity disorder. Symptoms that may overlap with depression are bipolar disorder, eating disorder, and conduct disorder. Gathering information including descriptions and context in which these symptoms affect daily functioning is important [16]. Also, sleep problems should be assessed as these are frequent comorbid symptoms to depressive disorders and may need to be targeted early in treatment to increase the youths' energy and ability to be engaged in the treatment. Importantly, the initial clinical interview not only provides essential diagnostic information but also offers an opportunity to establish a working alliance and build report with the young person and their family.

In addition, questionnaires are valuable for assessing depressive symptoms and monitoring treatment response. Self-report questionnaires can be completed prior to sessions and used throughout treatment to track progress and changes in symptoms.

The assessment is reviewed with input from the young person and the parents. Information from the assessment is used to tailor the treatment interventions, including psychoeducation (e.g., discussions on how the CBT model applies to the young person), case formulation (e.g., integrating information about their current behavior), goal setting, and the implementation of cognitive and behavioral strategies.

4.2 Therapeutic Alliance with Depressed Youth

During the assessment, the therapist should create a trusting and collaborative environment with the youth and their families (*see* also Chapter 3 of this volume). In the initial sessions, the foundation for the therapeutic alliance is established, which is essential for the success of treatment. Importantly, it has been suggested that an early therapeutic alliance is associated with early improvements in depressive symptoms [20].

Depressed youth are often referred to therapy because their parents are concerned about their well-being. However, the adolescent may not initially recognize symptoms of depression or may disagree with their parents about the need for help. This can make it challenging to engage adolescents in treatment. It may sometimes be necessary to spend extra time on creating a good therapeutic alliance with depressed youth to ensure that the youth feel understood and supported.

4.3 Psychoeducation

In the treatment of depressed youth, the psychoeducation should include information on depressive symptoms, the principles of CBT, and how behaviors and thoughts perpetuate depression (the CBT model).

The psychoeducation on depressive symptoms should explain common symptoms, that the origins of depression may stem from a combination of genetic, biological, psychological, and environmental factors, and that it's not always possible to pinpoint a single cause for the disorder. While it sometimes may be clear that the depression was triggered by particular situations or experiences (e.g., bullying at school, losing a friend), often the causes are less obvious. It should be explained that even if it might be difficult to pinpoint the exact cause, the treatment can still be effective in alleviating the symptoms. It may also be important to address common misconceptions about depression, such as beliefs that depression is merely a result of laziness or a lack of effort.

The psychoeducation should further explain the rationale behind the interventions and prepare the youth and their families for the work ahead. Providing a CBT model of depression helps to illustrate the role of cognitions and behavior in maintaining depression, as well as the interactions between emotions, thoughts, behavior and physical states. Figure 1 illustrates a common CBT model which may be useful in this regard.

Youth and caregivers need to be explained that the treatment will involve discussing difficult emotions and experiences, examining how the young person's thoughts and behaviors contribute to their depressive symptoms, working to change unhelpful thought patterns, and promoting positive behavioral changes to interrupt the "vicious cycle" of depression. CBT is a collaborative process, with both the clinician and the adolescent contributing to the work. It is important to emphasize that regular attendance and

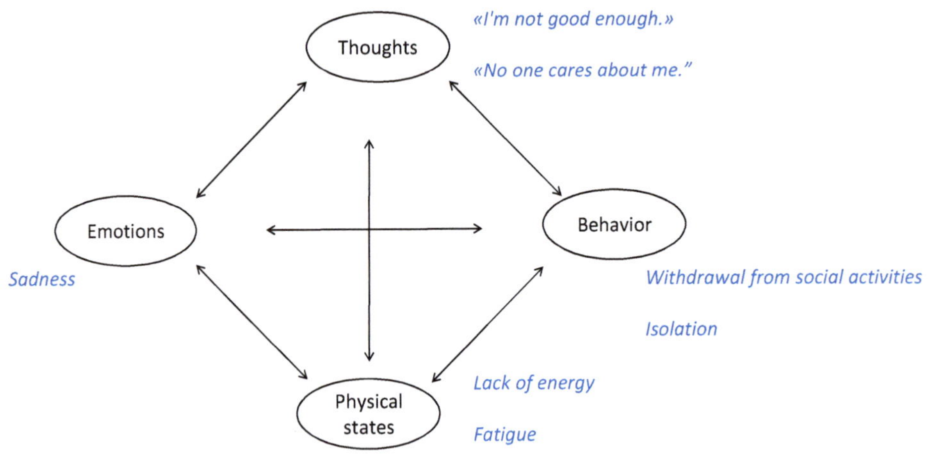

Fig. 1 CBT model

completion of homework tasks are part of this collaborative approach and essential for therapy. The young person should be encouraged to actively participate and to bring in any concerns they may have. Information about the treatment structure, the frequency of sessions, and the expected duration of treatment should also be covered. Providing clarity on the treatment approach and what to expect ensures predictability and transparency and further strengthens the treatment alliance.

Parents and family members may respond to their child's depression in a variety of ways, and their attitudes and behaviors can either support or hinder recovery. Psychoeducation should provide families with accurate information about depression, including its potential causes, available treatments, and expected outcomes. Addressing common but unhelpful reactions is essential. For example, parents may blame themselves for their child's illness, view the young person as lazy or not trying hard enough, believe the adolescent has conscious control over their symptoms, or assume that depression is purely a biological disorder with only biological treatments being effective. Families may need psychoeducation that specifically targets these concerns. For instance, when parents blame themselves, the therapist should emphasize that depression is typically the result of a combination of factors, not a single event, and that parents play an important role in their children's recovery. If parents feel their children are not trying hard enough, clinicians can help them understand that depression is an illness, and its symptoms and dysfunction are not within the young person's voluntary control. Lastly, if parents believe only biological treatments are effective, clinicians should explain that depression involves psychological, biological, and social factors, all of which contribute to both its onset and its effective treatment [14].

4.4 Case Conceptualization

A case conceptualization is created during the initial phase of the therapy (*see* also Chapter 2 of this volume). A thorough and thoughtful assessment lays the groundwork for a case conceptualization that resonates with the youth and is experienced as relevant and credible. Building on the information from the assessment, the therapist continues with identifying thoughts that maintain the depressed mood and behavioral avoidance, including behavior that leads to a decrease in naturally occurring reinforcement from the environment. Figure 2 illustrates a simple model for a case conceptualization for Mark, a 15-year-old boy who has developed depressive symptoms over the past year.

4.5 Setting Treatment Goals

Treatment goals are developed once a case conceptualization has been created. Goal setting helps break tasks into manageable steps. It is important to identify what outcomes matter most to the young person and the family and what they seek to change. Adolescents often start with broad or abstract goals, such as "feeling less depressed" or "becoming more popular." The therapist may need to guide the adolescent to ensure specific goals, like "spending more time with friends" or "building better confidence" to make the goals more achievable [21]. The goals help provide direction. Reviewing their goals throughout treatment enables adolescents to recognize their progress. It is important that adolescents feel ownership over their goals and that the goals are attractive for them, as this may promote engagement and motivate change [22].

Feelings of hopelessness, low self-esteem, lack of motivation, cognitive disorganization, or difficulty envisioning change can make it challenging to engage depressed youth in the goal-setting process [23]. For some, setting goals may feel overwhelming, especially when even small day-to-day tasks are challenging. Furthermore, adolescents referred to therapy by others may not fully acknowledge the symptoms, making it harder for them to identify meaningful goals. To navigate these challenges, therapists must adopt a flexible and collaborative approach. Taking the time to build trust and understand the adolescent's perspective is essential. The therapist's role is to motivate and guide the adolescent through this process by asking open-ended questions like "What would you like to change?" or "How do you want things to be different?" and helping them break down larger goals into smaller, manageable steps.

Therapists should also monitor progress in a thoughtful manner, avoiding overly frequent progress reviews [24]. It may be helpful to agree on a specific time to review each goal when setting them. Progress reviews should emphasize accomplishments and, when needed, adjust goals to align with the adolescent's evolving needs.

Example case formulation

- **Problem:** Mark was previously an enthusiastic student, but recently he has withdrawn from social activities. Mark feels sad and lacks motivation, especially when it comes to schoolwork and social interactions. He can't find joy in hobbies he used to enjoy, such as playing video games or hanging out with friends. Mark reports feeling tired and lacking energy. His thoughts are often dominated by self-criticism and a belief that he will fail or disappoint others, leading to feelings of worthlessness.
- **Development (including cognitive predispositions):** Mark has had a normal developmental history without major health problems. He has been a good student and has had a few close friendships. Over the past year, however, he has become increasingly withdrawn from social activities and less engaged in school. Mark has always been sensitive to criticism and has high standards for himself. He struggles with perfectionistic thinking, where he feels that anything less than his best is a failure. This has led him to believe that he is not good enough and that others have high expectations of him that he cannot meet.
- **Context:** At school, Mark feels disconnected. Despite having some friends, he compares himself to the popular boys and feels that he is not as fun or popular. This leads him to believe that nobody truly likes him, and he often feels like a failure. Mark avoids socializing with peers, further isolating himself. At home, his parents are supportive but often question why he isn't spending more time with friends, which adds to his guilt and shame. Mark feels torn between his parents' differing approaches to handling his struggles, intensifying his guilt. He feels unable to meet expectations from his parents, school, or socially, worsening his sense of failure and hopelessness.
- **Previous Experiences:** Mark has had several negative experiences that have reinforced his feelings of failure and low self-worth. He has struggled with poor academic performance and has experienced moments of feeling isolated from his peers, especially after a close friend moved.

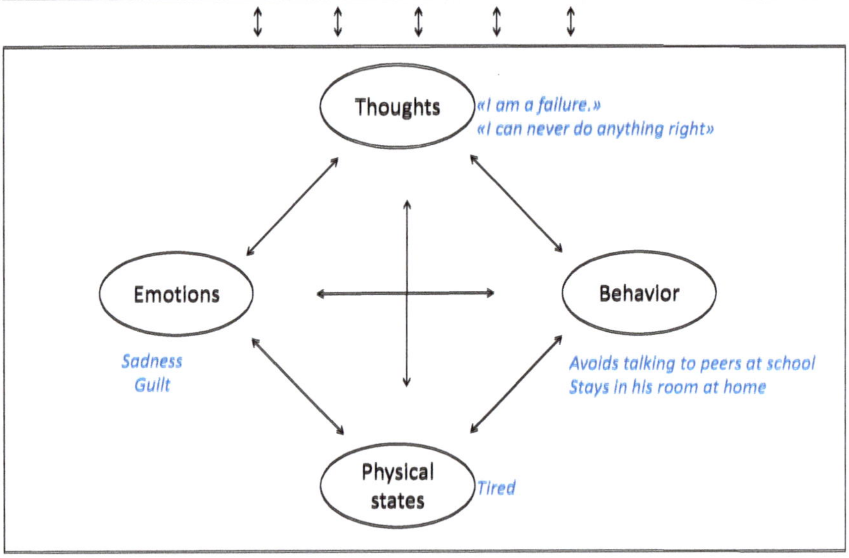

Fig. 2 Example case formulation

4.6 Self-Monitoring

Self-monitoring aims to increase the youth's awareness of his/her own thoughts, physical sensations, emotions, and behaviors—an awareness which is essential for change to occur. While self-monitoring is typically introduced early in therapy, it also serves to track depressive symptoms and behaviors throughout the treatment process. By tracking their own experience, like the frequency of negative thoughts or passive behavior, clients can gain a clearer understanding of the problem's scope, motivating them to engage in therapy. It can also reveal patterns of rumination or self-criticism that the youth may not have been aware of. Collecting data through self-monitoring makes it harder to dismiss facts and encourages youth to reconsider their beliefs. Interestingly, self-monitoring itself has been found to increase positive behaviors [25].

A common way of recording thoughts is to ask the youth to write down specific situations where they have felt sad or disappointed. The youth might be provided with a thought record sheet to write down what happened. There are several versions of thought recording sheets for youth with depression, but the majority of them are typically divided into columns and follow a traditional A-B-C model, i.e., antecedents, beliefs (thoughts), and consequences (emotions and behaviors) [14]. Alternatively, the youth can be provided with a less complex sheet to record the negative thought and the activating event (situation) [26].

In addition, activity monitoring is particularly relevant for youth depression. Activity monitoring helps adolescents with depression identify how daily activities affect their mood and energy, often revealing patterns of inactivity or withdrawal. It increases awareness of how reduced activity worsens mood and highlights opportunities to reintroduce enjoyable or meaningful tasks. Activity monitoring is typically recorded hour by hour across days of the week, providing a detailed overview of how time is spent. Figure 3 illustrates an example of an activity monitoring sheet filled out by a depressed adolescent.

5 Behavioral and Cognitive Interventions

Core CBT interventions for youth depression comprise addressing behaviors that contribute to depressive symptoms, increasing engagement in rewarding activities, and challenging maladaptive thought patterns. Youth are guided in practicing and applying these strategies to real-life situations to promote meaningful change.

5.1 Behavioral Interventions

When treating depression in children and adolescents through CBT, behavioral interventions are essential. These practices are grounded in the idea that thoughts, feelings, and behaviors are interconnected, emphasizing the impact of behavioral changes on emotions and thoughts.

5.1.1 Behavioral Activation

Inactivity and withdrawal are core symptoms of depression and behavior that tends to maintain the depressed mood. Behavioral activation is carried out in order to create experiences that increase positive experiences and social interaction. Depressed youth often find previously enjoyable activities less rewarding and lose motivation to participate. Since their motivation rarely returns on its own, a structured approach is crucial for reengaging them. Success requires identifying activities that have previously been enjoyable or given a sense of coping, and integrating these into their activity plan.

	Sun	Mon	Tue	Wed	Thurs	Fri	Sat
06:00 - 07:00	Sleeping	Sleeping	Sleeping	Sleeping	Sleeping	Sleeping	Sleeping
07:00 - 08:00	Sleeping	Getting ready for school	Getting ready for school	Sleeping	Getting ready for school	Getting ready for school	Sleeping
08:00 - 09:00	Sleeping	School	School	Stayed home watching TV /sleeping (headache)	School	School	Sleeping
09:00 - 10:00	Sleeping	School	School	Stayed home watching TV /sleeping (headache)	School	School	Sleeping
10:00 – 11:00	Sleeping	School	School	Stayed home watching TV /sleeping (headache)	School	School	Sleeping
11:00 – 12:00	Sleeping	School	School	Stayed home watching TV /sleeping (headache)	School	School	Sleeping
12:00 - 13:00	In my room watching TikTok videos	School lunch	School lunch	Stayed home watching TV /sleeping (headache)	School lunch	School lunch	On my phone
13:00 - 14:00	In my room watching TikTok videos	School	School	Stayed home watching TV /sleeping (headache)	School	School	Did nothing
14:00 - 15:00	In my room watching TikTok videos	School	School	Stayed home watching TV /sleeping (headache)	School	School	Did nothing
15:00 - 16:00	In my room watching TikTok videos	In my room trying to do maths	Watch TV	Watch TV	Watch TV	Watch TV	Did nothing
16:00 - 17:00	Eat	Sleeping	Watch TV	On iPhone, TikTok	Sleeping		Eat
18:00 – 19:00	Watch Tv	Sleeping	Sleeping	On iPhone, TikTok	Sleeping	Sleeping	Did nothing
19:00 – 20:00	Watch TV	On iPhone, TikTok	Sleeping	On iPhone, TikTok	On iPhone, TikTok	Sleeping	Did nothing
20:00 – 21:00	In my room watching TikTok videos	On iPhone, TikTok	On iPhone, TikTok	On iPhone, TikTok	On iPhone, TikTok	On iPhone, TikTok	In my room watching TikTok videos
21:00 – 22:00	In my room watching TikTok videos	Eat	On iPhone, TikTok	Eat	Eat	On iPhone, TikTok	In my room watching TikTok videos
22:00 – 23:00	In my room watching TikTok videos	Played candy crush	Tried to do homework, stared into wall thinking about what a failure I am	Tried to do homework but couldn't concentrate, did nothing.	Tried to do homework but couldn't concentrate, did nothing.		In my room watching TikTok videos
23:00- 00:00	Made ready for bed	Made ready for bed	Made ready for bed	Made ready for bed	Made ready for bed	Made ready for bed	Made ready for bed

Fig. 3 Example activity monitoring sheet

Activity monitoring, as previously described, is typically the first step of behavioral activation. It provides information on baseline activity levels to guide specific activation assignments and illustrates the treatment rationale by demonstrating the relationship between activity and mood [27]. A common strategy involves creating an activity schedule organized by the days of the week. During the session, a list of enjoyable activities is compiled, and the adolescent selects the ones they wish to engage in more often. These activities are then written down on the activity schedule and completed as homework between sessions and reviewed in the following session. The process usually begins with small, manageable steps to prevent feelings of being overwhelmed and avoid setting unrealistic or overly ambitious goals.

In addition to encouraging young individuals to engage in activities, another key advantage of activity scheduling is its ability to promote a sense of accomplishment and mastery, even with simple routine tasks. By incorporating simple routine tasks such as eating breakfast, taking a shower, preparing meals, or preparing for bed, alongside enjoyable activities, young people are gradually encouraged to build a structured and balanced daily routine. This structure can help disrupt behaviors that perpetuate depression, such as poor sleep patterns, excessive screen use, rumination, and irregular eating habits.

Behavioral activation can be challenging for youth with depression, as they often experience reduced energy, diminished motivation, and have negative expectations, such as believing the intervention will not be effective. Therapists may find it difficult to engage the adolescents in suggesting activities. If this occurs, therapists should focus on activities the patient previously found enjoyable and explain that engaging in activities will probably give more energy and improve his/her mood over time. Providing a clear and compelling rationale for behavioral activation can also be helpful. Repetition of psychoeducation about the interconnectedness of behavior, thoughts, and emotions may be useful, emphasizing that more helpful behaviors can change the negative cycle. Several additional strategies can enhance the effectiveness of behavioral activation. These include involving parents as supportive allies, stressing the importance of sticking to the plan, and using Socratic dialogue to encourage reflection on the pros and cons of passivity and withdrawal. Contingency management can also be beneficial for increasing motivation and enhance the likelihood of success in engaging the youth in activities. Effective planning is also crucial, making the plan as specific as possible (e.g., specifying when, where, for how long, and with whom the activity will occur), anticipating potential obstacles and finding strategies for overcoming them.

The following dialogue demonstrates how the activity monitoring sheet can serve as a starting point for developing an activity schedule.

Therapist (T): *Thanks for bringing back your activity monitoring sheet. Let's take a look at it together. How did you find the process of recording your activities?*

Youth (Y): *It was okay, I guess. But looking at it now… it doesn't seem very productive.*

T: *When you look at your week now, you don't find it very productive. Let's explore this further. When you look at your sheet, what stands out to you?*

Y: *I spend way too much time on TikTok and not enough time doing what I should, like homework or getting out of bed.*

T: *I see that too. How do you feel about the amount of time you're spending on your phone or watching TV?*

Y: *It feels like I waste so much time, but I also don't have the energy to do anything else.*

T: *That makes sense. When you're feeling low or tired, it's tempting to stick to things like TikTok because they're easy and distracting. But let's think about this—how do you usually feel after spending hours on your phone?*

Y: *Honestly, not great. Sometimes I feel worse, like I've just wasted the whole day.*

T: *That's a good point. Let's take a closer look at your activity sheet. Can you spot any patterns in how certain activities, like using your phone, affect your mood compared to others?*

Y: *Yeah, when I'm just scrolling or watching TV, I feel empty afterward. But when I actually manage to do something, like even eating or trying to do homework, I feel a little better—at least I don't feel as bad.*

T: *That's a valuable observation. It shows that even small actions can make a difference in how you feel. What do you think might happen if you started adding a few more structured or productive activities to your day?*

Y: *I guess it could help, but I don't know if I'd have the energy to actually do it.*

T: *That's a common concern, and it's why we'll start small—no overwhelming changes, just manageable steps. Let's brainstorm some activities together. Which ones from your current routine do you think you could build on?*

Y: *Maybe eating breakfast. I skip it most days, but it might help me feel more awake in the morning.*

T: *Great idea. Breakfast is a good way to start your day. What about something you used to enjoy but haven't been doing lately?*

Y: *I used to hang out with my friends at the park, but I haven't done that in a while.*

T: *That sounds like a great activity to add back. How about we create a schedule for next week where you meet up with a friend at the park for a short time?*

Y: *I think I could try that.*

T: *Perfect. Let's also look at your sleep schedule—it seems like you're staying up late scrolling. How do you think changing your bedtime routine might help?*

Y: *I guess if I wasn't on my phone so much at night, I'd fall asleep faster.*

T: *Exactly. Maybe we could set a goal to reduce phone time before bed. For example, putting your phone away 30 minutes before you go to sleep. How does that sound?*

Y: *It sounds hard, but I can try.*

T: *That's all we're asking—just to try. We'll review how it went next session and adjust the schedule if needed. Let's write down your activity plan for next week.*

5.1.2 Behavioral Experiments

The purpose of behavioral experiments is to evaluate a specific cognition through a real-life experience. With depressed youth, this often involves testing negative predictions about how certain situations will turn out. For example, consider a youth who believes, "If I try to do anything, it won't help, and I'll just feel worse." To challenge this prediction, the therapist might design a simple experiment. The youth could be encouraged to engage in a small, enjoyable activity, such as listening to his/her favorite music for 10 min or inviting a friend to a movie. In the next session, the therapist would review the results by asking the adolescent to reflect on the experience. Together they would explore how the activity impacted the youth's mood and whether the initial prediction—that doing something would make them feel worse—held true. The therapist would guide the youth in assessing the evidence: did the activity lead to any improvement, no change, or did they feel worse? This discussion allows the therapist to highlight any discrepancies between the youth's negative predictions and the reality of the situation. By examining the evidence together, the therapist can help youth recognize the unhelpfulness of negative thinking and begin to see that small actions may, in fact, lead to positive changes. This may further reinforce future engagement in similar or other activities.

5.1.3 Contingency Management

Contingency management can be an important part of treating depression, especially when positive behaviors are not recognized or rewarded by the youth's environment. For example, an adolescent who spends more time on homework may not necessarily get better grades at school due to underlying learning difficulties. Despite putting in extra effort, they may feel discouraged if their performance does not improve. This can lead to frustration and a sense of failure. The lack of positive reinforcement for their hard work can make it difficult for them to stay motivated or engaged in their studies. In such cases, the therapist might introduce a system of rewards or incentives to help motivate the adolescent. For instance, the parents and the youth might agree on a specific reward, like watching a favorite show or having some extra time for a hobby, after completing homework [13]. The therapist might also set up similar rewards during therapy sessions, such as allowing the youth to choose an activity after completing a therapeutic exercise.

The goal of contingency management is to create a system where positive actions, even small ones, are acknowledged and rewarded, helping the youth to engage in more positive behaviors.

5.2 Cognitive Interventions

5.2.1 Identifying and Restructuring Cognitive Distortions

A good starting point to help youth recognize that their own thoughts may not be entirely accurate or true is to explain common cognitive distortions in depression as described by Beck and colleagues [28]. Among others these include all-or-none thinking ("I am a failure because I made a mistake"), discounting the positives ("the date was a disaster even though she seemed to enjoy herself"), or emotional reasoning ("I feel worthless; therefore, I know that nobody will accept me"). For youth, cognitive distortions are often best explained using simpler language, referring to them as "thinking mistakes" [29] or "unhelpful thoughts" [14].

Furthermore, in addition to cognitive distortions that are identified in dialogue in sessions, thought recording sheets may be used to help adolescents identify their own thinking mistakes. Reflecting on thoughts they have had in specific situations retrospectively, can help them create a sense of distance from their thoughts, which often makes it easier to challenge them.

A common approach to arrive at more accurate and helpful thoughts is a test of evidence [30]. This is done by looking for evidence that support and evidence that contradict the negative thoughts. Once a list of supporting and disconfirming evidence is created, a more accurate and nuanced conclusion is generated. Figure 4 provides an example of a test of evidence for the thought "nobody likes me."

Another common way of restructuring cognitive distortions is through Socratic dialogue. In the Socratic dialogue the therapist coaches the child or adolescent to evaluate the accuracy of his or her

Thought or belief:
Nobody likes me.

Evidence that support the thought or belief	**Evidence against the thought or belief**
I don't get invited to hang out with my friends as often as others do.	I still have a few friends who talk to me regularly and invite me to do things, even if it's not as much as before.
Sometimes I feel like I'm left out in class or at lunch.	There are people at school who smile at me or say hello when they see me.
A close friend of mine moved away recently, and I haven't made new close friends.	My parents, although sometimes worried, always tell me that they care and are there for me.
	I am often told that I'm a good listener, which suggests that people do value me.

Conclusion
While I may not always feel connected, there are people who care about me and enjoy spending time with me.

Fig. 4 Example test of evidence

beliefs by asking a series of open-ended questions to guide the adolescent's attention toward alternative explanations. Examples of questions may be "What has happened previously in similar situations?", or "What would you tell a friend who had similar thoughts/feelings as you are experiencing now?". The therapist may use a series of questions to help young people reflect on and explore how their thoughts might be inaccurate. Alternative explanations may be written down and their plausibility rated. The goal is to encourage the youth to recognize and explain, in their own words, why their beliefs may not be fully correct. The following dialogue illustrates how a Socratic dialogue can facilitate cognitive restructuring and change cognitions in a depressed youth:

Therapist (T): *Mark, you mentioned that you feel like a failure because you missed school on Wednesday. I'd like to explore that thought with you. What does it mean to you to be a "failure"?*

Youth (Y): *Well, if I mess up or don't do what I'm supposed to, then it means I'm not good enough, like I'm failing at everything.*

T: *So, missing one day of school leads you to think that you're failing at everything. Have I understood you correctly? Let's try to break that down a bit. Is it possible that there are other factors contributing to you missing school that don't have anything to do with being a failure?*

Y: *I guess it was because I was feeling really tired, and I didn't feel like going. And I had a headache. I've been stressed, so maybe that's why.*

T: *Okay, so you weren't feeling your best that day, and that might have influenced your decision. If this had happened to a friend, would you say he was a failure for not going to school one day?*

Y: *I'd probably say that it was okay. Everyone has bad days sometimes.*

T: *So, you'd be more understanding with a friend who missed school, right? Now, what if you think about this from a different angle—do you think missing one day of school means you are a failure in other areas of your life too, like being a good friend or a brother?*

Y*: I don't know... I mean, I've done well in school before, I have good friends, and I think my sister cares for me. I don't think missing one day means I'm bad at everything.*

5.2.2 Problem-Solving

Problem-solving with depressed youth is done to solve problems that seem overwhelming or problems that perpetuate the depressed mood. A common method involves evaluating the advantages and disadvantages of a particular choice or decision. For example, if a depressed adolescent chooses to withdraw to their room when feeling down, the therapist and youth may explore the benefits and drawbacks of this choice and see if other solutions can be better. In other cases, depressed youth may feel overwhelmed by specific tasks, such as preparing for an upcoming exam. The therapist can guide the youth to break the tasks into smaller, more manageable steps [30]. This process helps to see the problem as less daunting, with each step bringing the youth closer to a solution [31]. Some CBT manuals outline specific stepwise problem-solving interventions [13], where the child or the family identifies the problem—for example "we would like to quarrel less with each other"—then brainstorms a range of possible solutions such as "spending more time together doing things we enjoy," "arranging family time each week where we can discuss our conflicts," "avoiding talking to each other when we disagree or get angry". Next each solution is evaluated, the best options are selected and implemented and the outcome is assessed in the next session by asking "Did it work or should we try other solutions?.

6 Skills Training

Skills training is typically used when a youth lacks the necessary ability to perform positive behaviors in ways that will be helpful. For example, engaging adolescents in social activities is an important part of treatment, but some youth might struggle with social skills and end up feeling worse after social activities. In these cases, it may be necessary to practice interpersonal skills in the sessions (e.g., making eye contact, speaking in a clear and loud voice, showing interest in others, respecting the personal space of others, or giving compliments). Others may have difficulty asserting their needs and may benefit from learning how to express themselves clearly and confidently. Some youth may struggle to solve interpersonal problems and could benefit from skills training focused on communication.

Depressed individuals often spend considerable time on depressive rumination. They may focus on their perceived flaws or shortcomings and dwell on upsetting events, problems, and the symptoms of depression, along with their causes, meanings, and consequences [32, 33]. Many of the core interventions of CBT—such as self-monitoring/behavioral monitoring, behavioral activation, problem-solving, and cognitive restructuring—can help reduce rumination by increasing awareness of negative thought patterns, providing strategies to modify these patterns, and engaging in behaviors that exacerbate rumination. However, some individuals may still experience excessive and persistent rumination. In some cases, where rumination is especially time-consuming and disturbing, therapists may have to focus on making the youth become aware of their ruminative thoughts and intentionally shift their attention away from them. Exercises often involve redirecting attention toward their breath or the environment around them, guiding them to focus on neutral sensory experiences [13, 14].

Finally, relaxation training may also be a useful technique to reduce physiological arousal for some youth with depression [14, 34, 35]. Relaxation techniques include controlled deep breathing, guided imagery, and progressive muscle relaxation.

7 Preventing Relapse

In the final phase of treatment, the therapist prepares the youth and the family for the termination of treatment, by focusing on relapse prevention. The therapist, the youth, and the caregivers review the strategies and skills learned during CBT, to help maintain progress. Youth are encouraged to continue using these skills to manage any future depressive symptoms. Families are advised on the importance of booster sessions or returning to therapy if depressive symptoms reemerge.

8 Empirical Support for CBT with Depressed Youth

Different psychotherapy approaches are used in treating youth with depression. Both CBT and interpersonal therapy (IPT) have the strongest evidence and are well-established treatments for adolescent depression, with CBT having the strongest evidence of effectiveness [36]. For example, both group and individual CBT and CBT in combination with medication are considered well-established for adolescents with depression, whereas technology-assisted and group- and family-based CBT are considered possibly efficacious for children [36].

A recent meta-analysis examining the efficacy of psychosocial treatments for subthreshold depression in children and adolescents found small to moderate effect sizes when comparing psychotherapy to usual care, attention control, or supportive treatment [37]. Of the 12 trials (and 13 comparisons between psychotherapy and a control condition), all but 1 were CBT or IPT, with 8 of 11 trials investigating CBT. Furthermore, only two trials included treatments for children (mean age <12 years). Importantly, results indicated a 48% reduction in risk of a major depressive episode at 1-year follow-up in youth receiving CBT and IPT compared to the control condition, indicating a potential to prevent the development of MDD at short-term follow-up. However, studies in children under 12 years of age and studies examining long-term effects are needed.

For clinically depressed youth, another meta-analysis examined the efficacy of CBT across 24 studies. This study also examined long-term effects [38]. Large pre-post effect sizes were found, and results were maintained or even increased after 1 year as well as at longer time follow-up. These findings support the evidence of long-term effects of CBT targeting youth depression, although only a few studies had assessments after a 12-month follow-up. A third meta-analysis that evaluated the effectiveness of CBT for clinically depressed youth [39] (i.e., CBT delivered in ordinary clinical practice where youths and therapists represent the conditions typical of that treatment setting) found a large pre-post effect size and a remission rate of 53% at post-treatment across the eight trials included. The results were maintained at follow-up. Importantly, the outcomes were comparable to those found in efficacy studies from research clinic settings, meaning that treatment effects are not lost when the treatments are implemented in routine clinical care.

Across various evidence-based treatment guidelines for depression in young persons, some minor differences exist in the recommendation of moderate to severe depression. CBT or IPT are recommended as the first-line treatment in some guidelines (e.g., [40, 41]), whereas CBT, IPT, or selective serotonin reuptake inhibitors (SSRI) are recommended by others [19, 42, 43].

Regarding combination treatment, data from two large, multisite studies provide solid support for the combination of CBT and antidepressant medication in the treatment of young persons with moderate to severe depression [44, 45]. Results from a third study suggest that adding CBT to medication could be especially valuable for those with higher levels of clinical severity [46]. Regarding pharmacological treatment in young persons with depression, SSRI is the recommended first-line treatment in moderate to severe depression in youth [47]. A network meta-analyses found fluoxetine to be associated with the best outcomes for youth with depression, including a better side effect profile. A discussion of SSRI in treating youth depression is beyond the scope of the current chapter and has been commented on extensively by others. The reader is referred to these for further information [44, 45, 47, 48].

In summary, CBT is considered a well-established treatment for depression in adolescents and children. At the same time, the effect sizes for depression treatments are modest to moderate in magnitude and among the lowest among interventions for common youth mental health problems. This indicates that a substantial number of youths do not respond to the treatments currently available [36, 49, 50]. These results suggest there is a need to improve treatment strategies to further enhance the effectiveness of CBT interventions for youth with depression.

9 Conclusion

In CBT for depressed youth, both cognitive and behavioral interventions work together to reduce symptoms and promote recovery. Cognitive interventions primarily aim to challenge the youth's negative thinking patterns, while behavioral interventions focus on changing actions and increasing positive engagement with the world around the youth. By addressing both cognitive and behavioral aspects of depression, these techniques help children and adolescents develop more flexible and adaptive thinking patterns as well as healthier coping strategies, leading to improvements in mood and functioning. Empirical research consistently supports the efficacy of CBT in reducing depression symptoms in youth, demonstrating its effectiveness in both clinical and real-world settings.

Conflicts of Interest

The authors have no conflicts of interest to declare that are relevant to the content of this chapter.

References

1. Merikangas KR, He JP, Burstein M, Swanson SA, Avenevoli S, Cui L et al (2010) Lifetime prevalence of mental disorders in U.-S. adolescents: results from the National Comorbidity Survey Replication—Adolescent Supplement (NCS-A). J Am Acad Child Adolesc Psychiatry 49(10):980–989
2. Shorey S, Ng ED, Wong CHJ (2022) Global prevalence of depression and elevated depressive symptoms among adolescents: a systematic review and meta-analysis. Br J Clin Psychol 61(2):287–305
3. Ghandour RM, Sherman LJ, Vladutiu CJ, Ali MM, Lynch SE, Bitsko RH, Blumberg SJ (2019) Prevalence and treatment of depression, anxiety, and conduct problems in US children. J Pediatr 206:256–67.e3
4. Morken IS, Viddal KR, Ranum B, Wichstrøm L (2021) Depression from preschool to adolescence—five faces of stability. J Child Psychol Psychiatry 62(8):1000–1009
5. Balázs J, Miklósi M, Keresztény Á, Hoven CW, Carli V, Wasserman C et al (2013) Adolescent subthreshold-depression and anxiety: psychopathology, functional impairment and increased suicide risk. J Child Psychol Psychiatry 54(6):670–677
6. Bor W, Dean AJ, Najman J, Hayatbakhsh R (2014) Are child and adolescent mental health problems increasing in the 21st century? A systematic review. Aust N Z J Psychiatry 48(7):606–616
7. Georgiades K, Lewinsohn PM, Monroe SM, Seeley JR (2006) Major depressive disorder in adolescence: the role of subthreshold symptoms. J Am Acad Child Adolesc Psychiatry 45(8):936–944
8. Essau CA, Chang WC (2009) Epidemiology, comorbidity, and course of adolescent depression. In: Essau C (ed) Treatments for adolescent depression: theory and practice. Oxford University Press, Oxford, pp 3–25
9. Epidemiology of depression in children and adolescents [press release] (2008). The Guilford Press, New York
10. Axelson DA, Birmaher B (2001) Relation between anxiety and depressive disorders in childhood and adolescence. Depress Anxiety 14(2):67–78
11. Comorbidity of unipolar depression: II. Comorbidity with other mental disorders in adolescents and adults [press release] (1991). American Psychological Association, Washington, DC
12. Beck AT (1979) Cognitive therapy and the emotional disorders. Penguin, London
13. Curry JF, Wells KC, Brent DA, Clarke GN, Rohde P, Albano AM (2005) Treatment for adolescents with depression study (TADS) cognitive behavior therapy manual: introduction, rationale, and adolescent sessions. Duke University Medical Center, Durham
14. Hetrick S, Parker A, Bailey A, Cahill S, Rice S, Garvin T et al (2015) Cognitive-behavioural therapy for depression in young people: a modular treatment manual. Orygen, The National Centre of Excellence in Youth Mental Health, Parkville
15. Clarke G, Hops H, Grossen B, Lewinshon PM (1990) Leader's manual for adolescent groups: adolescent coping with depression course. Castalia, Eugene
16. Thapar A, Eyre O, Patel V, Brent D (2022) Depression in young people. Lancet 400(10352):617–631
17. Maughan B, Collishaw S, Stringaris A (2013) Depression in childhood and adolescence. J Can Acad Child Adolesc Psychiatry 22(1): 35–40
18. Egger HL, Angold A (2006) Common emotional and behavioral disorders in preschool children: presentation, nosology, and epidemiology. J Child Psychol Psychiatry 47(3–4): 313–337
19. Walter HJ, Abright AR, Bukstein OG, Diamond J, Keable H, Ripperger-Suhler J, Rockhill C (2023) Clinical practice guideline for the assessment and treatment of children and adolescents with major and persistent depressive disorders. J Am Acad Child Adolesc Psychiatry 62(5):479–502
20. Labouliere CD, Reyes JP, Shirk S, Karver M (2017) Therapeutic alliance with depressed adolescents: predictor or outcome? Disentangling temporal confounds to understand early improvement. J Clin Child Adolesc Psychol 46(4):600–610

21. Grossoehme DH, Gerbetz L (2004) Adolescent perceptions of meaningfulness of psychiatric hospitalization. Clin Child Psychol Psychiatry 9(4):589–596
22. Cairns A, Kavanagh D, Dark F, McPhail SM (2015) Setting measurable goals with young people: qualitative feedback from the Goal Attainment Scale in youth mental health. Br J Occup Ther 78(4):253–259
23. Jacob J, Stankovic M, Spuerck I, Shokraneh F (2022) Goal setting with young people for anxiety and depression: what works for whom in therapeutic relationships? A literature review and insight analysis. BMC Psychol 10(1):171
24. Law D, Wolpert M (2014) Guide to using outcomes and feedback tools. CAMHS Press, London
25. Korotitsch WJ, Nelson-Gray RO (1999) An overview of self-monitoring research in assessment and treatment. Psychol Assess 11(4): 415–425
26. Clarke G, Lewinsohn P, Hops H (1990) Student workbook: adolescent coping with depression course. Castalia, Eugene
27. Kanter JW, Manos RC, Bowe WM, Baruch DE, Busch AM, Rusch LC (2010) What is behavioral activation?: a review of the empirical literature. Clin Psychol Rev 30(6):608–620
28. Beck AT, Rush AJ, Shaw BF, Emery G, DeRubeis RJ, Hollon SD (2024) Cognitive therapy of depression. Guilford Publications, New York
29. Lawrence HR, Buffie ML, Schwartz-Mette RA, Nangle DW (2020) Cognitive behavioral therapy with depressed youth. In: Friedberg RD, Nakamura BJ (eds) Cognitive behavioral therapy in youth: tradition and innovation. Humana Press, New York
30. Friedberg RD, McClure JM (2015) Clinical practice of cognitive therapy with children and adolescents: the nuts and bolts. Guilford Publications, New York
31. Stallard P. Think good, feel good: a cognitive behavioural therapy workbook for children and young people 2nd ed. Newark: Wiley; 2019
32. Nolen-Hoeksema S (1991) Responses to depression and their effects on the duration of depressive episodes. J Abnorm Psychol 100: 569–582
33. Watkins ER (2008) Constructive and unconstructive repetitive thought. Psychol Bull 134(2):163–206
34. Barrett P (2005) Friends for life: group leaders' manual for children, 4th edn. Australian Academic Press, Brisbane
35. Rapee RM, Lyneham HJ, Hudson JL, Wuthrich VM, Kangas M, Schniering CA, Wignall A (2019) Cool kids anxiety program, 2nd edn. Centre for Emotional Health, Macquarie University, Sydney
36. Weersing VR, Goger P, Schwartz KTG, Baca SA, Angulo F, Kado-Walton M (2024) Evidence-base update of psychosocial and combination treatments for child and adolescent depression. J Clin Child Adolesc Psychol 54:1–51
37. Cuijpers P, Pineda BS, Ng MY, Weisz JR, Muñoz RF, Gentili C et al (2021) A meta-analytic review: psychological treatment of subthreshold depression in children and adolescents. J Am Acad Child Adolesc Psychiatry 60(9):1072–1084
38. Rith-Najarian LR, Mesri B, Park AL, Sun M, Chavira DA, Chorpita BF (2019) Durability of cognitive behavioral therapy effects for youth and adolescents with anxiety, depression, or traumatic stress: a meta-analysis on long-term follow-ups. Behav Ther 50(1):225–240
39. Wergeland GJH, Riise EN, Ost LG (2021) Cognitive behavior therapy for internalizing disorders in children and adolescents in routine clinical care: a systematic review and meta-analysis. Clin Psychol Rev 83:101918
40. MacQueen GM, Frey BN, Ismail Z, Jaworska N, Steiner M, Lieshout RJV et al (2016) Canadian Network for Mood and Anxiety Treatments (CANMAT) 2016 clinical guidelines for the management of adults with major depressive disorder: section 6. Special populations: youth, women, and the elderly. Can J Psychiatry 61(9):588–603
41. American Psychological Association. Clinical practice guideline for the treatment of depression across three age cohorts (2019). Available from: https://www.apa.org/depression-guideline
42. Malhi GS, Bell E, Bassett D, Boyce P, Bryant R, Hazell P et al (2021) The 2020 Royal Australian and New Zealand College of Psychiatrists clinical practice guidelines for mood disorders. Aust N Z J Psychiatry 55(1):7–117
43. Pringsheim T, Gaudet LA, Pillay J, Hartling L, Fletcher J, Wilcox G et al (2024) Evidence-based guidelines for anxiety, depression, obsessive-compulsive disorder and Tourette syndrome in children and youth: a rapid review. J Can Acad Child Adolesc Psychiatry 33(3): 171–205
44. Treatment for Adolescents with Depression Study Team (2004) Fluoxetine, cognitive-behavioral therapy, and their combination for adolescents with depression treatment for adolescents with depression study (TADS) randomized controlled trial. JAMA 292(7):807–820

45. Brent D, Emslie G, Clarke G, Wagner KD, Asarnow JR, Keller M et al (2008) Switching to another SSRI or to venlafaxine with or without cognitive behavioral therapy for adolescents with SSRI-resistant depression: the TORDIA randomized controlled trial. JAMA 299(8):901–913
46. Asarnow JR, Emslie G, Clarke G, Wagner KD, Spirito A, Vitiello B et al (2009) Treatment of selective serotonin reuptake inhibitor—resistant depression in adolescents: predictors and moderators of treatment response. J Am Acad Child Adolesc Psychiatry 48(3):330–339
47. Cipriani A, Zhou X, Del Giovane C, Hetrick SE, Qin B, Whittington C et al (2016) Comparative efficacy and tolerability of antidepressants for major depressive disorder in children and adolescents: a network meta-analysis. Lancet 388(10047):881–890
48. Locher C, Koechlin H, Zion SR, Werner C, Pine DS, Kirsch I et al (2017) Efficacy and safety of selective serotonin reuptake inhibitors, serotonin-norepinephrine reuptake inhibitors, and placebo for common psychiatric disorders among children and adolescents: a systematic review and meta-analysis. JAMA Psychiatry 74(10):1011–1020
49. Cuijpers P, Miguel C, Harrer M, Plessen CY, Ciharova M, Ebert D, Karyotaki E (2023) Cognitive behavior therapy vs. control conditions, other psychotherapies, pharmacotherapies and combined treatment for depression: a comprehensive meta-analysis including 409 trials with 52,702 patients. World Psychiatry 22(1):105–115
50. Weisz JR, Venturo-Conerly KE, Fitzpatrick OM, Frederick JA, Ng MY (2023) What four decades of meta-analysis have taught us about youth psychotherapy and the science of research synthesis. Annu Rev Clin Psychol 19: 79–105

Chapter 5

Assessment and Treatment of Suicidal Thoughts and Behaviors in Youth

Melissa Padron, Dorothy H. Balser, and Jeremy W. Pettit

Abstract

Suicide in children and adolescents (youth) is a major public health crisis, with rates of suicide and suicide attempts showing alarming increases over the past two decades. This chapter describes the cognitive-behavioral approach to treatment of suicidal thoughts and behaviors in youth. We first review data on the prevalence of suicidal thoughts and behaviors in youth and describe prominent theoretical models of suicide. We then briefly summarize research on treatments for suicidal thoughts and behaviors in youth, including cognitive-behavioral therapy (CBT), and strategies for assessing and monitoring risk. We conclude with an overview of the basic principles of CBT for suicidal thoughts and behaviors in youth, followed by the structure and goals of treatment.

CBT leverages a collaborative partnership between therapist and client, with the initial phase (psychoeducation and stabilization) prioritizing assessment and management of risk, including the use of safety planning. The initial phase also emphasizes psychoeducation on the cognitive-behavioral conceptualization of suicidal thoughts and behaviors as maladaptive coping strategies, setting up the goal of developing alternative, adaptive strategies for coping with distress. Information gathered during assessment is used to devise a treatment plan. During the second phase of treatment (application), the emphasis shifts to implementing the treatment plan with an eye toward building adaptive coping skills. Strategies and tasks in the application phase include the identification of reasons to live and creation of a hope box to promote motivation and connect youth with their value system, as well as the development of more adaptive thinking patterns and emotion regulation skills for managing difficult situations. Family dynamics are also a focus of the application phase, enhancing communication strategies, active listening, conflict resolution, and emotional validation to strengthen the family support system. In the final phase (skills consolidation and relapse prevention), the emphasis shifts to helping youth solidify skills learned during treatment to prepare for termination.

Key words Suicidal ideation, Suicidal behaviors, Suicide attempt, Safety planning, Treatment, Children, Adolescents

1 Introduction and Overview

Suicide in children and adolescents (youth) is a major public health crisis in the United States, ranking as the fifth leading cause of death for children and fluctuating between the second and third leading

Robert D. Friedberg and Micaela A. Thordarson (eds.), *Cognitive Behavioral Therapy with Youth: Tradition and Innovation*, Neuromethods, vol. 231, https://doi.org/10.1007/978-1-0716-5134-6_5,

cause of death for adolescents [1]. Recent data reveal a troubling rise in suicidal thoughts and behaviors among youth [2–4]. Suicide rates rose by 14.7% annually between 2012 and 2017 in children aged 5 to 11, tripled in youth aged 10 to 14 from 2007 to 2018, and increased 57% in adolescents aged 15 to 19 from 2009 to 2017 [2, 5]. Emergency department visits for youth experiencing suicidal thoughts and behaviors doubled from 2007 to 2015 [6]. Additionally, the percentage of high school students who seriously considered attempting suicide increased from 16% in 2011 to 22% in 2021, with corresponding increases in the percentages who made a suicide plan (from 13% to 18%) and attempted suicide (from 8% to 10%) [7].

As these alarming rising rates suggest, many mental health clinicians who work with youth will encounter clients who experience suicidal thoughts or behaviors, highlighting the importance of accurate risk assessment and efficacious treatments for this population. In the following sections of this chapter, we review prominent theoretical models of suicide and summarize research on cognitive-behavioral and related treatments for addressing suicidal thoughts and behaviors in youth. We then describe specific assessment and treatment strategies for clinicians when working with youth.

2 Brief Review of Prominent Theoretical Models of Suicide

Theoretical models provide foundations upon which assessment and treatment approaches are built. Cognitive-behavioral theories of suicide broadly propose that information processing biases, especially distorted interpretations, and poor coping and problem-solving skills contribute to the development and maintenance of suicidal thoughts and behaviors [8–12]. For example, cognitive theories posit that individuals who attribute negative life situations to personal flaws and struggle to find solutions may become hopeless and see suicide as an escape from distressing situations [13–16]. Such theories informed the development of treatments that targeted cognitive distortions and problem-solving skills.

In recent years, researchers have expanded upon earlier work and proposed new theories of suicide grounded in the ideation-to-action framework. These theories include the interpersonal theory of suicide [17], the three-step theory [18, 19], and the integrated motivational-volitional model [20]. At the core of the ideation-to-action framework is the distinction between the desire for suicide and the capability to act on that desire. Suicide desire refers to a wish not to live, while suicide action refers to behaviors aimed at self-inflicted death. According to the ideation-to-action framework, suicide attempts occur when both desire and the capability to act on it are present [21]. This framework assumes a sequential progression from thoughts about suicide to actions toward suicide. This progression implies a window of opportunity to identify risk and intervene on suicidal thoughts before an attempt is made.

Ideation-to-action theories identify key drivers of suicidal ideation that can be targeted in treatment. For example, the interpersonal theory of suicide emphasizes a sense of social disconnection, referred to as thwarted belongingness, and the belief that one is a burden on others as key drivers of suicide desire, especially when individuals feel hopeless that these two painful states will change [17]. The three-step theory emphasizes pain and hopelessness as key drivers of suicidal ideation and further posits that the severity of suicidal ideation escalates when pain exceeds or overwhelms connectedness [22]. The integrated motivational-volitional model emphasizes feelings of defeat and entrapment that arise in the context of negative life circumstances as key drivers of suicidal ideation [20]. When combined with feelings of disconnection, burdensomeness, or hopelessness, entrapment leads to the belief that suicide provides a solution to negative life circumstances. Drivers of suicidal ideation common to ideation-to-action theories thus include feelings of hopelessness, social disconnection, and psychological pain, all potential targets for treatment [23].

3 Summary of Research on Treatments for Suicidal Thoughts and Behaviors

As noted, theoretical models inform assessment and treatment approaches that aim to ensure the safety of youth who experience suicidal thoughts and behaviors while addressing the drivers of such thoughts and behaviors. Historically, suicidal thoughts and behaviors were targeted indirectly in the context of treatments focused on other psychopathologies such as depression, anxiety, and borderline personality disorder, with the assumption that effective treatment of such psychopathologies would produce concomitant reductions in suicidal thoughts and behaviors. In light of data questioning the validity of that assumption [24], researchers in recent years have focused on developing treatments that directly target suicidal thoughts and behaviors in youth. We briefly summarize findings on the efficacy of these treatments, drawing upon recent reviews and meta-analyses.

In a meta-analysis published in 2015, Ougrin and colleagues evaluated 19 randomized controlled trials (RCTs) of treatments aimed at reducing suicidal behaviors and non-suicidal self-injury (NSSI) in youth aged 10 to 18 years [25]. The meta-analysis included treatments from diverse theoretical backgrounds such as cognitive-behavioral therapy (CBT), dialectical behavioral therapy (DBT), family therapy, and mentalization-based therapy (MBT). Findings demonstrated that multiple treatments were superior to treatment as usual in reducing self-harm in general, with large effect sizes for DBT, medium to large effect sizes for CBT, and small to medium effect sizes for MBT. However, effect sizes for suicide attempts specifically were smaller compared to effect sizes for NSSI, highlighting a need to develop more effective strategies for preventing suicide attempts.

In 2019, Glenn and colleagues conducted an evidence-based update for the treatment of self-injurious thoughts and behaviors in youth aged 10 to 19 years, reviewing 26 RCTs [26]. The evidence supporting each treatment was classified on five levels based on the number and quality of studies, ranging from well-established (Level 1) to questionable efficacy (Level 5) [27]. Using these criteria, only DBT was classified as Level 1 (well-established), indicating the presence of strong, replicable evidence for reducing deliberate self-harm and suicidal ideation in multiple independent RCTs. CBT, interpersonal psychotherapy, and family-based therapies were classified as Level 2 (probably efficacious), indicating the presence of moderate evidence of efficacy, though more support from independent RCTs was needed before these treatments could be considered well-established.

In 2022, Itzhaky and colleagues published a meta-analysis evaluating psychosocial treatments for suicide prevention in adolescents aged 10 to 18 years [28]. This study included 30 RCTs, with a focus on interventions that targeted self-harming behaviors (SHB) and suicidal ideation. The results indicated an overall small but statistically significant decrease in suicidal ideation from pre- to post-treatment across treatments collectively. However, the overall effect size for suicide attempts was very small and not statistically significant. Of the 30 studies reviewed, 10 showed superiority over comparator control arms. It is important to note that comparator control arms typically received active treatment as usual, providing a stringent benchmark for detecting superior effects in the experimental treatment arm. While the treatment and comparator control arms typically showed reductions in suicidal ideation by post-treatment, no treatment demonstrated clear superiority over others.

In 2022, Wu and colleagues published an umbrella review of the efficacy of CBT for suicide behaviors that synthesized nine systematic reviews and meta-analyses, of which seven included youth aged 11 years and older [29]. They concluded that CBT offers benefits for treating suicidal thoughts and behaviors, such as flexibility, cost-effectiveness, and feasibility to integrate suicide prevention within a larger treatment plan for psychopathologies. They further concluded that CBT yielded statistically significant but small effect sizes in the reduction of suicidal ideation and prevention of suicide attempts.

Based on the conclusions of these reviews and meta-analyses, it is clear that considerable progress has been made in the development and testing of treatments that target suicidal thoughts and behaviors in youth. It is also clear that considerable work remains to enhance the potency of treatments, especially for preventing suicide attempts. In the following section, we elaborate on the cognitive-behavioral conceptualization of suicide and describe the principles, structure, and implementation of CBT for suicidal thoughts and behaviors in youth.

4 CBT for Suicidal Behavior: In Practice

4.1 Cognitive-Behavioral Conceptualization of Suicide

The cognitive-behavioral conceptualization views suicidal thoughts and behaviors as problematic coping behaviors [30]. That is, suicidal thoughts and behaviors emerge from negative thought patterns and maladaptive coping mechanisms in response to distress. This framework suggests that youth who experience persistent negative beliefs about themselves, their future, and their relationships with others may feel trapped and hopeless and view suicide as a means of escaping emotional pain. Stressors in the context of interpersonal relations, together with cognitive distortions, may foster feelings of burdensomeness and social disconnection [31]. Coupled with limited problem-solving skills and impaired emotion regulation, these distorted beliefs can drive youth toward suicidal ideation and/or behavior. CBT aims to identify and change these harmful thought patterns, improve coping strategies, and promote adaptive thinking and behaviors to reduce suicide risk.

4.2 Principles, Structure, and Goals of CBT for Suicidal Thoughts and Behaviors

CBT is a structured, time-limited, and problem-focused approach that requires active participation from the therapist and client, and often from a caregiver, too. A typical course of treatment lasts 10 to 16 sessions [32]. The overarching goal of treatment is to reduce suicidal thoughts and prevent suicidal behaviors. Tasks used to achieve this goal include identifying and restructuring harmful thought patterns, enhancing problem-solving skills, and improving emotion regulation skills [33–35].

The initial phase of treatment, typically the first three sessions, involves stabilizing the client, providing psychoeducation about suicidal thoughts and behaviors, and establishing treatment goals. The therapist works to achieve consensus with the client and caregiver about the key issues that will be targeted in treatment sessions. Achieving consensus strengthens the alliance and fosters motivation, enhancing adherence to treatment. The middle phase of treatment, the application phase, involves developing, applying, and practicing coping and problem-solving skills to reduce suicidal thoughts and behaviors. The final phase of treatment focuses on consolidating skills learned during the middle phase and preparing the youth to continue applying adaptive coping and problem-solving skills independently.

Phase I: Psychoeducation and Stabilization The primary goal of Phase I is to stabilize the client and reduce immediate suicide risk. Therapists focus on building a strong therapeutic alliance, conducting a thorough risk assessment, and creating a plan to ensure the client's short-term safety while they develop skills to manage suicidal thoughts independently in the long term.

Assessment. Ensuring the client's safety must always remain at the forefront of treatment for suicidal thoughts and behaviors. For this reason, accurate assessment and determination of suicide risk, as well as steps to maintain the client's safety, precedes skill-building and continues regularly throughout treatment. Several scales exist for screening and monitoring suicidal ideation and/or behaviors in adolescents, each with strengths and limitations. Self-rating screening tools can be efficient and cost-effective and also may reduce respondents' concerns about social evaluation, potentially leading to more openness to disclose suicidal thoughts and behaviors. Examples of self-rating screening tools include the following: the Ask Suicide-Screening Questions (ASQ), a 4-item screening tool for youth aged 10 years and older [36]; the Risk of Suicide Questionnaire (RSQ), a 4-item screening tool for youth aged 8 years and older [37]; the Patient Health Questionnaire-9 Adolescent Version (PHQ-9A), a 9-item depression screening tool for youth aged 12 to 18 years that includes a single item on suicidal ideation [38]; the Suicidal Ideation Questionnaire (SIQ) [39], a 30-item tool for high school students, and the corresponding Suicidal Ideation Questionnaire-Junior (SIQ-JR), a 15-item tool for middle school students [40]; and the Suicide Behaviors Questionnaire-Revised (SBQ-R), a 4-item tool for youth aged 13 years and older [41].

Interviewer-administered scales can allow for probing, clarifying, and expanding on topics, potentially leading to deeper understanding of a client's thoughts and behaviors. Examples of interviewer-administered scales include the following: the Patient Safety Screener (PSS-3), a 3-item screening tool for youth aged 12 years and older [42]; the Modified Scale for Suicidal Ideation (MSSI), an 18-item scale originally developed for adults that has been used with youth aged 13 years and older [43]; the Self-Injurious Thoughts and Behaviors Interview-Revised (SITBI-R), a lengthy interview that covers multiple topics relevant to self-injury originally developed for adults that has been used with youth aged 12 years and older [44]; and the Columbia-Suicide Severity Rating Scale (C-SSRS), which offers several versions that assess recent and lifetime suicidal thoughts and behaviors [45]. Developed for adolescents, the C-SSRS has primarily been used with youth aged 12 years and older [46], but has also been adapted for children as young as 4 years [47].

Because most assessment tools were designed for adolescents or adults, there is little guidance on evaluating suicidal thoughts and behaviors in preadolescent children. In 2022, the National Institute of Mental Health (NOT-MH-22-085) highlighted a need to adapt measures used with older adolescents and adults for younger children. To address this need, researchers are beginning to develop and validate suicide screening tools specifically for children. Examples of emerging scales designed for children include the Suicide Risk and Ideation Scale for Kids-Child and Parent Versions (S-RISK-C/P) [48], a 22-item rating scale based on ideation-to-action frameworks.

Selection of assessment tools may be guided by the purpose, setting, resources, and provider and client burden. Brief tools, such as the ASQ, may be optimal for the purpose of initial screening especially when time and resources are limited. Following a positive screen, lengthier tools that provide more in-depth information about suicidal thoughts and behaviors, such as the S-RISK-C/P, and/or allow for flexibility and probing into risk areas, such as the SITBI-R, provide a more comprehensive assessment of risk areas and potential treatment targets.

Safety Planning. Following assessment, the therapist works with the client and family to agree upon a primary treatment goal of managing suicidal thoughts and behaviors. Once such agreement is achieved, the therapist collaborates with the client and family to devise a safety plan. Therapists can present safety planning to a client and family with the rationale that suicidal crises are typically time limited. Coming up with a plan to prevent youth from acting on suicidal thoughts allows time for the crisis to diminish in intensity while skills needed to prevent future crises are being developed in treatment [49].

Safety planning consists of a brief, sequential plan in the event that suicidal thoughts emerge in a triggering situation [24, 49]. The first step is identifying triggers that lead to suicidal ideation or urges (e.g., family/peer conflict, academic demands). This can be achieved through chain analysis, wherein the therapist and client dissect the events and emotional states leading to suicidal thoughts and behaviors [50]. The second step involves identification of activities that the client enjoys doing on their own, being careful to ensure the client has the means and capacity to engage in these activities (e.g., listening to music, watching a television show). The third step involves identification of people or places that offer distractions from suicidal thoughts, including a list of contact information and locations. The primary goal of steps 2 and 3 is distraction from negative emotions and suicidal thoughts. As such, the therapist can assist the client in ensuring that selected activities, people, and places are incongruent with the negative emotions of the client at the time of triggering event and promote more neutral or positive emotions. When working with children, therapists are encouraged to provide preplanned activities for the child, rather than relying on the child to generate activities [51].

The fourth step involves identification of trusted adults, typically family members, to whom the client can disclose the crisis and seek support. The adults' contact information should be included for accessibility in a crisis. When working with children, therapists can identify at least one trusted adult in all the settings in which the child spends time (e.g., home, school) [51]. The fifth step involves generating a list of professional resources the client can contact for help, including emergency numbers for the therapist, the National Suicide Prevention Lifeline (988), and the nearest emergency department.

Therapists use judgment in deciding how much to involve caregivers in steps 1–5. These steps are completed with caregivers present when working with children, but typically not with adolescents. When caregivers are not involved, the therapist should discuss with the client how caregivers or other supportive adults can best assist them in using the safety plan. Once this is determined, the therapist should guide the client through role-plays or planning discussions to prepare them for sharing this information and explaining the safety plan when caregivers join the session for step 6 [49].

In step 6, the therapist works with the client and caregivers to make the environment safe by limiting access to lethal means of self-harm, such as firearms, medications, and sharp objects. Therapists provide psychoeducation on the importance of means safety and confirm with caregivers that youths lack access to means of suicide. Caregivers are coached to implement safety measures, such as locking up medications, securing sharp objects, and using monitoring strategies like open-door policies and periodic check-ins. For a client at high risk, alternatives like hospitalization should be considered if caregivers cannot provide adequate monitoring and ensure a safe environment. A day after the safety planning session, therapists may call caregivers to confirm that agreed-upon safety measures have been implemented.

Setting the Stage for Application. Following appropriate risk management steps, therapists explore collaboratively with the client antecedents to the client's reported suicidal thoughts and/or behaviors. This includes mapping a client's history of suicidal behaviors to identify triggers and inform treatment plans and tasks [52, 53]. Identification of triggers alerts the client to when they should use safety plans and sets the stage for problem-solving, wherein the client and therapist collaboratively generate alternative, adaptive ways of responding to triggers. Such alternatives are added to the client's safety plan. Therapists can also present their client with a list of skills (i.e., a "treatment menu") that may be covered later in treatment and discuss which skills they believe would be most helpful to them [54]. This allows the client to participate in shaping the treatment plan in ways that will benefit them.

Phase II: Application In Phase II, the emphasis shifts to identifying and restructuring maladaptive thought patterns and beliefs associated with suicidal thoughts and behaviors. Therapists facilitate a client's identification of negative thoughts and provide skills used to challenge such thoughts and develop more adaptive thinking patterns. Informed by recent ideation-to-action theories [17, 18, 20], therapists can be particularly attuned to thought patterns indicative of social disconnection, hopelessness, and burdensomeness, which may exacerbate feelings of entrapment. Therapists also work with the client to develop emotion regulation strategies and practice skills for managing difficult situations. Clients are encouraged to engage in behavioral experiments and problem-solving exercises aimed at restructuring negative thought patterns, challenging beliefs around hopelessness, and building resilience [55]. This phase is critical for helping a client develop a toolkit of adaptive coping mechanisms to use when experiencing distress and/or suicidal thoughts, supporting a move toward stability and safety.

A strategy often used early in the application phase is developing a list of reasons to live. The purpose of developing this list is to explore and strengthen the reasons that have kept the client from attempting suicide and why those values or motivations are meaningful to them. A secondary aim of this strategy is to increase the client's sense of social connectedness, reduce psychological pain, and increase optimism. Common reasons for living identified by youth include concerns about the impact on loved ones, enjoyment of specific activities, religious or spiritual beliefs, and social connections [56–58]. Helping the youth articulate tangible reasons for living can foster a more positive perspective and counter feelings of hopelessness [59]. For clients who struggle to identify reasons, therapists can shift the emphasis to help them create a hope box or "hope kit," which is a collection of memories and objects associated with positive experiences or important people in their life [60]. These tangible items provide a visual reminder of their commitment to life and can be a useful reference during times of distress. Virtual hope boxes, delivered through smartphone apps, may increase the accessibility and effectiveness of this coping strategy [61].

Additional strategies used in CBT for suicidal thoughts and behaviors include cognitive restructuring, behavioral experiments, managing negative emotions, and facilitating improved family communication [50, 54, 62]. Therapists can implement cognitive restructuring by helping clients identify and challenge negative or distorted thinking patterns that contribute to feelings of hopelessness and despair. This process begins with psychoeducation, teaching the client how thoughts influence emotions and behaviors, and guiding them to recognize common cognitive distortions such as catastrophizing or black-and-white thinking, and how these can maintain the client's psychological pain. Therapists then collaborate with the client to evaluate the accuracy of these thoughts and replace them with more balanced and realistic perspectives. Techniques such as thought records, evidence-based questioning, and behavioral experiments can be used to reinforce these new thought patterns ([55]; *see* also Chapter 1 of this text).

Therapists can also plan behavioral experiments by collaboratively designing activities that test the validity of the client's negative beliefs or assumptions. These experiments are tailored to the client's specific thoughts, such as "No one cares about me" or "I always fail," and aim to provide real-world evidence that challenges these perceptions. The therapist works with the client to identify a manageable, low-risk action, like reaching out to a friend, participating in an enjoyable activity, or attempting a small task they perceive as difficult. Afterward, the client and therapist reflect on the outcome, discussing how the results align with or contradict the initial belief. These experiments help the client build confidence, foster hope, and develop alternative, more balanced ways of thinking, reducing the intensity of suicidal ideation.

Additionally, therapists can help a client develop emotional awareness and regulation skills. This process begins helping the client identify and label their emotions and understand how these emotions influence their thoughts and behaviors. Therapists may use techniques like mindfulness to increase awareness of emotional triggers and encourage acceptance of emotions without judgment [63]. Skills such as deep breathing, grounding exercises, and progressive muscle relaxation can also help the client reduce emotional intensity in the moment. Through these techniques, the client seeks to move into a state of emotional equilibrium and reduce feelings of hopelessness. Therapists may also explore with their client opportunities to develop new and healthy coping strategies, such as journaling, physical activity, or talking to a friend, to manage distress constructively. By fostering emotion regulation, therapists empower the client to feel more in control of their emotions, reducing the likelihood of suicidal thoughts [64].

Therapists can target family communication skills by teaching active listening techniques, encouraging both parents and clients to focus on understanding each other's perspectives without interrupting or judging. Therapists can introduce "I" statements to express feelings and needs constructively, reducing blame and defensiveness. Role-playing exercises can be used to practice these skills in a safe setting, allowing families to address common conflicts more effectively. Further, therapists may work with caregivers to validate their child's emotions, reinforcing a sense of trust and support. By enhancing communication skills, therapists help families build stronger connections, resolve misunderstandings, and create a more supportive home environment.

Phase III: Skills Consolidation and Relapse Prevention In the final phase, the focus shifts to consolidating skills built during Phase II and developing plans to prevent relapses into suicidal thought patterns and behaviors. The therapist and client work on consolidating the skills they have learned, identifying warning signs for future suicidal thoughts, and creating a personalized plan for managing future challenges. This phase empowers the client to apply their skills independently, reinforcing a sense of hope and confidence in their ability to manage distress.

Reviewing and reinforcing coping mechanisms and skills learned during treatment is vital for maintaining progress after treatment ends. During this phase, it is important to emphasize continued practice of these skills, recognizing that suicidal thoughts may recur. Before concluding treatment, therapists should update safety plans to ensure they reflect the most practical and effective strategies for the client. Encouraging continued social engagement and behavioral activation can also serve as powerful preventative tools.

Now that suicide risk has stabilized and clients have developed skills for managing suicidal thoughts and behaviors, this phase also provides an opportunity for therapists to (re)assess potential comorbid concerns, such as trauma or anxiety, that might not have been addressed in treatment. If such concerns are present, therapists can help the client and family evaluate their options for addressing them in continued treatment or through appropriate referrals.

As treatment concludes, therapists can normalize setbacks and negative emotions, helping the client view these experiences as part of a normal growth pattern rather than as failures. Framing the client as an expert on their own warning signs and coping strategies fosters a sense of empowerment and self-efficacy, equipping them to proactively manage future challenges. This approach helps the youth leave treatment feeling prepared, confident, and resilient.

5 Conclusion

Suicidal thoughts and behaviors are prevalent and alarmingly on the rise among youth. Informed by theoretical models of suicide, CBT conceptualizes suicidal thoughts and behaviors as maladaptive ways of coping with distress. Following a thorough risk assessment using developmentally appropriate tools, CBT therapists engage in safety planning with the client to collaboratively identify triggers and develop actionable steps to manage distress, including specific activities for distraction and use of emergency contacts. Assessment and identification of triggers also guides the selection of specific treatment targets that will help the client develop alternative, more adaptive ways of coping with distress.

As therapists shift to implementing a treatment plan, they leverage activities like identifying reasons to live and creating a hope box to help the client identify tangible motivations and shift toward positive thinking. Additional strategies include cognitive restructuring to identify and change maladaptive thinking patterns and strengthening social support networks, particularly through improving family communication, to facilitate adaptive emotion regulation and foster a sense of social connection.

In conclusion, CBT is an evidenced-based treatment for suicidal thoughts and behaviors in youth. Advances in theoretical models, informed by the ideation-to-action framework, have provided novel cognitive and emotional treatment targets that can be addressed in CBT [65]. Building on the promising foundational evidence base and theoretical advancements, further research will be needed to move CBT toward classification as "well-established" and enhance understanding of mechanisms by which positive treatment response is achieved.

Conflicts of Interest

The authors have no conflicts of interest to declare that are relevant to the content of this chapter.

References

1. National Center for Injury Prevention and Control. Web-based injury statistics query and reporting system. Centers for Disease Control and Prevention. https://wisqars.cdc.gov/. Accessed 1 Dec 2024
2. Curtin SC, Garnett MF (2023) Suicide and homicide death rates among youth and young adults aged 10–24: United States, 2001–2021. NCHS Data Brief 471:1–8
3. Ruch DA, Sheftall AH, Schlagbaum P, Rausch J, Campo JV, Bridge JA (2019) Trends in suicide among youth aged 10 to 19 years in the United States, 1975 to 2016. JAMA Netw Open 2(5):e193886. https://doi.org/10.1001/jamanetworkopen.2019.3886
4. Xiao Y, Cerel J, Mann JJ (2021) Temporal trends in suicidal ideation and attempts among US adolescents by sex and race/ethnicity, 1991–2019. JAMA Netw Open 4(6): e2113513. https://doi.org/10.1001/jamanetworkopen.2021.13513
5. Stijelja S, Mishara BL (2020) COVID-19 and psychological distress-changes in internet searches for mental health issues in New York during the pandemic. JAMA Intern Med 180(12):1703–1706. https://doi.org/10.1001/jamainternmed.2020.3271
6. Burstein B, Agostino H, Greenfield B (2019) Suicidal attempts and ideation among children and adolescents in US emergency departments, 2007–2015. JAMA Pediatr 173(6):598–600. https://doi.org/10.1001/jamapediatrics.2019.0464
7. Centers for Disease Control and Prevention (2023) Youth risk behavior survey: data summary and trends report. National Center for HIV, Viral Statistics, STD, and TB Prevention, Division of Adolescent and School Health, Atlanta, pp 60–69. Published online
8. Brüdern J, Spangenberg L, Stein M et al (2024) Implicit measures of suicide vulnerability: investigating suicide-related information-processing biases and a deficit in behavioral impulse control in a high-risk sample and healthy controls. Behav Res Ther 180: 104601. https://doi.org/10.1016/j.brat.2024.104601

9. Wenzel A, Spokas M (2014) Cognitive and information processing approaches to understanding suicidal behaviors. In: The Oxford handbook of suicide and self-injury. Oxford library of psychology, Oxford University Press, pp 235–254. https://doi.org/10.1093/oxfordhb/9780195388565.001.0001
10. Weismoore JT, Esposito-Smythers C (2010) The role of cognitive distortion in the relationship between abuse, assault, and non-suicidal self-injury. J Youth Adolesc 39(3):281–290. https://doi.org/10.1007/s10964-009-9452-6
11. Sadowski C, Kelley ML (1993) Social problem solving in suicidal adolescents. J Consult Clin Psychol 61(1):121–127. https://doi.org/10.1037/0022-006X.61.1.121
12. Xu L, Zhang H, Zhou C et al (2023) Deficits in psychological resilience and problem-solving ability in adolescents with suicidal ideation. Child Adolesc Psychiatry Ment Health 17(1): 31. https://doi.org/10.1186/s13034-023-00577-z
13. Baumeister RF (1990) Suicide as escape from self. Psychol Rev 97(1):90–113. https://doi.org/10.1037/0033-295X.97.1.90
14. Beck AT, Weissman A, Lester D, Trexler L (1974) The measurement of pessimism: the hopelessness scale. J Consult Clin Psychol 42(6):861–865. https://doi.org/10.1037/h0037562
15. Joiner T (2005) Why people die by suicide. Harvard University Press, Cambridge, p 276
16. Nock MK (2010) Self-injury. Annu Rev Clin Psychol 6:339–363. https://doi.org/10.1146/annurev.clinpsy.121208.131258
17. Van Orden KA, Witte TK, Cukrowicz KC, Braithwaite SR, Selby EA, Joiner TE (2010) The interpersonal theory of suicide. Psychol Rev 117(2):575–600. https://doi.org/10.1037/a0018697
18. Anderson AM, Happ MB (2021) The three-step theory of suicide: analysis and evaluation. ANS Adv Nurs Sci 44(1):89–100. https://doi.org/10.1097/ANS.0000000000000337
19. Meerwijk EL, Jones GA, Shotqara AS et al (2024) Development of a 3-step theory of suicide ontology to facilitate 3ST factor extraction from clinical progress notes. J Biomed Inform 150:104582. https://doi.org/10.1016/j.jbi.2023.104582
20. O'Connor RC, Kirtley OJ (2018) The integrated motivational–volitional model of suicidal behaviour. Philos Trans R Soc B Biol Sci 373(1754):20170268. https://doi.org/10.1098/rstb.2017.0268
21. Ribeiro JD, Joiner TE (2009) The interpersonal-psychological theory of suicidal behavior: current status and future directions. J Clin Psychol 65(12):1291–1299. https://doi.org/10.1002/jclp.20621
22. Klonsky ED, May AM (2015) The three-step theory (3ST): a new theory of suicide rooted in the "ideation-to-action" framework. Int J Cogn Ther 8(2):114–129. https://doi.org/10.1521/ijct.2015.8.2.114
23. Kirshenbaum JS, Pagliaccio D, Bitran A, Xu E, Auerbach RP (2024) Why do adolescents attempt suicide? Insights from leading ideation-to-action suicide theories: a systematic review. Transl Psychiatry 14(1):1–22. https://doi.org/10.1038/s41398-024-02914-y
24. Asarnow JR, Porta G, Spirito A et al (2011) Suicide attempts and nonsuicidal self-injury in the treatment of resistant depression in adolescents: findings from the TORDIA study. J Am Acad Child Adolesc Psychiatry 50(8): 772–781. https://doi.org/10.1016/j.jaac.2011.04.003
25. Ougrin D, Tranah T, Stahl D, Moran P, Asarnow JR (2015) Therapeutic interventions for suicide attempts and self-harm in adolescents: systematic review and meta-analysis. J Am Acad Child Adolesc Psychiatry 54(2):97–107.e2. https://doi.org/10.1016/j.jaac.2014.10.009
26. Glenn CR, Esposito EC, Porter AC, Robinson DJ (2019) Evidence base update of psychosocial treatments for self-injurious thoughts and behaviors in youth. J Clin Child Adolesc Psychol 48(3):357–392. https://doi.org/10.1080/15374416.2019.1591281
27. Southam-Gerow MA, Prinstein MJ (2014) Evidence base updates: the evolution of the evaluation of psychological treatments for children and adolescents. J Clin Child Adolesc Psychol 43(1):1–6. https://doi.org/10.1080/15374416.2013.855128
28. Itzhaky L, Davaasambuu S, Ellis SP et al (2022) Twenty-six years of psychosocial interventions to reduce suicide risk in adolescents: systematic review and meta-analysis. J Affect Disord 300:511–531. https://doi.org/10.1016/j.jad.2021.12.094
29. Wu H, Lu L, Qian Y et al (2022) The significance of cognitive-behavioral therapy on suicide: an umbrella review. J Affect Disord 317: 142–148. https://doi.org/10.1016/j.jad.2022.08.067
30. Stanley B, Brown GK (2012) Safety planning intervention: a brief intervention to mitigate suicide risk. Cogn Behav Pract 19(2): 256–264. https://doi.org/10.1016/j.cbpra.2011.01.001

31. Wolff J, Esposito-Smythers C, Becker S et al (2014) Social-cognitive moderators of the relationship between peer victimization and suicidal ideation among psychiatrically hospitalized adolescents. J Aggress Maltreat Trauma 23(3):268–285. https://doi.org/10.1080/10926771.2014.883458
32. Kagan ER, Frank HE, Knepley MJ, Kendall PC (2021) Beyond 16 sessions: extending manualized treatment of anxious youth. J Child Fam Stud 30(2):493–501. https://doi.org/10.1007/s10826-020-01872-5
33. Dobson KS, Poole JC, Beck JS (2018) The fundamental cognitive model. In: Science and practice in cognitive therapy: foundations, mechanisms, and applications. The Guilford Press, New York, pp 29–47
34. Kellogg SH, Young JE (2008) Cognitive therapy. In: Twenty-first century psychotherapies: contemporary approaches to theory and practice. Wiley, New York, pp 43–79
35. Knapp P, Beck AT (2008) Cognitive therapy: foundations, conceptual models, applications and research. Rev Bras Psiquiatr Sao Paulo Braz 1999 30(Suppl 2):s54–s64. https://doi.org/10.1590/s1516-44462008000600002
36. Horowitz LM, Bridge JA, Teach SJ et al (2012) Ask suicide-screening questions (ASQ): a brief instrument for the pediatric emergency department. Arch Pediatr Adolesc Med 166(12): 1170–1176. https://doi.org/10.1001/archpediatrics.2012.1276
37. Horowitz LM, Wang PS, Koocher GP et al (2001) Detecting suicide risk in a pediatric emergency department: development of a brief screening tool. Pediatrics 107(5): 1133–1137. https://doi.org/10.1542/peds.107.5.1133
38. Leslie KR, Chike-Harris K (2018) Patient-administered screening tool may improve detection and diagnosis of depression among adolescents. Clin Pediatr (Phila) 57(4): 457–460. https://doi.org/10.1177/0009922817730343
39. Pinto A, Whisman MA, McCoy KJM (1997) Suicidal ideation in adolescents: psychometric properties of the suicidal ideation questionnaire in a clinical sample. Psychol Assess 9(1): 63–66. https://doi.org/10.1037/1040-3590.9.1.63
40. Courtney DB, Iseyas N, Monga S et al (2024) Systematic review: the measurement properties of the suicidal ideation questionnaire and suicidal ideation questionnaire-Jr. J Am Acad Child Adolesc Psychiatry 63(9):870–887. https://doi.org/10.1016/j.jaac.2023.10.013
41. Osman A, Bagge CL, Gutierrez PM, Konick LC, Kopper BA, Barrios FX (2001) The Suicidal Behaviors Questionnaire-Revised (SBQ-R): validation with clinical and nonclinical samples. Assessment 8(4):443–454. https://doi.org/10.1177/107319110100800409
42. Boudreaux ED, Jaques ML, Brady KM, Matson A, Allen MH (2015) The patient safety screener: validation of a brief suicide risk screener for emergency department settings. Arch Suicide Res 19(2):151–160. https://doi.org/10.1080/13811118.2015.1034604
43. Pettit JW, Garza MJ, Grover KE et al (2009) Factor structure and psychometric properties of the modified scale for suicidal ideation among suicidal youth. Depress Anxiety 26(8): 769–774. https://doi.org/10.1002/da.20575
44. Gratch I, Fernandes SN, Bell KA et al (2022) Self-Injurious Thoughts and Behaviors Interview-Revised (SITBI-R): reliability, validity, and inter-informant agreement in an adolescent sample. J Clin Child Adolesc Psychol 51(4):484–494. https://doi.org/10.1080/15374416.2021.1901229
45. Posner K, Brent D, Lucas C, et al (2008) Columbia-suicide severity rating scale (C-SSRS). Published online. https://cssrs.columbia.edu/wp-content/uploads/C-SSRS_Pediatric-SLC_11.14.16.pdf. Accessed 27 Oct 2024
46. Jakobsen IS, Larsen KJ, Horwood JL (2017) Suicide risk assessment in adolescents – C-SSRS, K10, and READ. Crisis 38(4): 247–254. https://doi.org/10.1027/0227-5910/a000450
47. Cooper D (2017) A study of the validity of the C-SSRS in assessing risk of suicidality in children and adolescents. Dr.H.Sc. The University of Tennessee. https://www.proquest.com/docview/2447888948/abstract/ED7141CC4FB548E3PQ/1. Accessed 1 Dec 2024
48. Pettit JW, Yeguez CE, Budagzad-Jacobson R et al (2024) Using cognitive interviewing to develop rating scales of suicide thoughts and behaviors in preadolescent children. J Appl Res Child Inform Policy Child Risk. https://doi.org/10.58464/2155-5834.1544. Published online
49. Pettit JW, Buitron V, Green KL (2018) Assessment and management of suicide risk in children and adolescents. Cogn Behav Pract 25(4): 460–472. https://doi.org/10.1016/j.cbpra.2018.04.001

50. Stanley B, Brown G, Brent D et al (2009) Cognitive behavior therapy for suicide prevention (CBT-SP): treatment model, feasibility and acceptability. J Am Acad Child Adolesc Psychiatry 48(10):1005–1013. https://doi.org/10.1097/CHI.0b013e3181b5dbfe
51. Itzhaky L, Stanley B (2024) The safety planning intervention for children (C-SPI): rationale and case illustration. Cogn Behav Pract 31(2):250–258. https://doi.org/10.1016/j.cbpra.2022.10.001
52. Mann JJ, Apter A, Bertolote J et al (2005) Suicide prevention strategies a systematic review. JAMA 294(16):2064–2074. https://doi.org/10.1001/jama.294.16.2064
53. Chu C, Buchman-Schmitt JM, Stanley IH et al (2017) The interpersonal theory of suicide: a systematic review and meta-analysis of a decade of cross-national research. Psychol Bull 143(12):1313–1345. https://doi.org/10.1037/bul0000123
54. Esposito-Smythers C, Walsh A, Spirito A, Rizzo C, Goldston DB, Kaminer Y (2012) Working with the suicidal client who also abuses substances. Cogn Behav Pract 19(2):245–255. https://doi.org/10.1016/j.cbpra.2010.11.004
55. Curtiss JE, Levine DS, Ander I, Baker AW (2021) Cognitive-behavioral treatments for anxiety and stress-related disorders. Focus 19(2):184–189. https://doi.org/10.1176/appi.focus.20200045
56. Blum RW, Lai J, Martinez M, Jessee C (2022) Adolescent connectedness: cornerstone for health and wellbeing. BMJ 379:e069213. https://doi.org/10.1136/bmj-2021-069213
57. Pirani S, Kulhanek C, Wainwright K, Osman A (2021) The reasons for living inventory for young adults (RFL-YA-II). Assessment 28(3):942–954. https://doi.org/10.1177/1073191119900242
58. Gorraiz G, Porta G, McMakin DL et al (2024) Factors associated with reasons for living among suicidal adolescents. Arch Suicide Res 28(2):471–481. https://doi.org/10.1080/13811118.2023.2190367
59. Brent DA, Greenhill LL, Compton S et al (2009) The treatment of adolescent suicide attempters study (TASA): predictors of suicidal events in an open treatment trial. J Am Acad Child Adolesc Psychiatry 48(10):987–996. https://doi.org/10.1097/CHI.0b013e3181b5dbe4
60. Berk MS, Henriques GR, Warman DM, Brown GK, Beck AT (2004) A cognitive therapy intervention for suicide attempters: an overview of the treatment and case examples. Cogn Behav Pract 11(3):265–277. https://doi.org/10.1016/S1077-7229(04)80041-5
61. Bush NE, Dobscha SK, Crumpton R et al (2015) A virtual hope box smartphone app as an accessory to therapy: proof-of-concept in a clinical sample of veterans. Suicide Life Threat Behav 45(1):1–9. https://doi.org/10.1111/sltb.12103
62. Brent DA, Poling KD, Goldstein TR (2011) Treating depressed and suicidal adolescents: a clinician's guide. Guilford Press, New York
63. Schuman-Olivier Z, Trombka M, Lovas DA et al (2020) Mindfulness and behavior change. Harv Rev Psychiatry 28(6):371–394. https://doi.org/10.1097/HRP.0000000000000277
64. Asarnow JR, Berk MS, Bedics J et al (2021) Dialectical behavior therapy for suicidal self-harming youth: emotion regulation, mechanisms, and mediators. J Am Acad Child Adolesc Psychiatry 60(9):1105–1115.e4. https://doi.org/10.1016/j.jaac.2021.01.016
65. Buitron V, Hill RM, Cabrera V, Pettit JW (2022) Development and initial validation of a parent report measure of youth belongingness and burdensomeness. Suicide Life Threat Behav 52(5):857–865. https://doi.org/10.1111/sltb.12869

Chapter 6

Coping When the Struggle Is Real: Cognitive Behavioral Therapy Strategies for Youth Anxiety

Brian C. Chu, Emily Badin, and Sandra S. Pimentel

Abstract

Anxiety disorders and related problems affect a large number of children and adolescents, are impairing, and can affect long-term functioning if left untreated. Recent years, beset by the COVID-19 pandemic and reckoning with social injustice, have only amplified the everyday challenges that evoke anxiety in our youth. Youth can be encouraged to actively cope with environmental and societal challenges using cognitive behavioral therapy (CBT), an efficacious set of interventions that teach anxious youth skills to help manage anxiety symptoms, adjust unconstructive thinking and behavioral patterns, and build on strengths. This chapter describes and illustrates CBT principles and strategies that can help children and adolescents address the anxiety they experience in light of trials grounded in real-world challenges where barriers extend beyond their immediate control. Worksheet templates and case illustrations are provided to demonstrate how a clinician turns principles into practice using a flexible, principles-based approach.

Key words Child, Adolescent, Anxiety, Cognitive behavioral therapy (CBT)

1 Introduction

Anxiety disorders reflect a broad category of socio-emotional disorders characterized by intense fear reactions, pervasive worries, avoidant behaviors, and self-critical and threat-related thinking. Concerns that bring youth to treatment include a diverse set of disorders, including generalized anxiety, social anxiety, separation anxiety, panic and agoraphobia, specific phobias, and selective mutism. Traditionally, anxiety disorders have been cited as the most common psychological disorder among children and adolescents (hereafter referred to as youth), affecting nearly 32% of 13- to 18-year-olds, followed by behavior disorders (19.1%), mood disorders (14.3%), and substance use disorders (11.4%) [1]. These disorders come with a host of impairing consequences across social, occupational, academic, and family functioning that may extend into adulthood if left untreated [2]. The degree of comorbidity

Robert D. Friedberg and Micaela A. Thordarson (eds.), *Cognitive Behavioral Therapy with Youth: Tradition and Innovation*, Neuromethods, vol. 231, https://doi.org/10.1007/978-1-0716-5134-6_6,

between anxiety and depression is also significant and reflected on dimensional measures [3–5].

Youth populations were particularly impacted by the COVID-19 pandemic and its sequelae as initial response to the pandemic severely interrupted in-person learning routines and social and community networks, limited recreational activities, and strained limited societal resources and supports [6]. Estimates suggest that the prevalence of anxiety and mood problems may have doubled during the most acute pandemic periods [7, 8]. Globally, these outcomes were distributed unevenly with females, younger people, and individuals of lower socioeconomic status experiencing the greatest impact [9]. In the United States, younger individuals and ethno-racial minorities reported disproportionately worse mental health outcomes, highlighting the elevated risk to vulnerable communities [10].

At the same time, multiple social justice issues came to the forefront that forced a national/international reckoning with social-cultural inequities and a growing awareness of environmental catastrophe. The past decade has seen a rise in youth exposure to violence, poor mental health, and suicidal thoughts and behaviors [11]. In particular, Black youth and adults experience more discriminatory practices [12] and mental health problems related to racial violence than others [13]. Following the police killings of George Floyd and Breonna Taylor in 2020, 50% of Black teens reported thinking about racism and police brutality as a key cause of distress [14]. As a result of pandemic-related misattributions to China, violence and discrimination against Asian-born and Asian-American individuals escalated in the early 2020s [15]. More broadly, discrimination and anti-immigration policies, particularly in the current political climate, have been linked to mental health concerns for immigrant and undocumented families [16, 17]. Faith-based discrimination continues to have detrimental impact in America, as Jewish youth and young adults reported sharp increases in anti-Semitic incidents, particularly after the October 2023 attack on Israel which has been connected with increased depressed symptoms [18–21]. Anti-Muslim and anti-Arab sentiments remain high in the United States and abroad and has been associated with distress and hypervigilance [22–24].

Focused attention on gender, sex, and sexuality inequalities has also intensified. Teen girls are experiencing record high levels of violence, sadness, and suicide risk with rates rising faster than teen boys over the past decade [25]. Oppressive legislation (and its associated public debates) related to LGBTQ rights have stressed youth mental health, in which 86% of transgender and nonbinary youth connect emotional distress to anti-trans policies, 45% of trans youth experience cyberbullying, and a third feel unsafe to go to a doctor or hospital when feeling sick [26]. And in national and international polls, at least 85% of teens and young people report feeling at least moderately worried about climate change, and over 40% report a current impact on mental health [27, 28]. Importantly, the degree to which youth feel ignored or dismissed when expressing concern is linked to increased disparagement. Our kids are dealing with a lot, and anxiety is a normative reaction to world events.

The encouraging news is that exposure itself does not equate to harm. Youth activism and social engagement can help buffer some of the effects of social injustice. For example, teens who engaged with social media related to Black Lives Matter reported greater positive feelings (e.g., inspired, hopeful) than negative feelings (e.g., angry, afraid), and in-person engagement (e.g., demonstrations) engendered more positive feelings than online engagement [29]. However, it is important to note that Black teens reported both higher negative and positive emotions than other teens. Among undocumented Latinx college students, social justice activism (actions to address social injustice) and critical agency (belief one can make a difference) have been found to mediate (weaken) the impact of discrimination on student depression [30]. In the context of climate anxiety, a combination of problem-focused and meaning-focused (trust in leaders, optimism for solutions) coping has been found to buffer emotional distress in teens [31]. Thus, the challenge for any cognitive behavioral therapist is to help youth navigate a complex world where anxiety is an understandable reaction to the stressors and barriers they face, while also promoting action and self-efficacy.

This chapter extends work we introduced in an earlier edition [32] that described how to conceptualize cognitive behavioral therapy (CBT) in a "nutshell" for anxious youth, including a review of classic assessment and intervention strategies. This current chapter develops our conceptualization model to place an increasing importance on understanding the multiple systems that impact youth and to recognize the local, regional, and cultural supports and barriers that promote or impede youth well-being. This chapter further highlights advances in CBT strategies that have been integrated into traditional approaches (e.g., values assessment, mindfulness) to help address contemporary youth and family concerns. We provide case illustrations that demonstrate how therapists can apply CBT in these complicated times when the struggle often feels real and in reality can reflect very real challenges.

2 Theoretical and Empirical Foundations

2.1 Evidence Base for Cognitive Behavioral Therapy

A substantial body of research has demonstrated the efficacy of cognitive behavioral therapy (CBT) for treating youth anxiety [33, 34]. The largest randomized clinical trial, the Child Anxiety Multimodal Study (CAMS), evaluated the efficacy of individual cognitive behavioral therapy (Coping Cat; CBT), medication (sertraline; SRT), a combination of the two treatments, and a pill placebo among 488 youth aged 7–17 for treating anxiety [35]. Response rates indicated notable symptom reductions of 80.7% among participants in the combined treatments, 59.7% among those in CBT, and 54.9% in the pill placebo condition. Long-term outcomes of youth who received CBT as part of CAMS (4–12 years after randomization) showed that rates of remission ranged between 40% and 60% across the 4 years of the uncontrolled follow-up and that successful response during the acute period predicted remission at follow-up [36].

Other trials support efficacy for individual, family, and group versions of CBT. Kendall and colleagues [37] conducted a randomized clinical trial comparing the efficacy of individual cognitive behavioral therapy (ICBT), family cognitive behavioral therapy (FCBT), and a family-based education/support/attention (FESA) active control for treating youth ages 7–14 years old with an anxiety disorder. FCBT and ICBT conditions were superior to FESA in remission of principal anxiety diagnosis (57%, 55%, and 37%). Another RCT conducted by Hudson and colleagues [38] randomized 112 children aged 7–16 with a principal anxiety disorder diagnosis to either a group cognitive behavioral treatment (CBT) or control group (group support and attention [GSA]). Remission rates at 6-month follow-up were significantly higher for CBT compared to GSA condition (68.6% versus 45.5%). CBT has also been extended to community settings and has shown modest results relative to usual care [39]. More recently, transdiagnostic adaptations of CBT, which are designed to simultaneously target shared features across disorder classes, have efficacious results for youth presenting for either anxiety or depression [40–43].

2.2 Anxiety from the CBT Perspective

The cognitive behavioral model helps conceptualize anxiety in the context of three main components (thoughts, feelings, and actions) that are interrelated and influence each other reciprocally [44]. Set within a socio-ecological model, we can understand how an individual youth "moves" through the world as they navigate challenges and successes. For example, a child who is anxious about an upcoming language arts project may have the thought, "I'm no good at writing; I'm going to fail," and procrastinate until the day before to start the assignment. Observing this pattern, a CBT therapist can plan to use cognitive restructuring to identify the assumptions implicit in the thoughts and devise more adaptive coping thoughts. The therapist could also teach the child relaxation exercises to address the physiological arousal and muscle tension and promote a sense of control over bodily feelings. Avoidance (procrastination) and other escape behaviors (e.g., calling out sick on the due date) could be addressed through incremental activation around homework completion and processing ambiguous feedback.

At the same time, we might notice that the child is particularly nervous in their language arts class. When asked, the youth reports that the teacher seems to be more critical of students who are learning English as a second language, which this child is, as well. The therapist might include parent advocacy, assertiveness, or problem-solving into the plan. Alternatively, the child might report that their parents set high expectations that they feel insecure about achieving. The therapist might include parent education or family sessions to address communication and family goals. The individual child will likely still benefit from adjusting their reaction to the anxiety of a project; helping address systemic challenges around the child will make any solutions more robust and sustainable.

The CBT model benefits from two principal traditions: cognitive science and learning theory. Cognitive science has shown that anxious youth tend to demonstrate threat-focused attentional biases, negative automatic thinking, and ingrained negative schema [45, 46]. Specifically, anxious children will selectively focus on threat-related information, assume ambiguous situations are threatening, conclude threat-related attributions, expect negative outcomes, and make behavioral choices prioritizing personal safety and comfort over goal achievement or mastery [47]. As these processes interact, anxious youth overemphasize threat in various situations and make decisions based on the perceived threat.

Behavioral science helps explain the functions that maintain maladaptive patterns. Within the framework of classical conditioning, youth who repeatedly experience intense anxiety in specific situations (e.g., giving a class presentation) will learn over time to associate the experiential stimulus (class presentation) with the cognitive, behavioral, and/or physiological response (developing anxiety symptoms). As this process is repeated, operant conditioning reinforces the child's avoidance of the stimuli, making them more likely to experience higher degrees of anxiety with each exposure. Youth are particularly susceptible to long-term increases in anxiety through avoidant or safety behaviors, which are actions performed to prevent, escape, or minimize feared outcomes or potential associated distress [48–50].

Avoidance interferes with effective learning and processing of emotions because it prevents new anti-anxiety information from being learned. When individuals avoid or escape (i.e., remove themselves from the presence of the feared situation or thing), it renders it impossible to learn that the stimulus lacked inherent danger in the first place. In our previous example, the youth avoids working on their writing assignment because they fear it will con firm their inadequate writing skills. The procrastination puts them in a position to execute at a suboptimal level for their own capabilities. Starting their assignment earlier would likely improve their chances of learning (gaining evidence) that their writing is adequate. The short-term relief that the procrastination provides functions to perpetuate the cycle even as it comes with long-term impairment. The goal of behavior therapy is to expose individuals to these challenges and illustrate the youth's natural abilities and/or capacity to learn skills that can help them feel successful in these scenarios. The theory of inhibitory learning encapsulates this process of learning new lessons that compete with old knowledge [51]. As the youth experiences success and benefit from small, incremental approach steps (e.g., doing homework in small parts, tolerating the distress of engaging work, getting praise for handing in work), the new lessons compete (i.e., inhibit) the older nonconstructive patterns (e.g., choosing avoidance to manage distressing anxiety).

Recent work in values-based assessment and mindfulness interventions show consistent impact for individuals managing anxiety [52, 53]. "Values" within a behavioral context refers to life directions and processes that inform an individual's choices [54]. Values are distinguished from concrete goals which orient a person toward achievement of a particular task or outcome. For teens, values-based assessment and concepts have been incorporated to enhance work with behavioral health (e.g., chronic pain, cystic fibrosis), habit reversal (e.g., trichotillomania), attention-deficit-hyperactivity disorders, and emotional distress (e.g., anger, depression, trauma) [55–57]. Mindfulness is a complementary set of strategies designed to provide psychological distance (defusion) from ruminative and intrusive thought processes. It is often used to help teens and adults work toward observing the difference between values and goals and to notice when they are working toward one or the other [58].

2.3 Cognitive Behavioral Approach to Youth Anxiety: A Case Illustration

The remainder of the chapter makes use of a case vignette to illustrate how evidence-based clinical evaluation and values assessment can be integrated into a multisystemic CBT conceptualization and how to apply advanced strategies, like mindfulness, social communication, exposure, and cognitive reframing in real-world examples. Our case example features a 14-year-old, Asian-American, cisgender female, Nicole (fictional name; the case reflects a compilation of relevant clients), with a clinical profile that includes generalized anxiety disorder and depressive disorder not otherwise specified. After observing withdrawal and isolation from friends and activities, Nicole's mother and father sought outpatient care for her. Traditionally an engaged and successful student, teachers reported that Nicole often seems distracted in class and hands in incomplete assignments. Her parents disclosed that Nicole has always been a "sensitive" child and worries about people, animals, and world events far more than her peers. In recent years, Nicole has frequently expressed anxiety about upsetting historical events that she learned about in school as well as difficult current global events (e.g., climate change, poverty), and seeks frequent reassurance. However, Nicole often becomes discouraged when adults do not have clear solutions. While Nicole used to be able to function well despite worries, she has become increasingly preoccupied by her thoughts and exhibits disrupted sleep, muscle tension, and retraction from activities and peer connections. After phone screening, the therapist agreed to conduct an initial assessment and present a treatment plan.

2.3.1 Initial Assessment

During the initial sessions, the therapist conducts a comprehensive assessment that permits a diagnostic evaluation and a functional assessment of Nicole's thoughts, feelings, and behaviors. Additional assessment of family history, social/cultural environment, parenting behaviors, and youth values and goals are also essential elements to guide a case conceptualization and treatment plan.

Diagnosis, Symptom, Coping, and Family Assessment

Diagnostic assessment helps clarify overlapping and distinctive clinical syndromes that can inform which specific CBT strategies would be useful in a particular case (e.g., panic disorder may indicate more psychoeducation and greater exposures over a shorter period of time compared to treating social anxiety disorder). We recommend using the Anxiety and Related Disorders Interview Schedule for DSM-5, Child and Parent Versions (ADIS-5) for evaluating youth diagnosis [59, 60]. The ADIS-5 is a semi-structured interview consisting of independent but congruent parent and child interviews. Typically, the parent and child interviews are conducted separately, allowing the clinician to collect data from each reporter to compose a composite diagnosis. Psychometric studies for the previous version of the ADIS (ADIS-IV-TR) have demonstrated good interviewer reliability (e.g., $k = 0.98$, parent interview; $k = 0.93$, child interview), retest reliability (e.g., $r = 0.76$, parent interview), and sensitivity to treatment effects [35, 61, 62]. Three core principles distinguish assessment with youth from adults: (1) collecting information from multiple informants, (2) using multiple measures to assess various constructs and repeating assessment measures over time, and (3) incorporating developmentally appropriate measures [63, 64]. The ADIS-5 possesses several advantages over other diagnostic measures (e.g., Kiddie Schedule for Affective Disorders and Schizophrenia (K-SADS) and Diagnostic Interview Schedule for Children (DISC)). First, it includes a separate parent interview, which allows the therapist to collect and integrate critical information that would not typically be disclosed by the child. Additionally, the ADIS-5 enables the clinician to assign both a diagnosis and interference rating (e.g., rating from 0 to 8, 0 = no interference, 8 = significant interference). Ranking interference within various life domains (e.g., school, home, friendships) helps the clinician better understand the degree of negative impact within specific contexts of the child's life.

Use of short, objective measures can augment assessment and allow clinicians to capture more detail of a child's symptom and interference profile. The Screen for Child Anxiety Related Emotional Disorders (SCARED) is a short, psychometrically sound measure that can be used to measure anxiety symptoms periodically (e.g., weekly or monthly) throughout treatment. The SCARED has demonstrated good internal consistency ($\alpha = 0.78$ to 0.87), test-retest reliability (intraclass correlation coefficients = 0.6 to 0.9), and discriminative validity between anxiety disorders [65]. It has also been validated for use across cultures [66]. The Multidimensional Anxiety Scale for Children-Child and Parent Forms (MASC) is another empirically validated measure designed for assessments of four domains of anxiety psychopathology (e.g., physical symptoms, social anxiety, harm avoidance, and separation anxiety) [67]. The 39-item form (MASC-39) has been normed on large samples and is one of the most widely used anxiety symptom measures. Additionally, the Revised Children's Anxiety and Depression Scale-Parent/Youth (RCADS) is a 47-item scale, rated on a 4-point (Never to Always) scale [68]. Subscales demonstrate good factorial validity (e.g., association with common anxiety and depressive disorders), 1-week test-retest reliability, internal consistency, and good convergent and discriminant validity with other validated internalizing and externalizing symptom measures. Additional free valid and reliable measures are also available online [69, 70].

Assessment of emotion regulation and other psychological and behavioral patterns can be useful in identifying high-risk behaviors and vulnerability [71]. Measures like the Difficulties in Emotion Regulation Scale [72] help identify youth regulation strategies, including difficulties engaging in goal-directed behaviors, impulse control difficulties, and nonacceptance of emotions. The Responses to Stress Questionnaire [73] helps identify youth use of 19 coping styles across 3 broad dimensions (primary control, secondary control, and involuntary experiences). The Behavioral Activation for Depression Scale [74, 75] can help identify level of youth activity in terms of activation and avoidance across work, school, and social domains.

Assessment of family environment and parenting styles helps a therapist understand the family context of youth clients and identify patterns worth addressing [76]. The Family Environment Scale [77] assesses family climate across ten domains, including cohesion, conflict, independence, and control, which have been linked with internalizing problems (anxiety/depression) in youth. The Family Accommodation Scale [78] for Anxiety assesses the frequency and extent of parental accommodation to youth anxiety (e.g., "How often do you reassure your child?"), parental distress when accommodating (e.g., "Does helping your child in these ways cause you distress?"), and consequences of not accommodating (e.g., "Has your child become distressed/anxious when you have not provided assistance?). The Parental Acceptance and Action Questionnaire [79] is a parent-report measure designed to assess parental experiential avoidance, focusing on a parent's ability to tolerate their child's negative emotions and manage their own responses to their child's emotional experiences. Family accommodation and parent acceptance of internal distress have been linked with elevated levels of anxiety in youth [80].

Functional Assessment

While diagnosis and symptom severity help to classify a youth's concerns into recognizable diagnostic profiles, a functional assessment helps identify observable behavioral patterns that call for intervention. Functional assessments identify antecedent-behavior-consequence (ABC) sequences or behavioral chains as they relate to target problem behaviors [44, 81]. An *antecedent* is considered to be any trigger, cue, or circumstance (e.g., person, place, object, event, thought, feeling, action) that exhibits discriminating properties. We define the youth's reaction to the antecedent (e.g., thoughts, feelings, action, physiological response) as their *behavior*. *Consequences* are any outcomes (internally, externally, or interpersonally) that follow the youth's initial response. Identifying the consequences that follow problematic behaviors helps the therapist recognize the reinforcers that maintain those behaviors. A functional assessment gives the client (and other family members) an opportunity to disclose specific details of problematic behavior while gaining insight into the triggers and emotional reactions that precede them and the consequences that follow. Likewise, a functional assessment acts as an essential assessment tool for helping the client, family, and therapist determine treatment priorities and target the behavioral chains that lead to the most concerning problem behaviors.

In Nicole's case, her parents were concerned with her withdrawal and isolation when confronted with stressors. They noted that she has always been "sensitive" and would worry about things beyond her years. These patterns bore out in her functional assessment (Fig. 1: Functional assessment worksheet) in which Nicole often avoids strong emotional reactions with withdrawal, cognitive rumination, or reassurance seeking. For example, fears and worry (emotional response) about the Los Angeles wildfires (antecedent) sent her to seek news coverage and images of the fires on social media (behavioral response), which ultimately caused more stress and intrusive thoughts (long-term results). Oftentimes, a client (or their parents) will report this sequence, dumbfounded that the youth would respond this way: "Doesn't she see how looking at more images only makes it worse?" The key to this assessment is to help the youth identify the function of her unhelpful response by homing in on the immediate consequences that follow her

Functional Assessment Worksheet. Tell us about your triggers and how you reacted. Describe your feelings, what you did (action), what happened right away (immediate outcome), and then what happened later (long-term outcome).				
Trigger	**Emotional Response (feelings, thoughts)**	**Action (Behavioral Response)**	**Immediate Results (What keeps it going?)**	**Long-term Results (What gets you in trouble?)**
Watching a kid at school be mean to another kid	Sad - "I wish someone would stop that."	Look away, walk away	Don't have to see it; pretend I didn't see it	Feel sad for the kid; guilty I didn't help
Watch news about LA forest fires	Worried, scared - "Can this happen to us?"	Look up more news/images on phone	Trying to reassure myself - find anything I can on it	More stressed, can't stop thinking about it. Spending too much time online.
Volunteering at an animal shelter	Sad, scared - "Will all these animals find a home?"	Freeze up; get lost in thoughts about what will happen	Something I can't help; it makes me so anxious	I leave without telling supervisor - still anxious
Mom and dad argue; dad leaves	Scared, sad - "What if they break up?"	Hide in my room; get on my phone	Get out of there. Don't have to listen or see mom upset.	Stuck with all this worry.

Fig. 1 Functional assessment worksheet. (Worksheet adapted from Chu and Pimentel [44])

behavioral reaction. Here, Nicole immerses herself in wildfire news in a misdirected attempt to reassure herself—if she knows everything about it, maybe she will know how to avoid it to stay safe. While this reaction provides a temporary sense of psychological control ("I know everything about fire prevention!"), it is short-lived because the scenario is predominantly uncontrollable. This insight provides a direction for intervention—introducing alternative methods to cope with unpredictability and minimal control will be key.

3 Values and Goals Assessment

Coinciding with a functional assessment, therapists may wish to conduct a holistic values and goals assessment with youth. Goals and values differ in that goals reflect concrete, achievable end-states that the youth would like to achieve, whereas values reflect an ongoing set of principles or processes that inform the specific goals [54]. Goals are necessary to set a direction for behavior, but an awareness of values can enhance motivation and sustain interest in goals, even as challenges arise. Marriage offers a classic example: Getting married is a concrete goal that can be achieved. The *purpose* of, or *meaning* to, getting married differs from person to person. For some, getting married represents being a part of a loving and reciprocal relationship. For others, the value is in growing a family and feeling a greater extension of oneself. The critical features of values are that they are (a) ongoing and cannot be completed (e.g., there is always room to demonstrate more loving behavior to a partner) and (b) personal, idiosyncratic, and chosen by the individual themselves (e.g., no one else has the authority to dismiss ones declaration of a value as incorrect) [54].

To help assess this admittedly abstract construct with youth, we use an exercise called the "Life Buckets Assessment," which helps youth conduct a 360-degree assessment of the key domains in life [82]. Each life domain (e.g., family, peers, school/work, activities, etc.) is conceived as a life bucket. Youth start by describing how they currently fill each life bucket. How do they categorize their daily activities? For the peers bucket, do they see friends at school? In clubs? After school or on the weekends? Who are their closest friends? Next, youth are asked to describe how full each bucket is. Similar to carrying a bucket of water, one would not want a bucket that is completely empty (creating a feeling of lacking or wanting), nor would want the bucket to be overflowing or overfilled (creating an overwhelming feeling). Instead, we aim for a bucket that is full but manageable (creating a sense of "just right"). Finally, youth are asked to identify any ways in which they would like things to be different in those buckets. For example, in the peers domain, would they like more or less friends, pursue different kinds of relationships, or engage in different activities with friends. This assessment helps identify both concrete goals and a sense of what is most important to the youth. A key tenet of behavioral therapy is that a person's actions speak louder than their words; the activities that fill a youth's day say more about what's important to them than what they could describe.

For teens, we can also use metaphors to elicit long-term goals and the meaning behind them. In one example, we ask teens to envision a high school or college graduation party where everyone they wish to attend has gathered. They envision each guest giving a toast to the youth, asking the youth to imagine what they would want each guest to say, rather than what they believe the guest might actually say. For example, "I think my mother would be there, and I would want her to say that I looked out for my brothers and sisters. And that I put my best foot forward..." The list of attendees gives some sense about *who* is important to them and the toast hints at *what* is important to them. Rarely do youth imagine a list of achievable goals like "I'd want her to say that I got all A's through high school." Nor do they describe toasts that center around their emotional insecurities (e.g., "My friends would say they never saw me nervous").

During Nicole's values assessment, common themes emerged across domains that reflected the kind of person she wants to be and how she wants to approach life. Nicole's sensitivity and worry stemmed from her deep empathy for both people and animals, along with a strong desire to contribute to solving social issues. Consistent with our notion of values, these principles can be continuously applied and inform *how* Nicole wishes to live life and work toward goals. For example, showing empathy is a behavior she can perform whether her goal is to volunteer at the animal shelter, comfort a fellow student who is being harassed, or call her local politicians to advocate for climate change policy reform. In this way, the therapist can help Nicole set specific goals and reinforce success as she pursues these goals in a valued way. A values-based approach helps the teen learn a sustainable way to maintain motivation and interest in the face of temporary setbacks and challenges.

3.1 CBT Case Formulation: A Multisystemic Approach to Capturing Client Stories

Based on the information collected during the diagnosis and functional assessment, the clinician begins to develop a case formulation. A case formulation (or conceptualization) refers to a working hypothesis of the internal (e.g., physiological arousal or maladaptive beliefs) and external (e.g., environmental reinforcers and punishers) factors that contribute to the development and maintenance of a problem behavior [44, 83, 84]. In previous editions, we have described capturing a client's functioning "within a nutshell." [32] To recognize the impact of local and broader cultural influences on youth and families, we have extended this model to incorporate individual and family factors (e.g., youth strengths and resources) that support growth, and also include the local, regional, and systemic cultural factors that may impede client progress and wellness.

To start, a CBT therapist works to identify relevant patterns of thoughts, feelings, and behaviors that typify the youth's unconstructive response patterns. Data from diagnostic interviews, self- and other-report questionnaires, and functional assessment are all used to inform initial theories. Once a clinician identifies problematic response patterns, they begin to build an initial case formulation that serves as a template for understanding what motivates and

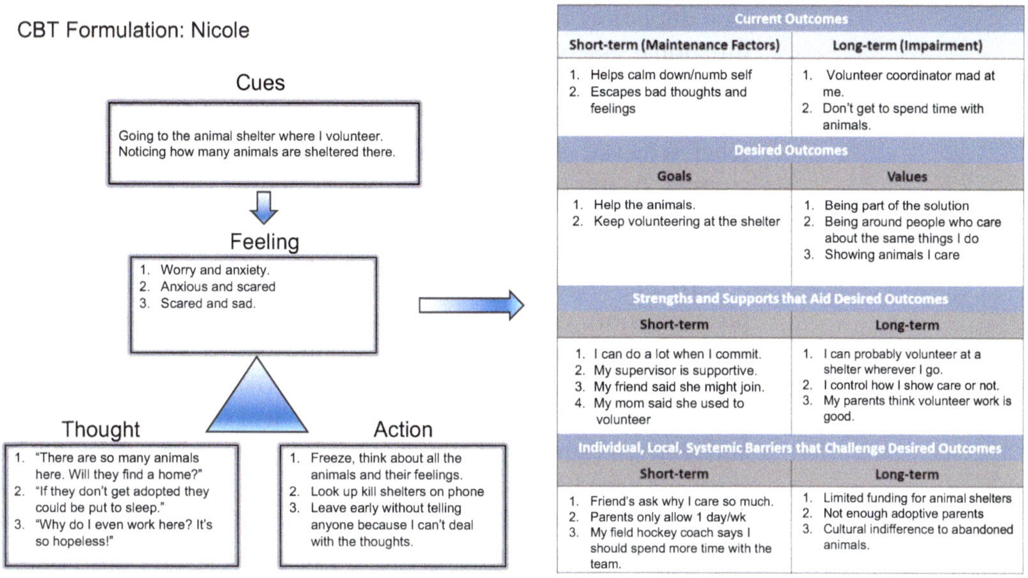

Fig. 2 Multisystemic case formulation. (Figure adapted from Badin et al. [32])

impedes a client's progress. From this, an initial treatment plan can be created. The case formulation is not meant to be static but continuously refined. As the therapist identifies new behavioral patterns, or observes the success/failure of planned interventions, the conceptualization can be updated with new information.

Figure 2 illustrates the key components to a multisystemic case formulation. On the left side, the clinician fills out a prototypical scenario for the client that illustrates a problematic response pattern. In this example, Nicole's reaction to a volunteer day at the animal shelter is described. When Nicole walks into the animal shelter, she notices that the shelter is overcrowded that day (initial cue). This observation produces a rapid sequence of reactions including emotions (worry and anxiety), thoughts ("There are so many animals here"; "Will they find a home?"), and actions (freezing, ruminating about the animals and their experience). The form provides room to consider subsequent cycles to illustrate how initial reactions spiral into continuous emotional cascades. For example: scared → "If they don't get adopted, they could be put to sleep" → start researching what happens to abandoned animals in kill shelters. Ultimately, this cascade leads Nicole to leave early without informing a supervisor.

The right side of the conceptualization worksheet highlights the key motivators, supports, and barriers to client change. The first row reflects the short- and long-term outcomes that currently result from Nicole's behavioral patterns (Current Outcomes) and the second row provides room to list the youth's preferred responses and results (Desired Outcomes). The Current Outcomes row duplicates the short- and long-term results identified during functional assessment (*see* Functional Assessment above). In this case, Nicole tolerates the long-term impairment (e.g., poor review by supervisor, missing the animals) in exchange for the short-term relief achieved through escape and distance from the distressing context. As this pattern persists, Nicole deprives herself of time with the animals and her supervisor may become increasingly disappointed in her performance.

To her credit, Nicole envisions alternative goals that would produce more constructive outcomes for her. Nicole's goals are concretely operationalized. She would like to keep helping the animals by volunteering at the shelter. To make goals optimally attainable, the therapist works with the client to make goals specific and concrete, such as "I aim to (a) arrive on time, (b) clean the dog cages, (c) feed the dogs, (d) take the dogs for walks, etc." If necessary, goals can be itemized into smaller steps to ensure they are achievable (e.g., getting leash, signing dog out, walking dog around block, etc.).

Based on her earlier values assessment, Nicole expressed her vision of approaching all living beings with empathy and playing an active role in addressing social issues. These values serve as psychological "guardrails" that the therapist and client can use to evaluate how the client is approaching life. When she has a future choice point to make, [[54]] we can ask, "How did your response show empathy?" "In what way were you part of the solution for this particular issue?" In Nicole's case, did she show care and warmth for the animals; did she show patience for behaviorally challenged dogs as she walked them? Does she consider volunteering at a publicly funded shelter that has a no-kill policy as being "part of the solution"? If she can answer affirmatively to these queries, then she is living within her preferred values. In this way, values provide self-evaluative benchmarks that can be used universally across contexts and goals.

The conceptualization worksheet also prompts the clinician to consider strengths and supports that are available to the youth, as well as any individual, local, or systemic barriers that can impede change. Youth are embedded in multiple systems (e.g., family, school, community, national, etc.) and the youth's relationship with, and across, each system can support or impede their growth. Acknowledging a youth's individual strengths and interests, in addition to their local supports, helps the therapist apply a strength-based approach that mobilizes a youth's natural resources for greater well-being [85]. Reviewing Nicole's history of academic and social success reminds us that she is capable of accomplishing meaningful goals when she commits; she can acknowledge that her supervisor at the animal shelter has been a support and role model in the past and her parents support her volunteer focus. She can also appreciate how her strengths and supports will be present, or further develop, in the future. She acknowledges that volunteering is a life-long activity which she can build upon across time and contexts and at the level of involvement that works for her. She also imagines her parents will continue to model community action through their own deeds. The youth and therapist can draw upon these resources as challenges arise.

Challenges and barriers should be acknowledged too. Are there practical, resource, or attitudinal barriers for Nicole? Does she have transportation to the animal shelter? Does the timing work with her demanding school and sports schedule? Do her friends, family, and other key figures support her efforts? The answers to these questions raise opportunities for the array of CBT interventions, including problem-solving, assertiveness/communication, behavioral activation (e.g., activity scheduling), and cognitive reframing. In Nicole's case, her parents are supportive of her efforts in principle but have also expressed concerns about the volunteer work interfering with school responsibilities. Nicole's friends do not dedicate time to volunteer efforts, which leaves her feeling disconnected from her friends on this topic. Additionally, her field hockey coach has suggested she needs to practice with the team more, which typically occurs during after-school times Nicole is scheduled to volunteer. The therapist can build with the youth a challenge hierarchy that addresses each of these barriers, practicing communication to assert her interests and problem-solving to overcome conflicts and practical hurdles.

Systemic barriers, including cultural attitudes toward race, gender, and sexuality, as well as the challenges posed by limited resources, significantly affect youth development and well-being [86]. Recognizing the general impact of broad cultural factors is one important aspect of a client's context, while discerning their specific influence on the psychology and life of an individual is another. One way to approach these questions includes a focused assessment of the youth's cultural identity in the formulation of the problem. Several frameworks have been developed to help clinicians identify relevant cultural factors, some created to reflect the influence of specific cultural groups, while others broadly encourage the clinician to consider how culture can affect symptom presentation and treatment [87–92]. For example, Hays' ADDRESSING model [93] prompts consideration of age, disability (developmental or acquired later in life), religion/spiritual orientation, ethnicity/racial identity, socioeconomic status, sexual orientation, indigenous heritage, national origin, and gender during assessment and treatment. The American Psychiatric Association's Cultural Formulation Interview (CFI) [94–97] helps clients (youth or family) to define their concerns in light of their self-endorsed cultural identities (e.g., "Sometimes people have different ways of describing their problems to their family, friends, or others in the community. How would you describe your problem to them?"). The interview provides space to consider cultural perceptions of the cause, the sociopolitical context in which the problems occur, and sources of support or challenge. The therapist queried Nicole about these topics who did not perceive links between her personal identities (cisgender girl, Asian-American) and her volunteer work, but she did acknowledge that older generations (e.g., parents, coach) struck her as dismissive, even callous, when discussing animal welfare. For example, she considered restricting her consumption of animal products and worried that adults might dismiss or criticize her beliefs. Through increasing awareness of generational differences in cultural-political issues, the therapist helped Nicole depersonalize her worries to some extent ("It's not just me who feels this way") and mobilize her to commit more to her values.

Although a clinician begins to develop a case formulation as soon as he/she meets the client, they are continuously revising it as they gather new information. The case conceptualization should be continually refined as treatment progresses and as the clinician tests out their hypotheses in treatment. This process of routinely integrating new information into conceptualization, and therefore informing treatment planning, has been called the "assessment-treatment dialectic" [98].

3.2 Treatment Planning

Armed with a multidimensional assessment, case conceptualization, and list of target problems, the therapist can then plan interventions to strategically address the youth's concerns. Research has documented which practices qualify as evidence-based for broad-based diagnostic profiles [99]. Selecting which strategies to teach, and in what order, should follow the client's prioritization of target problems and clinical needs [44]. Youth who are primarily presenting with anhedonia, isolation, and limited activation would likely benefit from behavioral activation and exposure. Youth who suffer from impairing rumination and intrusive thoughts might benefit from mindfulness exercises. Youth with overly negative views of themselves and others may benefit from cognitive restructuring. Significant interpersonal or family conflict may call for increased social support, family-based interventions, and effective communication skills.

The therapist shares their ideas for a treatment plan with the family, including a general sequence and time frame. Research suggests that involving youth and caregivers in this process improves attendance, treatment engagement, and ultimate outcomes [100]. Thus, making treatment selection a collaborative decision can lay a strong foundation for the therapeutic process.

3.3 Interventions

3.3.1 Establishing Dialogue Through Reciprocal Education

When a youth presents with identifiable forms of anxiety, it can be helpful for the therapist to provide psychoeducation on the nature of anxiety and how treatment will intend to target anxious symptoms. Interested clinicians are referred to an excellent set of free online FAQ sheets on specific anxiety disorders, produced by the

Association for Behavioral and Cognitive Therapies [http://www.abct.org/Information/?m=mInformation&fa=fs_ANXIETY], and to information regarding effective therapies for anxiety, produced by the Society of Clinical Child and Adolescent Psychology [www.effectivechildtherapy.com]. By emphasizing that anxiety, like all emotions, is natural and serves an important function, the clinician is also communicating that experiences of intense emotion, particularly anxiety, are not dangerous or intolerable.

In cases where clients are presenting with stress reactions to realistic challenges (e.g., family conflict, peer harassment, adverse sociopolitical climate, natural disasters), the therapist is wise to leave space for these discussions whenever it seems helpful. Youth surveys indicate that dismissive adult responses can have a significant negative impact on mental health as they wrestle with complicated social issues (e.g., climate anxiety, social injustice) [27, 28, 31]. It is critical for the clinician to leave space to hear client concerns and how they connect with their presenting problems to engender trust and foster hope. When helping youth understand complex cultural or social issues, experts recommend taking an open stance by asking youth for their understanding of the issue, focusing on the youth's individual concerns, and creating a sense of safety [101]. It is important to tailor the discussion to the youth's age and temperament, recognizing that some youth are more activated by communal events than others. Emphasizing that youth have agency in their everyday life, even as big issues mount, can also be grounding. Much of the anxiety stems from not knowing how to process the frequent and sometimes contradictory information presented in modern news and social media outlets [102]. The therapist can help the youth determine which sources are reliable and examine their thinking, which can promote a sense of grounding. For further help, the Society of Clinical Child and Adolescent Psychology has collated resources for therapists and parents to address youth reactions to crisis events (https://effectivechildtherapy.org/concerns-symptoms-disorders/concerns/natural-disasters-terrorism/).

If the therapist had conducted the Cultural Formulation Interview, they may have a sense of how a youth's identities impact their sense of the problem and preferences for intervention. Hays [91, 93] lists ten strategies that foster "cultural competency" to help therapy be more culturally salient. Some examples of these strategies include identifying cultural strengths and supports, validating client's self-reported experiences of oppression, emphasizing collaboration over confrontation, avoiding challenging core cultural beliefs, and developing homework assignments that are culturally congruent. Often cultural themes manifest as intergenerational differences in values or practices. In these cases, therapists can work with the youth in finding the right balance between acceptance and communicating their needs to important stakeholders; adjunctive work with parents in session may also be useful.

3.3.2 Mindfulness Training: Getting Distance from Intrusive Thoughts

Youth with persistent intrusive thoughts and worry can benefit from mindfulness training to help a client achieve psychological distance from the emotional attachment of negative thinking. Mindfulness exercises help interrupt ruminative cycles by providing an alternative to thought suppression, distraction, or cognitive restructuring [58, 103]. They start with just *noticing* the thought. For example, one mindfulness exercise is to envision placing each intrusive thought on a leaf floating gradually down a stream. Another is to envision your thoughts being placed on pieces of luggage as they circle around a conveyor belt at the airport [44]. This metaphor acknowledges that thoughts will frequently resurface, encouraging the client to observe them and gently return their focus to the present moment. You can also ask the youth to imagine the thought as a cloud slowly drifting by. The goal of mindfulness exercises is to accept that a range of thoughts (negative and positive in valence) enter our minds continuously throughout the day. Over-attending to any one thought gives that thought too much "power" over one's mood, choices, and beliefs.

In Nicole's case, intrusive thoughts dovetail many of her emotional reactions to upsetting events. When she sees a fellow student being harassed or sees news about the LA forest fires, she experiences intrusive thoughts that imply threatened safety or personal responsibility (e.g., "What if I'm next? What if I can't protect myself or others?"). These momentary thoughts become increasingly distressing as she progressively ruminates. The therapist encourages her to notice these thoughts, recognize that it is natural to have these thoughts in the present context, and also realize that over-attending to them may not serve her personal interests (e.g., help her meet her goals or live within valued directions). Instead, they practice mindfulness exercises to help Nicole create some distance between herself and the thought—"I am having the thought that I need to do something now!"—and instead allow herself to decide when and how to address such thoughts. She practices re-attending to her present activity (e.g., classwork, talking with a friend) while committing to address the thoughts later.

3.3.3 Behavioral Activation Promotes Opportunity

Where rumination and intrusive thoughts persist, behavioral activation (BA) can help [104, 105]. Advances in BA approaches have applied functional assessment and exposure philosophies to address entrenched stuck points for clients [40, 82, 106–108]. Avoidance in all its forms (procrastination, delay, escape, distraction) is evaluated using a functional assessment. Popular versions use the acronym TRAP to help youth identify the trigger, response, and avoidance patterns that keep a youth stuck in a pattern of avoidance that prioritizes short-term relief over longer-term impairment [109]. This approach is consistent with the functional assessment described earlier. As the therapist identifies a client's TRAPs, they help the youth use it as a "cue for activation" and work to enact a proactive, approach-oriented strategy to address the concern more directly. In Nicole's situation, she could act more proactively by confiding in her supervisor or a friend about her frequent worries for the animals. Nicole could look up the shelter's policies about animal care or volunteer to conduct outreach to prospective donors. BA acknowledges a bias in its approach that taking some kind of action puts an individual in a better situation than inaction. Even if perfect outcomes cannot be guaranteed, activation generally produces more opportunities and options than choosing to do nothing or to avoid ("You miss 100% of shots you don't take").

3.3.4 Cognitive Restructuring When the Struggle Is Real

Cognitive restructuring (or cognitive reframing) involves techniques aimed at helping youth recognize how unhelpful or unrealistic negative thoughts affect their emotions and subsequently learn how to modify these thought patterns. When the youth is dealing with concerns rooted in real-life challenges and impediments, it can be a struggle to know when and how to challenge thoughts contributing to emotional cascades and stuck points. In these cases, it is incumbent on the therapist to spend adequate time assessing the thought most responsible for impairment and to pinpoint if the thought is indeed overly rigid, personal, overgeneralized, or unfair to the situation. After comprehensive assessment, identifying thinking traps and generating more flexible and fair coping thoughts can help.

In Nicole's case, she displays a tendency to label scenarios, which are genuinely upsetting, with assumptions and generalizations that compound her negative emotions (*see* Fig. 3: Coping thoughts tracker). For example, when she witnesses a group of kids pick on another student at school, her automatic thought, "People just don't care for anyone else," might be understandable given what she just observed. However, it reflects an "overgeneralization" thinking trap that increases her disparagement and hopelessness. The therapist can take a variety of approaches to this situation, including brainstorming active solutions that help Nicole feel efficacious in the moment (e.g., sticking up for the victim or giving them comfort, reporting the incident to the school, getting emotional support from parents). Sometimes those active solutions seem feasible and functional for the child. In the times when they are not, helping the youth generate evidence for and against their negative assumptions can be helpful too. Nicole can reflect on the facts that people demonstrate both good and nice acts and accept a coping thought that "People can be both nice and mean to others." This does not change her judgment of the poor actors, nor does it mean she condones their misbehavior, but it can help stave off a generalized helplessness that festers. In so doing, it may give impetus to try out her proactive solutions or to solicit help from others.

Coping Thoughts Tracker. Brainstorm coping thoughts that could answer your thinking trap! Try and come up with coping statements that are more realistic and ask, "How am I not seeing the whole picture?"				
Trigger	**Thought**	**Thinking TRAP**	**Coping Thought**	**Result?**
Watching a kid at school be mean to another kid	"People just can't care for anyone else."	Over-generalizing	"People can be both nice and mean to others"	Open to looking for others
Watch news about LA forest fires	"Why don't people care at all? There' no reason to even bother"	All-or-nothing thinking; catastrophizing	"People are doing what they can. It's a long-term fight"	Looking for optimism.
Volunteering at an animal shelter	"There are so many animals here who will never find a home"	?? Jumping to conclusions	"I'm here to help as much as I can"	Still sad
Mom and dad argue; dad leaves	"Why can't we get along; if they fight, they'll split up. And then, I'll never have a good relationship with dad."	Jumping to conclusions; over-generalizing	"I know my parents love me even if they argue."	A little less worried and sad

Fig. 3 Coping thoughts tracker. (Adapted worksheet from Chu and Pimentel [44])

4 Exposures and Behavioral Experiments to Build on Strengths

Exposures are behavioral experiments that promote approach behaviors. Exposures help youth practice new skills, challenge thinking traps, and learn to persist through effortful and uncomfortable experiences (distress tolerance) [44, 49–51]. The first step in designing an exposure is to brainstorm relevant challenges that align with the goals of treatment. These challenges should activate some level of fear or distress, which should be rated by the client and ranked in order of difficulty. When working with younger clients, first categorizing the situations from easy to medium and then hard may provide an important first step in ranking each situation individually. The challenges in the hierarchy should vary in order to reflect increased difficulty and promote generalization across contexts. Exposures are then designed to be conducted in session, with the therapist coaching the youth through the challenge and encouraging practice of the skills they have learned in therapy. Exposures can be conducted in vivo, meaning "in life," or imaginally. Imaginal exposures involve the client recalling or envisioning themselves in the challenging situations. These are typically used when a situation listed cannot be replicated in session. During both in vivo and imaginal exposures, the therapist acts as both a coach and cheerleader.

Traditional examples include a youth giving a speech in front of similar-aged peers to practice speaking in front of a group or giving a school presentation. Furthermore, the youth may engage in an interoceptive exposure, such as hyperventilating, to violate expectancies that they will die during a panic attack. Another value of exposures is to provide the youth with an opportunity to develop skills, build confidence, and leverage available strengths. When realistic challenges exist (e.g., interpersonal conflict, resource limitations), the goal of an exposure may focus less on helping the youth achieve the concrete goal that they desire and more on demonstrating that the youth has the capacity to weather the storm or ask for what is needed.

One of the challenges that maintain doubt and evoke sadness for Nicole is that some of her closest friends do not care about animal welfare in the same way that she does. Some of her friends have questioned her interest in becoming vegetarian. Some of her field hockey teammates have sounded critical when she chooses to volunteer at the animal shelter instead of attending an additional voluntary workout. Nicole does not consider herself a confrontational person and does not believe that friends have to hold all the same beliefs to be close. However, she does value open listening between friends and signs that each person cares about and respects each other's interests. To help Nicole protect her own integrity and feel like she's given her friendship a fair shot, the therapist helps devise a behavioral experiment where she discloses her hurt feelings to one particular friend who recently made fun of vegetarians (*see* Fig. 4: Exposure worksheet).

Nicole's behavioral goals for the exposure are to convey her beliefs about the values of vegetarianism and to hear her friends' viewpoints. She uses cognitive techniques to challenge her assumptions that the friend will dismiss her again and pick another fight. She also uses mindfulness strategies to briefly compartmentalize remaining intrusive thoughts so she can initiate the experiment. In this challenge, Nicole texts her friend to talk and then holds a FaceTime session in front of her therapist. The therapist coaches her through the conversation (with silent cheering and encouragement) and allows the experiment to unfold naturally. At completion, Nicole reveals that it was difficult and that she wanted to

Worksheet 6. In vivo Exposure/Behavioral Experiments

Complete this worksheet with the youth as you are preparing for a behavioral experiment.

1. **Situation: (What's the situation?)** *Telling friend that it hurt my feeling when she made fun of vegetarians.*
2. **Feelings:** *Hurt, disappointed* **Distress Rating:** __7______
3. **Anxious/Negative Thoughts** | **Thinking Traps (See list below)**

Anxious/Negative Thoughts	Thinking Traps (See list below)
She's going to deny she did that and we'll get into a big fight.	*Jumping to conclusions; catastrophizing*
She'll make more fun of it and make me feel more bad.	*Mind Reading, catastrophizing*

Thinking Traps: Mind Reading, Fortune-telling, Catastrophizing, Jumping to Conclusions, What if's, Discounting the positives, Looking for the Negatives, Overgeneralizing, All-or-Nothing, Should Statements, Self-Critic

4. **Coping Thoughts: (How do you respond to your anxious thoughts?)**

"She didn't know how important it was to me. I'm sure she'll listen if I tell her."
"We don't have to agree on everything to be friends."

Challenge Questions: Do I know for certain that ________? Am I 100% sure that ________? What evidence do I have that ________? What is the worse that could happen? How bad is that? Do I have a crystal ball?

5. **Achievable Behavioral Goals: (What do you want to accomplish?)**

Goal	Accomplished?
a. *Tell her I was hurt.*	
b. *Finish my thoughts about vegetarianism*	
c. *Re-connect about something we both like*	

6. **Rewards**

Reward	Earned?
a. *Go out for Acai bowls after*	
b. *Write positive affirmations in my journal to myself.*	
c.	

Fig. 4 Exposure worksheet. (Adapted worksheet from Chu and Pimentel [44])

appease her friend many times, but sticking with her goals helped her feel more confident about herself and secure in her beliefs. She was relieved to learn at the end of the conversation that her friend didn't take Nicole's assertions personally; rather she welcomed the conversation. As a reward for this difficult friendship building exercise, they met up for acai bowls later.

5 Conclusion/Discussion

Anxiety is a universal experience that has been amplified by recent global and societal issues. The impairment associated with anxiety can be exacerbated when youth get stuck in unconstructive thinking or behavioral patterns. Cognitive behavioral treatment that makes use of a multisystemic case formulation, multidomain assessment, and strategies that address cognitive reframing and behavioral approach in light of a youth's values can be useful. This chapter illustrated the use of CBT strategies in a flexible principles-based way to address complex scenarios that come up in a challenging world.

Conflicts of Interest

The authors have no conflicts of interest to declare that are relevant to the content of this chapter.

References

1. Merikangas KR, He JP, Burstein M et al (2010) Lifetime prevalence of mental disorders in U.S. adolescents: results from the National Comorbidity Survey Replication–Adolescent Supplement (NCS-A). J Am Acad Child Adolesc Psychiatry 49(10): 980–989. https://doi.org/10.1016/j.jaac.2010.05.017
2. Swan AJ, Kendall PC (2016) Fear and missing out: youth anxiety and functional outcomes. Clin Psychol Sci Pract 23(4): 417–435. https://doi.org/10.1111/cpsp.12169
3. Cole DA, Peeke LG, Martin JM, Truglio R, Seroczynski AD (1998) A longitudinal look at the relation between depression and anxiety in children and adolescents. J Consult Clin Psychol 66(3):451–460. https://doi.org/10.1037/0022-006X.66.3.451
4. Stark KD, Laurent J (2001) Joint factor analysis of the children's depression inventory and the revised children's manifest anxiety scale. J Clin Child Adolesc Psychol 30(4):552–567. https://doi.org/10.1207/S15374424JCCP3004_11
5. Kessler RC, Berglund P, Demler O et al (2003) The epidemiology of major depressive disorder results from the National Comorbidity Survey Replication (NCS-R). JAMA 289(23):3095–3105. https://doi.org/10.1001/jama.289.23.3095
6. Samji H, Wu J, Ladak A et al (2022) Review: mental health impacts of the COVID-19 pandemic on children and youth – a systematic review. Child Adolesc Ment Health 27(2): 173–189. https://doi.org/10.1111/camh.12501
7. Nearchou F, Flinn C, Niland R, Subramaniam SS, Hennessy E (2020) Exploring the impact of COVID-19 on mental health outcomes in children and adolescents: a systematic review. Int J Environ Res Public Health 17(22):8479. https://doi.org/10.3390/ijerph17228479

8. Racine N, McArthur BA, Cooke JE, Eirich R, Zhu J, Madigan S (2021) Global prevalence of depressive and anxiety symptoms in children and adolescents during COVID-19: a meta-analysis. JAMA Pediatr 175(11): 1142–1150. https://doi.org/10.1001/jamapediatrics.2021.2482
9. World Health Organization (2022) Mental health and COVID-19: early evidence of the pandemic's impact: scientific brief, 2 March 2022. In: Mental Health and COVID-19: early evidence of the pandemic's impact: scientific brief, 2 March 2022. https://iris.who.int/handle/10665/352189. Accessed 31 Mar 2025
10. Czeisler MÉ (2020) Mental health, substance use, and suicidal ideation during the COVID-19 pandemic – United States, June 24–30, 2020. MMWR Morb Mortal Wkly Rep. 69. https://doi.org/10.15585/mmwr.mm6932a1
11. Centers for Disease Control and Prevention (2024) Youth Risk Behavior Survey Data Summary & Trends Report: 2013–2023. U.-S. Department of Health and Human Services
12. Pierson E, Simoiu C, Overgoor J et al (2020) A large-scale analysis of racial disparities in police stops across the United States. Nat Hum Behav 4(7):736–745. https://doi.org/10.1038/s41562-020-0858-1
13. Curtis DS, Washburn T, Lee H et al (2021) Highly public anti-black violence is associated with poor mental health days for black Americans. Proc Natl Acad Sci U S A 118(17): e2019624118. https://doi.org/10.1073/pnas.2019624118
14. C.S. Mott Children's Hospital, Michigan Medicine (2020) Mott Poll, "Teen Involvement in Demonstrations against Police Brutality
15. Tian NGR Carolyne Im and Ziyao (2025) Asian Americans and discrimination during the COVID-19 pandemic. Pew Research Center. November 30, 2023. https://www.pewresearch.org/race-and-ethnicity/2023/11/30/asian-americans-and-discrimination-during-the-covid-19-pandemic/. Accessed 25 May 2025
16. Barajas-Gonzalez RG, Ayón C, Brabeck K, Rojas-Flores L, Valdez CR (2021) An ecological expansion of the adverse childhood experiences (ACEs) framework to include threat and deprivation associated with U.-S. immigration policies and enforcement practices: an examination of the Latinx immigrant experience. Soc Sci Med 282:114126. https://doi.org/10.1016/j.socscimed.2021.114126
17. Cobb CL, Xie D, Sanders GL (2016) Coping styles and depression among undocumented Hispanic immigrants. J Immigr Minor Health 18(4):864–870. https://doi.org/10.1007/s10903-015-0270-5
18. Anti-Defamation League (2024) Audit of Antisemitic Incidents 2023. April 16, 2024. https://www.adl.org/resources/report/audit-antisemitic-incidents-2023. Accessed 23 May 23 2025
19. Anti-Defamation League (2024) Antisemitism in Schools and Support for Holocaust Education | ADL. December 23, 2024. https://www.adl.org/resources/report/antisemitism-schools-and-support-holocaust-education. Accessed 23 May 2025
20. BeWell, Jewish Federations of North America (2025) Strength, Stress, and Support: A Portrait of American Jewish Teen Well-Being. Published online 2025. chrome-extension://efaidnbmnnnibpcajpcglclefindmkaj. https://cdn.fedweb.org/fed-42/2/BEWELL_STANFORD-RESEARCH-REPORT_FINAL.pdf
21. Morstead T, DeLongis A (2025) Antisemitism on campus in the wake of October 7: examining stress, coping, and depressive symptoms among Jewish students. Stress Health 41(1):e3529. https://doi.org/10.1002/smi.3529
22. Maqbool A, Parry J (2024) Religious hate crime at record high after Israel-Hamas conflict. https://www.bbc.com/news/articles/c0kjrr44mvdo. Accessed 25 May 2025
23. Scheitle C, Mabute-Louie B, Ferguson J, Hawkins E, Ecklund EH (2023) Fear of religious hate crime victimization and the residual effects of anti-semitism and islamophobia. Soc Forces 101(4):2059–2086. https://doi.org/10.1093/sf/soac100
24. Yang M (2023) Islamophobia and antisemitism on rise in US amid Israel-Hamas war. The Guardian. https://www.theguardian.com/us-news/2023/nov/10/us-islamophobia-antisemitism-hate-speech-israel-hamas-war-gaza. Accessed 25 May 2025
25. CDC (2023) U.S. Teen Girls Experiencing Increased Sadness and Violence. Centers for Disease Control and Prevention. https://www.cdc.gov/media/releases/2023/p0213-yrbs.html. Accessed 25 May 2025
26. The Trevor Project (2023) New poll emphasizes negative impacts of anti-LGBTQ policies on LGBTQ Youth. The Trevor Project. https://www.thetrevorproject.org/blog/new-poll-emphasizes-negative-impacts-of-anti-lgbtq-policies-on-lgbtq-youth/. Accessed 31 Mar 2025

27. Hickman C, Marks E, Pihkala P et al (2021) Climate anxiety in children and young people and their beliefs about government responses to climate change: a global survey. Lancet Planet Health 5(12):e863–e873. https://doi.org/10.1016/S2542-5196(21)00278-3
28. Lewandowski RE, Clayton SD, Olbrich L et al (2024) Climate emotions, thoughts, and plans among US adolescents and young adults: a cross-sectional descriptive survey and analysis by political party identification and self-reported exposure to severe weather events. Lancet Planet Health 8(11):e879–e893. https://doi.org/10.1016/S2542-5196(24)00229-8
29. Baskin-Sommers A, Simmons C, Conley M et al (2021) Adolescent civic engagement: lessons from black lives matter. Proc Natl Acad Sci 118(41):e2109860118. https://doi.org/10.1073/pnas.2109860118
30. Cadenas GA, Garcia E, Autin K (2025) Social justice activism and critical agency among undocumented students: coping mediators between discrimination and depression. J Divers High Educ 18(2):123–130. https://doi.org/10.1037/dhe0000495
31. Ojala M (2013) Coping with climate change among adolescents: implications for subjective well-being and environmental engagement. Sustainability 5(5):2191–2209. https://doi.org/10.3390/su5052191
32. Badin E, Alvarez E, Chu BC (2020) Cognitive behavioral therapy for child and adolescent anxiety: CBT in a nutshell. In: Friedberg RD, Nakamura BJ (eds) Cognitive behavioral therapy in youth: tradition and innovation. Springer US, pp 41–71. https://doi.org/10.1007/978-1-0716-0700-8_3
33. Higa-McMillan CK et al (2016) Evidence base update: 50 years of research on treatment for child and adolescent anxiety. J Clin Child Adolesc Psychol 45(2):91–113. https://doi.org/10.1080/15374416.2015.1046177
34. Comer JS et al (2019) Evidence base update on the treatment of early childhood anxiety and related problems. J Clin Child Adolesc Psychol 48(1):1–15. https://doi.org/10.1080/15374416.2018.1534208
35. Walkup JT, Albano AM, Piacentini J et al (2008) Cognitive behavioral therapy, sertraline, or a combination in childhood anxiety. N Engl J Med 359(26):2753–2766. https://doi.org/10.1056/NEJMoa0804633
36. Ginsburg GS, Becker-Haimes EM, Keeton C et al (2018) Results from the child/adolescent anxiety multimodal extended long-term study (CAMELS): primary anxiety outcomes. J Am Acad Child Adolesc Psychiatry 57(7): 471–480. https://doi.org/10.1016/j.jaac.2018.03.017
37. Kendall PC, Hudson JL, Gosch E, Flannery-Schroeder E, Suveg C (2008) Cognitive-behavioral therapy for anxiety disordered youth: a randomized clinical trial evaluating child and family modalities. J Consult Clin Psychol 76(2):282–297. https://doi.org/10.1037/0022-006X.76.2.282
38. Hudson JL, Rapee RM, Deveney C, Schniering CA, Lyneham HJ, Bovopoulos N (2009) Cognitive-behavioral treatment versus an active control for children and adolescents with anxiety disorders: a randomized trial. J Am Acad Child Adolesc Psychiatry 48(5): 533–544. https://doi.org/10.1097/CHI.0b013e31819c2401
39. Southam-Gerow MA, Weisz JR, Chu BC, McLeod BD, Gordis EB, Connor-Smith JK (2010) Does cognitive behavioral therapy for youth anxiety outperform usual care in community clinics? An initial effectiveness test. J Am Acad Child Adolesc Psychiatry 49(10): 1043–1052. https://doi.org/10.1016/j.jaac.2010.06.009
40. Chu BC, Crocco ST, Esseling P, Areizaga MJ, Lindner AM, Skriner LC (2016) Transdiagnostic group behavioral activation and exposure therapy for youth anxiety and depression: initial randomized controlled trial. Behav Res Ther 76:65–75. https://doi.org/10.1016/j.brat.2015.11.005
41. Weersing VR, Brent DA, Rozenman MS et al (2017) Brief behavioral therapy for pediatric anxiety and depression in primary care: a randomized clinical trial. JAMA Psychiatry 74(6):571–578. https://doi.org/10.1001/jamapsychiatry.2017.0429
42. Ehrenreich-May J, Rosenfield D, Queen AH, Kennedy SM, Remmes CS, Barlow DH (2017) An initial waitlist-controlled trial of the unified protocol for the treatment of emotional disorders in adolescents. J Anxiety Disord 46:46–55. https://doi.org/10.1016/j.janxdis.2016.10.006
43. Kennedy SM, Bilek EL, Ehrenreich-May J (2019) A randomized controlled pilot trial of the unified protocol for transdiagnostic treatment of emotional disorders in children. Behav Modif 43(3):330–360. https://doi.org/10.1177/0145445517753940
44. Chu BC, Pimentel SS (2023) CBT treatment plans and interventions for depression and anxiety disorders in youth. Guilford Publications
45. Aldao A, Nolen-Hoeksema S (2010) Specificity of cognitive emotion regulation strategies: a transdiagnostic examination. Behav Res

Ther 48(10):974–983. https://doi.org/10.1016/j.brat.2010.06.002

46. Chu BC, Chen J, Mele C, Temkin A, Xue J (2017) Transdiagnostic approaches to emotion regulation: basic mechanisms and treatment research. Emot Regul Psychopathol Child Adolesc 1:419–451
47. Puliafico AC, Kendall PC (2006) Threat-related attentional bias in anxious youth: a review. Clin Child Fam Psychol Rev 9(3): 162–180. https://doi.org/10.1007/s10567-006-0009-x
48. Chu BC, Skriner LC, Staples AM (2014) Behavioral avoidance across child and adolescent psychopathology. In: Transdiagnostic treatments for children and adolescents: principles and practice, pp 84–110
49. Blakey SM, Abramowitz JS (2016) The effects of safety behaviors during exposure therapy for anxiety: critical analysis from an inhibitory learning perspective. Clin Psychol Rev 49:1–15. https://doi.org/10.1016/j.cpr.2016.07.002
50. Hoffman LJ, Chu BC (2019) When is seeking safety functional? Taking a pragmatic approach to distinguishing coping from safety. Cogn Behav Pract 26(1):176–185. https://doi.org/10.1016/j.cbpra.2018.11.002
51. Craske MG, Treanor M, Conway CC, Zbozinek T, Vervliet B (2014) Maximizing exposure therapy: an inhibitory learning approach. Behav Res Ther 58:10–23. https://doi.org/10.1016/j.brat.2014.04.006
52. Twohig MP, Levin ME (2017) Acceptance and commitment therapy as a treatment for anxiety and depression: a review. Psychiatr Clin 40(4):751–770. https://doi.org/10.1016/j.psc.2017.08.009
53. Hofmann SG, Gómez AF (2017) Mindfulness-based interventions for anxiety and depression. Psychiatr Clin North Am 40(4):739–749. https://doi.org/10.1016/j.psc.2017.08.008
54. Harris R (2019) ACT made simple: an easy-to-read primer on acceptance and commitment therapy. New Harbinger Publications
55. Halliburton AE, Cooper LD (2015) Applications and adaptations of acceptance and commitment therapy (ACT) for adolescents. J Context Behav Sci 4(1):1–11. https://doi.org/10.1016/j.jcbs.2015.01.002
56. Coyne LW, McHugh L, Martinez ER (2011) Acceptance and commitment therapy (ACT): advances and applications with children, adolescents, and families. Child Adolesc Psychiatr Clin 20(2):379–399. https://doi.org/10.1016/j.chc.2011.01.010
57. Petersen JM, Zurita Ona P, Twohig MP (2024) A review of acceptance and commitment therapy for adolescents: developmental and contextual considerations. Cogn Behav Pract 31(1):72–89. https://doi.org/10.1016/j.cbpra.2022.08.002
58. Greco LA, Blackledge JT, Coyne LW, Ehrenreich J (2005) Integrating acceptance and mindfulness into treatments for child and adolescent anxiety disorders. In: Orsillo SM, Roemer L (eds) Acceptance and mindfulness-based approaches to anxiety: conceptualization and treatment. Springer, pp 301–322. https://doi.org/10.1007/0-387-25989-9_12
59. Silverman WK, Albano AM. Anxiety and related disorders interview schedule for DSM-5, child and parent version: clinician manual. Oxford University Press; 2024
60. Kerns CM, Albano AM, Silverman WK (2024) Anxiety and related disorders interview schedule for DSM-5, child and parent version, with autism Spectrum addendum (ADIS/ASA): parent interview schedule – 5 copy set. Oxford University Press
61. Silverman WK, Eisen AR (1992) Age differences in the reliability of parent and child reports of child anxious symptomatology using a structured interview. J Am Acad Child Adolesc Psychiatry 31(1):117–124. https://doi.org/10.1097/00004583-199201000-00018
62. Silverman WK, Nelles WB (1988) The anxiety disorders interview schedule for children. J Am Acad Child Adolesc Psychiatry 27(6): 772–778. https://doi.org/10.1097/00004583-198811000-00019
63. Chu BC (2008) Child and adolescent research methods in clinical psychology. In: Handbook of research methods in abnormal and clinical psychology. Sage, pp 405–426
64. Chu BC (2019) Evidence-based therapist flexibility. In: Treatment of disorders in childhood and adolescence, pp 27–44
65. Birmaher B, Brent DA, Chiappetta L, Bridge J, Monga S, Baugher M (1999) Psychometric properties of the screen for child anxiety related emotional disorders (SCARED): a replication study. J Am Acad Child Adolesc Psychiatry 38(10):1230–1236
66. Hale WW, Crocetti E, Raaijmakers QAW, Meeus WHJ (2011) A meta-analysis of the cross-cultural psychometric properties of the screen for child anxiety related emotional disorders (SCARED). J Child Psychol Psychiatry

52(1):80–90. https://doi.org/10.1111/j.1469-7610.2010.02285.x
67. March JS (2013) Multidimensional anxiety scale for children, 2nd edn. Multi-Health Systems, Inc
68. Chorpita BF, Yim L, Moffitt C, Umemoto LA, Francis SE (2000) Assessment of symptoms of DSM-IV anxiety and depression in children: a revised child anxiety and depression scale. Behav Res Ther 38(8):835–855. https://doi.org/10.1016/S0005-7967(99)00130-8
69. Beidas RS, Stewart RE, Walsh L et al (2015) Free, brief, and validated: standardized instruments for low-resource mental health settings. Cogn Behav Pract 22(1):5–19. https://doi.org/10.1016/j.cbpra.2014.02.002
70. Becker-Haimes EM et al (2020) Evidence base update for brief, free, and accessible youth mental health measures. J Clin Child Adolesc Psychol 49(1):1–17. https://doi.org/10.1080/15374416.2019.1689824
71. Yadlosky LB, Mowrey WB, Pimentel SS (2023) Risky business: considerations of emotion regulation and high-risk behaviors in anxious adolescents. J Anxiety Disord 99:102760. https://doi.org/10.1016/j.janxdis.2023.102760
72. Gratz KL, Roemer L (2004) Multidimensional assessment of emotion regulation and dysregulation: development, factor structure, and initial validation of the difficulties in emotion regulation scale. J Psychopathol Behav Assess 26(1):41–54. https://doi.org/10.1023/B:JOBA.0000007455.08539.94
73. Connor-Smith JK, Compas BE, Wadsworth ME, Thomsen AH, Saltzman H (2000) Responses to stress in adolescence: measurement of coping and involuntary stress responses. J Consult Clin Psychol 68(6):976–992. https://doi.org/10.1037/0022-006X.68.6.976
74. Kanter JW, Mulick PS, Busch AM, Berlin KS, Martell CR (2007) The behavioral activation for depression scale (BADS): psychometric properties and factor structure. J Psychopathol Behav Assess 29(3):191–202. https://doi.org/10.1007/s10862-006-9038-5
75. Kanter JW, Rusch LC, Busch AM, Sedivy SK (2009) Validation of the behavioral activation for depression scale (BADS) in a community sample with elevated depressive symptoms. J Psychopathol Behav Assess 31(1):36–42. https://doi.org/10.1007/s10862-008-9088-y
76. McLeod BD, Wood JJ, Weisz JR (2007) Examining the association between parenting and childhood anxiety: a meta-analysis. Clin Psychol Rev 27(2):155–172. https://doi.org/10.1016/j.cpr.2006.09.002
77. Google Scholar (2025) Moos and Moos family environment scale. https://scholar.google.com/scholar?hl=en&as_sdt=0%2C31&inst=1210600082925750810 3&q=moos+and+moos+family+environment+scale&oq=Moos+and+Moos%2C+2009. Accessed 3 Apr 2025
78. Lebowitz ER, Marin CE, Silverman WK (2020) Measuring family accommodation of childhood anxiety: confirmatory factor analysis, validity, and reliability of the parent and child family accommodation scale – anxiety. J Clin Child Adolesc Psychol 49(6):752–760. https://doi.org/10.1080/15374416.2019.1614002
79. Cheron DM, Ehrenreich JT, Pincus DB (2009) Assessment of parental experiential avoidance in a clinical sample of children with anxiety disorders. Child Psychiatry Hum Dev 40(3):383–403. https://doi.org/10.1007/s10578-009-0135-z
80. Lebowitz ER, Woolston J, Bar-Haim Y et al (2013) Family accommodation in pediatric anxiety disorders. Depress Anxiety 30(1):47–54. https://doi.org/10.1002/da.21998
81. Rizvi SL, Ritschel LA (2014) Mastering the art of chain analysis in dialectical behavior therapy. Cogn Behav Pract 21(3):335–349. https://doi.org/10.1016/j.cbpra.2013.09.002
82. Chu BC, Temkin A, Hoffman L, Skriner LC, Zendegui E (2025) Individual behavioral activation program: growth through action. Unpublished Workbook
83. Christon LM, McLeod BD, Jensen-Doss A (2015) Evidence-based assessment meets evidence-based treatment: an approach to science-informed case conceptualization. Cogn Behav Pract 22(1):36–48. https://doi.org/10.1016/j.cbpra.2013.12.004
84. Persons JB (2006) Case formulation-driven psychotherapy. Published online 2006. https://psycnet.apa.org/fulltext/2006-07167-007.html. Accessed 1 Apr 2025
85. Yuen E, Sadhu J, Pfeffer C et al (2020) Accentuate the positive: strengths-based therapy for adolescents. Adolesc Psychiatry 10(3):166–171. https://doi.org/10.2174/2210676610666200225105529
86. MacIntyre MM, Zare M, Williams MT (2023) Anxiety-related disorders in the context of racism. Curr Psychiatry Rep 25(2):31–43.

https://doi.org/10.1007/s11920-022-01408-2

87. Berke DS, Liautaud MM, Chen D, Sloan CA (2024) Applying cognitive behavioral principles to promote health in transgender and gender diverse individuals. Cogn Behav Pract 31(1):35–47. https://doi.org/10.1016/j.cbpra.2022.05.002
88. Burton CL, Wang K, Pachankis JE (2019) Psychotherapy for the spectrum of sexual minority stress: application and technique of the ESTEEM treatment model. Cogn Behav Pract 26(2):285–299. https://doi.org/10.1016/j.cbpra.2017.05.001
89. Sanchez AL, Comer JS, LaRoche M (2022) Enhancing the responsiveness of family-based CBT through culturally informed case conceptualization and treatment planning. Cogn Behav Pract 29(4):750–770. https://doi.org/10.1016/j.cbpra.2021.04.003
90. Bailin A, Burton S, Rego S, Alpert J, Pimentel S (2021) Integrating advocacy for marginalized children and families into evidence-based care during COVID-19: clinical vignettes. Cogn Behav Pract 28(4):701–715. https://doi.org/10.1016/j.cbpra.2021.04.004
91. Wenzel A, Dobson KS, Hays PA (2016) Culturally responsive cognitive behavioral therapy. In: Cognitive behavioral therapy techniques and strategies. American Psychological Association, pp 145–160. https://doi.org/10.1037/14936-008
92. Huey SJ, Park AL, Galán CA, Wang CX (2023) Culturally responsive cognitive behavioral therapy for ethnically diverse populations. Annu Rev Clin Psychol 19:51–78. https://doi.org/10.1146/annurev-clinpsy-080921-072750
93. Hays PA (2001) Addressing cultural complexities in practice: a framework for clinicians and counselors. American Psychological Association, pp vii–239. https://doi.org/10.1037/10411-000
94. DeSilva R, Aggarwal NK, Lewis-Fernández R (2018) The DSM-5 cultural formulation interview: bridging barriers toward a clinically integrated cultural assessment in psychiatry. Psychiatr Ann 48(3):154–159. https://doi.org/10.3928/00485713-20180214-01
95. American Psychiatric Association (2013) Cultural formulation interview (CFI). In: Diagnostic and statistical manual of mental disorders, 5th edn. American Psychiatric Publishing, Inc.
96. Alarcón RD, Vega-Dienstmaier J, Cruzado L (2016) Use of the cultural formulation interview in different clinical settings. In: DSM-5® handbook on the cultural formulation interview. American Psychiatric Publishing, Inc., pp 208–223
97. Jarvis GE, Kirmayer LJ, Gómez-Carrillo A, Aggarwal NK, Lewis-Fernández R (2020) Update on the cultural formulation interview. Focus 18(1):40–46. https://doi.org/10.1176/appi.focus.20190037
98. Weisz JR, Chorpita BF, Frye A et al (2011) Youth top problems: using idiographic, consumer-guided assessment to identify treatment needs and to track change during psychotherapy. J Consult Clin Psychol 79(3):369–380. https://doi.org/10.1037/a0023307
99. Chorpita BF, Daleiden EL (2009) Mapping evidence-based treatments for children and adolescents: application of the distillation and matching model to 615 treatments from 322 randomized trials. J Consult Clin Psychol 77(3):566–579. https://doi.org/10.1037/a0014565
100. Langer DA, Jensen-Doss A (2018) Shared decision-making in youth mental health care: using the evidence to plan treatments collaboratively. J Clin Child Adolesc Psychol 47(5):821–831. https://doi.org/10.1080/15374416.2016.1247358
101. Moyer MW (2023) How to Talk to Kids About Israel and Gaza. Now What. https://melindawmoyer.substack.com/p/how-to-talk-to-kids-about-israel. Accessed 2 Apr 2025
102. Caporino NE, Exley S, Latzman RD (2020) Youth anxiety about political news. Child Psychiatry Hum Dev 51(5):683–698. https://doi.org/10.1007/s10578-020-00972-z
103. Kallapiran K, Koo S, Kirubakaran R, Hancock K (2015) Review: effectiveness of mindfulness in improving mental health symptoms of children and adolescents: a meta-analysis. Child Adolesc Ment Health. 20(4):182–194. https://doi.org/10.1111/camh.12113
104. Martin F, Oliver T (2019) Behavioral activation for children and adolescents: a systematic review of progress and promise. Eur Child Adolesc Psychiatry 28(4):427–441. https://doi.org/10.1007/s00787-018-1126-z
105. Tindall L, Mikocka-Walus A, McMillan D, Wright B, Hewitt C, Gascoyne S (2017) Is behavioural activation effective in the treatment of depression in young people? A systematic review and meta-analysis. Psychol Psychother Theory Res Pract 90(4):770–796. https://doi.org/10.1111/papt.12121

106. McCauley E et al (2016) The adolescent behavioral activation program: adapting behavioral activation as a treatment for depression in adolescence. J Clin Child Adolesc Psychol 45(3):291–304. https://doi.org/10.1080/15374416.2014.979933
107. Chu BC, Colognori D, Weissman AS, Bannon K (2009) An initial description and pilot of group behavioral activation therapy for anxious and depressed youth. Cogn Behav Pract 16(4):408–419. https://doi.org/10.1016/j.cbpra.2009.04.003
108. Ritschel LA, Ramirez CL, Jones M, Craighead WE (2011) Behavioral activation for depressed teens: a pilot study. Cogn Behav Pract 18(2):281–299. https://doi.org/10.1016/j.cbpra.2010.07.002
109. Martell CR, Dimidjian S, Herman-Dunn R (2021) Behavioral activation for depression: a clinician's guide. Guilford Publications

Chapter 7

Cognitive-Behavioral Therapy for Pediatric Obsessive-Compulsive Disorder

Jonathan S. Abramowitz and Emily K. Juel

Abstract

Pediatric obsessive-compulsive disorder (OCD) is characterized by intrusive obsessions and compulsive rituals that significantly interfere with daily functioning. Cognitive-behavioral therapy (CBT), specifically exposure and response prevention (ERP), is the gold-standard treatment for pediatric OCD. This chapter provides an overview of the cognitive-behavioral model of OCD, highlighting the role of avoidance and compulsive rituals in maintaining symptoms. We discuss the implementation of ERP, including functional assessment, psychoeducation, exposure hierarchy development, and response prevention strategies. Additionally, we explore the importance of parental involvement, given the impact of symptom accommodation on treatment outcomes. Finally, we examine common barriers to treatment, strategies to enhance engagement, and emerging therapeutic augmentations such as acceptance and commitment therapy (ACT). By emphasizing evidence-based interventions and family-based strategies, this chapter aims to provide clinicians with a comprehensive guide to optimizing treatment outcomes for children with OCD.

Key words Obsessive-compulsive disorder, Exposure and response prevention, Acceptance and commitment therapy, Family accommodation, Functional analysis

1 Definition of OCD

Obsessive-compulsive disorder (OCD) is a mental health condition characterized by persistent unwanted intrusive thoughts, ideas, and images that are senseless, yet upsetting (obsessions; e.g., images of germs and illness) and urges to perform repetitive behaviors or mental acts in a ritualized way or to excess (compulsions; e.g., handwashing) to alleviate this distress. Patients may also go to great lengths to avoid stimuli that provoke obsessions and compulsions (e.g., public bathrooms). Although the late teens and early 20 s is the most common age of onset, symptoms may begin at any point in life, including during childhood [1]. Among children, the most common themes of obsessions include fears of contamination, harm, or making mistakes, and intrusive thoughts about symmetry, order, or taboo subjects such as violence, sex, and morality. The

Robert D. Friedberg and Micaela A. Thordarson (eds.), *Cognitive Behavioral Therapy with Youth: Tradition and Innovation*, Neuromethods, vol. 231, https://doi.org/10.1007/978-1-0716-5134-6_7,

most common compulsions include excessive washing, checking, counting, seeking reassurance, and arranging items in a specific manner [2]. Although symptoms can range in severity and may wax and wane across the course of the disorder, obsessions and compulsions often interfere with school performance, social interactions, and family life, leading to significant distress and functional impairment [3].

An important aspect of OCD symptoms that varies widely from individual to individual, as well as across time, is the level of insight into the senselessness of symptoms. Specifically, some children recognize that their obsessions and compulsions are irrational or excessive, while others have little to no insight into the irrational nature of their thoughts and behaviors. This *poor insight* can make OCD more difficult to manage, as children may resist the idea that their thoughts and behaviors are problematic and be less likely to engage in treatment [4]. Parental involvement may also play a role in the severity of OCD symptoms among children. In particular, as we discuss later in the chapter, parents often become involved in their child's avoidance and compulsive behaviors. This involvement, termed *symptom accommodation*, while well-intentioned, inadvertently reinforces OCD symptoms [5].

2 Prevalence and Clinical Course

Epidemiological studies around the world report OCD prevalence rates of approximately 1–3% among children, underscoring the widespread nature of this condition [6, 7]. However, the disorder may go unrecognized, especially among younger children, due to their limited verbal skills and inability to articulate their intrusive thoughts. Children may also not recognize their obsessional thoughts as irrational and thus not report them to adults. Finally, children may hide their OCD symptoms due to embarrassment or shame.

The course of OCD typically involves two peaks of incidence: one in preadolescent children and another in early adulthood, around the mean age of 21 years [7, 8]. This biphasic pattern of onset highlights the importance of early detection and intervention to mitigate the long-term impact of OCD on affected individuals. The natural course of pediatric OCD, if left untreated, is typically chronic and debilitating, often persisting into adulthood. Longitudinal studies show that early-onset OCD is associated with a higher risk of comorbid diagnoses—often including anxiety disorders, depression, and tic disorders—which can complicate the clinical picture and treatment outcomes [8, 9]. Therefore, recognizing and addressing OCD in children is critical to improving prognosis and quality of life.

3 Conceptualization and Treatment Implications

Whereas numerous theoretical approaches have been offered to explain the psychopathology of OCD, cognitive-behavioral models have the most consistent empirical support. This approach begins with the well-established fact that spontaneously occurring negative intrusive thoughts, ideas, and images (e.g., the thought of one's parents dying) are normal occurrences in the general population [10] including, of course, for children. Such thoughts may arise without warning or may be evoked by external stimuli (e.g., hearing about a car accident on the news). Yet, borrowing from Beck's [11] cognitive model of emotion, if such intrusions are catastrophically misappraised as significant or meaningful (e.g., "thinking about my parents having an accident could cause this to happen"), it leads to anxiety or guilt, as well as attempts to control or remove the unwanted thoughts, upsetting emotions, and reduce the perceived threat. This occurs in the form of avoidance and compulsive rituals, which often technically "work" to restore feelings of safety. However, any reduction in distress is temporary and this merely serves to negatively reinforce the compulsive behaviors. Moreover, when the feared consequences associated with the intrusive thoughts do not ensue, the child comes to believe that their compulsive behavior was the savior. These processes preserve the obsessional fear, misinterpretations, and rituals, contributing to a self-sustaining vicious cycle.

This conceptual model is supported by decades of empirical research [12] and provides the foundation for the successful treatment of pediatric OCD. Specifically, it implies the need for (a) the correction of maladaptive appraisals of intrusive thoughts, including reducing estimates of threat, and (b) the termination of avoidance and compulsive rituals that prevent the self-correction of maladaptive beliefs. In short, the task of cognitive-behavioral therapy (CBT) is to foster an evaluation of obsessional stimuli (including the experience of fear itself) as nonthreatening and therefore not demanding of compulsive rituals or avoidance.

To illustrate, consider a child who compulsively taps a certain number of times to negate recurrent upsetting thoughts of bad luck befalling his family. This child overestimates the likelihood of threat associated with the intrusive thoughts and believes he is responsible for preventing a tragedy. The most robust way to modify such beliefs is for the child to gradually approach his feared situations and thoughts (i.e., exposure therapy) without the use of ritualistic behavior (i.e., response prevention). When exposure and response prevention (ERP) is practiced repeatedly, the child will learn through a process of fear extinction that (a) his *thoughts* about danger are not actual danger *signals*, (b) that doing compulsive tapping is not necessary to keep his family safe, and (c) that obsessional distress is temporary and manageable. Over the past 40 years, numerous treatment trials and meta-analytic studies provide robust support for the use of ERP as a centerpiece in the treatment of OCD [13]. Further below, we discuss the implementation of this approach, along with augmentative procedures.

3.1 Interpersonal Considerations

The above conceptualization applies to the experience of the individual child with OCD. However, OCD also has an *interpersonal* component, which for children, typically involves their parents. One way this might be manifested is when parents inadvertently maintain OCD symptoms by "helping" with compulsive rituals and avoidance behavior out of love for their child (a.k.a. symptom accommodation) [14]. A second is when obsessional problems create interpersonal distress and conflict, which then exacerbates anxiety, uncertainty, and obsessional thinking for the child [15]. This often happens when a parent becomes frustrated and argues with their child over the senselessness of OCD symptoms or over how the disorder is interfering with family life.

Symptom accommodation occurs when a parent (or other caretaker) of a child with OCD participates in rituals, facilitates avoidance, assumes daily responsibilities for the child, or steps in and helps to resolve problems that have resulted from the child's obsessional fears and compulsive urges—problems that it is best for the child to learn to solve on their own (e.g., speaking with a high school teacher) [16]. The accommodation might occur at the request (or *demand*) of the child with OCD, who directs their parent to help with reducing anxiety. In other instances, parents or other loved ones may voluntarily accommodate without needing to be asked because they feel the need to show care and concern for their suffering child and do not wish to see them become highly anxious. Accordingly, accommodation may become part of the family system.

Since avoidance and compulsive rituals prevent the natural extinction of obsessional fear and ritualistic urges, accommodation to these symptoms by a parent also perpetuates OCD symptoms in the same way, contributing to the vicious cycle described previously. For instance, consider a child with obsessional fears of acting on unwanted impulses to harm her younger sibling. By accommodating the avoidance of the younger sibling, parents prevent the child with OCD from learning that her distress over the senseless obsessions is manageable and temporary and that she is unlikely to act on her unwanted obsessional thoughts. As a result, the child's fear will persist. Hence, parents might feel additionally frustrated that the more they try to "help," the more their "help" is needed (or demanded).

Accommodation has various additional negative consequences. First, it may decrease a child's incentive to participate in ERP since he or she might not recognize good reasons to change the status quo—especially given that ERP involves provoking distress through facing one's fears. Second, in some families, accommodation becomes the primary strategy by which parents express warmth, caring, and compassion for their child. Not only does such accommodation maintain obsessional fear in the ways discussed previously, it also begets additional accommodation as the family relationship develops around this sort of "supportive" behavior. Not surprisingly, accommodation is related to more severe OCD symptoms and poorer long-term treatment outcome [16, 17]. Accordingly, as we discuss later in this chapter, it is critical to address symptom accommodation in the treatment of pediatric OCD [18].

3.2 Implementing Treatment

Because OCD is such a heterogeneous condition, the implementation of CBT using ERP varies widely from child to child. In this chapter, we provide an overview of the course of treatment, and the nuts and bolts of this intervention. For additional details, we recommend the reader seek out one or more of the comprehensive treatment manuals available for clinicians.

Treatment begins with an assessment of the child's present complaints, mental status, treatment history, social and family history, and an evaluation of various aspects of functioning including school, family, and personal wellness and self-care. Next, a functional analysis of OCD symptoms is conducted and used as the basis for psychoeducation and the development of a treatment plan. Treatment itself entails the primary ingredients of ERP. Sometimes, therapists will use techniques derived from acceptance and commitment therapy (ACT). Finally, we discuss the role of the parents or family in CBT.

3.2.1 Assessment and Diagnosis

The initial step includes gathering detailed information about the child's OCD symptoms, their onset, duration, and impact on daily functioning. Clinicians should conduct interviews with both the child and parents to get various perspectives on the severity level and functional interference. We recommend the use of a semi-structured interview such as the Children's Yale-Brown Obsessive-Compulsive Scale (CY-BOCS) [1], which allows for the identification of specific obsessions and compulsions as well as providing an index of symptom severity. Understanding the child's developmental stage, family dynamics, and any co-occurring conditions is also essential for understanding the clinical picture and crafting an effective treatment plan.

3.2.2 Functional Analysis of OCD Symptoms

Functional analysis helps the clinician understand the specific patterns of cognition and behavior that maintain the child's OCD symptoms according to the cognitive-behavioral conceptualization. The goal is therefore to gather detailed information about the triggers (antecedents), the child's specific thoughts and emotions (obsessions), behavioral responses (avoidance, compulsions, and accommodation), and the consequences of those behaviors.

With the child and parents (as necessary and developmentally appropriate) present, the clinician begins by asking for a general description of the obsessions and compulsions the child experiences. After this general overview, the therapist identifies the cues, triggers, and situations that provoke the child's obsessions (e.g., "What are you doing or thinking about when you start to feel anxious?"). Parents can often provide valuable insights based on their observations, such as noticing that the child avoids certain places or gets particularly anxious in specific situations, like before bedtime or after watching a certain TV show. Next, the clinician creates a list of all avoidance strategies and compulsive behaviors and learns the specific function of these behaviors—i.e., what purpose do they serve for the child. Typically, compulsions are performed to reduce anxiety or discomfort caused by obsessions. Compulsive rituals can be overt (e.g., washing, asking for reassurance) but also subtle or mental (e.g., repeating phrases or numbers to oneself). They may also involve family members as we have already discussed.

The functional analysis is complete when the clinician is able to "map out" the connections between the child's thoughts, feelings, and actions, identifying the patterns that maintain the OCD symptoms. This information will be essential for developing a targeted and effective treatment plan, particularly for guiding the exposure and response prevention as we will see in the sections that follow.

3.2.3 Education and Treatment Planning

As the clinician collects functional analysis data and begins thinking about treatment planning, they are also engaging in psychoeducation, which is a critical part of treatment often overlooked by therapists. Indeed, the child (and family) must understand the nature of OCD and how it can be treated to get the most out of CBT. Psychoeducation therefore focuses on presenting the cognitive-behavioral model of OCD and the rationale for CBT in a way that everyone involved can grasp. Below is an example of a didactic presentation of such material. We, however, recommend the use of Socratic discussion as much as possible to keep the child engaged.

> *Now that we've spent a lot of time finding out about your obsessions and compulsions, let's talk about how OCD works. Our brains are designed to keep us safe and help us survive, so when we feel like something is dangerous, our brain sends out a warning signal—like an alarm going off. But with OCD, the brain's alarm system gets a little too sensitive. The alarm starts going off even when there's nothing really dangerous happening. For example, your brain might be telling you that touching a doorknob is dangerous because it could make you sick, even though most people touch doorknobs all the time without getting sick (use an example relevant to the child's obsessions). That's the obsession—the thought that something bad might happen. When the alarm goes off, it feels really uncomfortable, so you might do something to try to make it*

stop. That's the compulsion—like washing your hands over and over to try to get rid of the worry (insert the child's specific compulsion). The problem is, when you do the compulsion, it teaches your brain that the alarm was right, so the next time, the alarm gets even stronger.

This is where treatment comes in. We've got a really powerful way to help your brain learn that the things it's worried about aren't as dangerous as they seem, and that you don't have to do the compulsions to feel okay. Here's how it works: We'll start by making a list of the things that you try to avoid or that make you use compulsions. Then, beginning with easier items from the list, we'll work together to help you approach these things without doing the compulsion. For example, if you avoid doorknobs, we might practice touching one without washing your hands right away. This might feel uncomfortable at first, but over time the alarm in your brain will start to get quieter as you see that everything is OK.

I want to make sure you (and your family) know that this is a team effort. We're going to work together to come up with a plan that's just right for you. Remember that this process might feel a little challenging at times, but that's okay. We'll take things step by step, and I'll be here to support you the whole way. How does all that sound? Do you have any questions or thoughts about this?

Next, the child and therapist work together to develop an exposure list—a compilation of between 10 and 20 fear-provoking thoughts, situations, and cues that the child will practice approaching during therapy. Therapists work with clients to rate the distress each item is expected to evoke, often using a subjective units of distress (SUDS) scale from 0 (no distress) to 10 (maximum distress). Items on the exposure list should be flexible enough to allow for varied exposures across different contexts, enhancing the generalization of learning. Additionally, response prevention strategies are planned to help the child resist rituals.

3.2.4 Exposure and Response Prevention

By the third or fourth session, a treatment plan is usually complete, and ERP can begin. Contemporary ERP involves therapist-guided systematic repeated practice approaching and remaining in the presence of the fear cues identified during the functional assessment (exposure) along with abstaining from compulsive and ritualistic behaviors (response prevention). Exposure can occur in the form of repeated (a) actual confrontation with feared but objectively low-risk situations (i.e., in vivo exposure) and (b) imaginal confrontation with anxiety-provoking intrusive thoughts (imaginal exposure) or (c) deliberate provocation of distressing (but harmless) internal bodily sensations (interoceptive exposure; e.g., drinking caffeine to produce feelings of jitters, or squinting one's eyes to provoke a tension headache). In the case of a child with OCD who has contamination obsessions triggered by "floor germs," the client would practice touching floors, working their way up from less "dangerous" floors (e.g., her bedroom) to more "risky" ones (e.g., in the bathroom). She might also practice imaginal exposure to the idea of having germs on her hands and then spreading these germs and becoming ill.

Response prevention is a vital component of effective treatment because the performance of rituals to reduce fear would prematurely discontinue exposure and consequently prevent the client from learning that (a) the fear trigger is not realistically dangerous and (b) anxiety subsides on its own even if the ritual is not performed. Thus, successful ERP requires that the client remain in the exposure situation until either the anxiety or distress decreases naturally (i.e., habituation) or expectations of danger have been disconfirmed, and the client no longer expects that the feared outcome is likely to occur. Continuing on with the exposure examples above, the person with OCD would refrain from washing or showering and deliberately use their hands for various things, such as eating or toothbrushing.

The therapist typically supervises exposure practices during the treatment session and then assigns additional practices to be completed as homework, perhaps with parental supervision. Depending on the child's specific fear triggers and the practicality of confronting actual feared situations, treatment sessions might involve varying amounts of in vivo and imaginal exposure practice.

These exposure exercises typically begin with moderately distressing situations, stimuli, and images and escalate to increasingly distressing situations. Beginning with less anxiety-evoking exposure tasks increases the likelihood that the child will learn to manage their distress and complete the exposure exercise successfully. Moreover, having success with initial exposures increases the child's confidence in the treatment and reduces the likelihood of premature discontinuation. Some research, however, suggests that a varied approach—randomly selecting items from different fear levels—can be more effective [19]. This method enhances learning by helping clients to approach different anxiety levels and thereby develop confidence in managing varying degrees of fear. Regardless of the order of stimuli, the full list (or hierarchy) of exposure stimuli should be confronted during treatment.

ERP sessions typically continue until the child has completed the full list of feared stimuli, and there is significant reduction in distress, avoidance, and ritualistic behaviors across a range of situations. Treatment is usually considered complete when the child can consistently approach previously feared situations and thoughts without experiencing overwhelming anxiety and without resorting to compulsions or avoidance behaviors. This usually corresponds to new skills in tolerating discomfort and new confidence in the ability to manage anxiety independently. Importantly, while structured ERP sessions may conclude, ongoing practice and occasional "booster" sessions may be recommended to prevent relapse and help the child maintain their progress over time.

Incorporating Acceptance and Commitment Therapy (ACT)

ACT strategies can enhance ERP, particularly for older children who are able to think more abstractly [20]. Specifically, ACT emphasizes developing a more flexible relationship with thoughts and feelings, including the importance of accepting distressing thoughts and emotions rather than trying to eliminate them. The child is encouraged to focus on what matters in their life and commit to actions that align with those values, even when it is uncomfortable. This approach can be useful in promoting a mindset that is less focused on eliminating anxiety and more on living a meaningful life despite it [21].

When complementing ERP with ACT, a therapist first helps the child clarify what they value in life (e.g., doing well in school, maintaining friendships). Then, within the context of ERP, the therapist might use a combination of experiential exercises, mindfulness practices, and discussions that encourage child to become more willing to experience difficult thoughts and emotions during exposure without trying to fight or change them. For example, during an exposure session, a therapist may use metaphors to help clients learn to observe their thoughts and feelings without becoming entangled in them. They might set a specific goal for the exposure which is consistent with the child's values, fostering a sense of purpose and commitment to action in the face of the challenging thoughts and emotions provoked by the ERP practice. Research indicates that parents can be helped to facilitate the use of ACT between sessions [22].

4 The Involvement of Parents and Caregivers

Parents (and other caregivers) play an important role in the treatment of pediatric OCD, supporting the child throughout the therapeutic process and ensuring that treatment gains are maintained at home. During the first few sessions, parents are educated about the nature of OCD, the importance of supporting their child's ERP work, and how to avoid accommodating or reinforcing compulsive behaviors. Family-based interventions may also focus on improving communication, reducing conflict, and fostering an environment that supports the child's progress. Engaging parents as partners in treatment helps to create a consistent and supportive approach, which is vital for the success of therapy. This role is particularly important for younger children or for those whose symptoms are significantly interfering with family life.

One core aspect of involving parents is teaching them to refrain from accommodating the child's OCD symptoms. As discussed, accommodation includes participating in rituals, providing excessive reassurance, or altering routines to prevent the child from feeling anxious. While accommodation may seem helpful in the short term (e.g., reducing the child's distress), it reinforces OCD symptoms by preventing the child from learning to tolerate anxiety and from disconfirming their fears. The therapist can help parents recognize instances of accommodation and then gradually reduce these behaviors as part of the response prevention component of ERP. For example, instead of giving repeated reassurances about cleanliness, parents might be coached to say, "I know this feels hard right now, but you can handle this discomfort without washing."

Lebowitz et al. [23] developed the SPACE (Supportive Parenting for Anxious Childhood Emotions) program specifically to help parents reduce accommodation behaviors in a supportive and compassionate way. Parents learn to acknowledge their child's anxiety with empathetic statements (e.g., "I see that you're feeling really worried right now") while clearly communicating a new boundary (e.g., "But I won't be helping with the ritual because that won't be helpful for you in the long run"). When symptom accommodation is present in the affected family, we recommend that therapists introduce parents to SPACE strategies by first explaining the detrimental effects of accommodation on the child's progress in ERP and then guiding parents in implementing supportive, non-accommodating responses. For example, if a child with contamination fears demands that the parent sanitize their belongings, the parent might respond with, "I understand that this makes you feel anxious, and I'm here to support you as you face this fear, but I won't be cleaning your things for you." The therapist may also facilitate parent-child conversations in the session to practice these new responses, providing coaching and feedback to help the parent set limits calmly and empathetically.

In addition to reducing accommodation, therapists can involve parents in exposure practices as appropriate, ensuring they are aware of the exposure list and the rationale for each task to be approached. Parents can be trained to supervise and support at-home exposures, acting as "coaches" who encourage the child to face their fears rather than offering reassurance or helping them avoid distress. This approach not only enhances the child's progress but also empowers parents with tools they can use to maintain and reinforce the gains made in therapy after treatment ends.

5 Overview of Empirical Support for CBT

The efficacy and effectiveness of CBT is supported by extensive empirical research, making it the first-line treatment for pediatric OCD [13, 24, 25]. Meta-analyses consistently report large effect sizes, with Öst et al. [26] finding a pooled effect size of 1.43, indicating substantial symptom reduction compared to control conditions. About 60–80% of youth show significant symptom improvement, typically defined as a CY-BOCS score of 12 or lower at posttreatment [27]. These effects are not impacted by the mode of delivery (i.e., telehealth, in-person). Long-term studies suggest that these benefits are generally maintained for 1–3 years, though some children may experience symptom recurrence, especially during stressful periods. Booster sessions and continued ERP practice are effective in preventing relapse [17].

Still, not all youth show improvement [24]. Specifically, higher severity of OCD symptoms, initial levels of impairment, depressive symptoms, presence of a comorbid diagnosis, and greater family accommodation all predict higher levels of OCD symptoms even after a course of CBT [28–30]. Comparative research shows that CBT often outperforms pharmacotherapy alone, and while combined treatment (CBT + SSRIs) may offer additional benefits for severe cases, ERP remains the preferred standalone intervention [31].

6 Obstacles to Effective Treatment

Despite the demonstrated efficacy of CBT, several factors can limit its success. This section discusses the key barriers, including misdiagnosis, refusal to do ERP tasks, parental accommodation, and comorbidity.

6.1 Misdiagnosis and Challenges with Symptom Reporting

One obstacle to successful CBT is the difficulty in identifying OCD symptoms in youth since they can present with subtle or unusual symptoms [27]. Misdiagnosis or delayed diagnosis often leads to inappropriate treatment, delaying access to ERP and reducing its effectiveness [31]. Additionally, children may be reluctant or embarrassed to report their symptoms, particularly if the obsessional content is taboo or distressing (e.g., aggressive or sexual thoughts). Psychoeducation and rapport-building are crucial steps to help children feel comfortable disclosing their symptoms, allowing for accurate diagnosis and tailored treatment planning [17].

6.2 Refusal to Engage in ERP

Engagement in ERP can be challenging and refusal is a common barrier. Children may lack the incentive or insight needed to push through their discomfort, especially in the early stages of treatment before they have experienced its benefits. This can result in incomplete or inconsistent participation, which reduces treatment efficacy. To address this issue, therapists may use motivational interviewing techniques, ACT metaphors, or increase parent involvement to help the child understand the rationale for ERP and the importance of approaching what they fear [23]. Building a strong therapeutic alliance and setting achievable, small initial goals can also help reduce resistance and increase the child's willingness to engage in ERP tasks.

6.3 Accommodation

Greater levels of parental accommodation are associated with worse treatment outcomes [32]. Addressing accommodation in therapy involves educating parents about the negative impact of these behaviors and equipping them with strategies to respond supportively without enabling the child's OCD. Programs like SPACE provide structured interventions to help parents reduce accommodation while maintaining a supportive stance [23].

6.4 Comorbidity

The presence of comorbid disorders can also affect both the course and outcome of CBT. Common comorbidities among youth include anxiety disorders, attention-deficit/hyperactivity disorder (ADHD), depression, and tic disorders [33]. These disorders can interfere with the child's ability to participate fully in or consolidate what is to be learned from doing ERP, and may necessitate modifications to the standard process. For example, children with ADHD may struggle with the sustained focus needed for ERP tasks, while those with depression may lack motivation or energy to engage actively in therapy [17]. In some cases, it may be necessary to address the comorbid symptoms first, such as stabilizing mood or improving attention, before initiating ERP. Integrated treatment approaches, psychoeducation, and close collaboration with other healthcare providers (e.g., psychiatrists, school counselors) can help manage these challenges and enhance the effectiveness of CBT [4].

7 Case Illustration

Alex[1] was a 12-year-old child who was given a diagnosis of OCD after presenting with his parents to a local mental health clinic. Alex's primary concern was an extreme fear that bad luck would befall his loved ones. For example, he reported unwanted thoughts and images that his parents would be killed in a car accident, or that his grandfather would fall down the stairs and be seriously injured. These images might be cued by any stimuli related to accidents and injuries, as well as the number 13, which Alex believed represented bad luck. In response to these unwanted intrusions and stimuli, Alex had the urge to tap six times on any surface to "negate" the effects of his unwanted thoughts and prevent bad luck from occurring. In addition, he frequently asked his parents for reassurance that everything would be OK. These reassurance-seeking questions often turned into lengthy routines in which Alex demanded that his parents respond with specific phrases to alleviate his anxiety. Often, these routines devolved into arguments because Alex did not feel satisfied by his parents' attempts to reassure him.

[1] This case is fictional and represents a synthesis of different cases seen in our clinic.

7.1 Assessment and Psychoeducation

During the first session following the intake appointment, Alex's therapist met with Alex and his parents to conduct a functional assessment and collect as much information as possible about Alex's obsessional cues, avoidance patterns, and rituals. This included an administration of the CY-BOCS to generate information about the breadth and severity of Alex's obsessions and compulsions, as well as an assessment of how Alex' parents managed and participated in OCD symptoms. Alex's score on the CY-BOCS was 21, indicating a moderate degree of severity.

Session 2 focused on psychoeducation. The cognitive-behavioral model of OCD was discussed at length with Alex and his parents using metaphors from ACT to illustrate how obsessions provoke anxiety and compulsive rituals lead to a temporary reduction in distress. An emphasis was placed on how rituals—including when parents are involved—calm Alex's anxiety in the short term, but maintain his fears in the long run. This led to a discussion of the pros and cons of accommodation. Next, the therapist provided a description and rationale for the use of ERP, including how Alex's parents would play a role in treatment by coaching Alex and reducing their accommodation.

7.2 Treatment Planning

During session 3, an exposure hierarchy and response prevention plan was developed. Alex's exposure list (please *see* Fig. 1) was as follows (listed in order of SUDs):

Alex's response prevention plan included resisting his tapping rituals when unwanted thoughts came up, refraining from checking on family members to make sure they are safe, and abstaining from asking his parents for reassurance. Treatment also included a plan for Alex's parents to gradually end their accommodation of his reassurance-seeking rituals. Essentially, this involved no longer

Item	SUDS
Write the number 13 on paper	4
Write the number 13 on a picture of loved ones	5
Read information (e.g., statistics) about car accidents	6
Read/watch news stories about serious car accidents	6
Read/watch news stories about household accidents (e g., falls, electrocution, fires)	6
Watch videos about people who have had serious household accidents (e.g., on YouTube)	6
Write a story about grandfather falling and injuring himself	9
Write a story about parents dying	9

Fig. 1 Alex's exposure list

answering requests for reassurance, and instead, reminding Alex that he is strong and capable of facing his anxiety. Both Alex and his parents agreed on this plan during session 3.

The next sessions occurred on a weekly basis, and Alex's parents were present for the first two exposure sessions so they could observe the therapist's role and coach Alex between sessions.

During the first exposure session, Alex practiced writing the number 13 on a piece of paper. Before beginning the exposure, Alex was worried that doing this would cause bad luck. When his therapist asked about what specific types of bad luck would occur, Alex said that he was worried his grandparents would die if he wrote the number 13. After some discussion about the likelihood of this occurring, Alex was willing to use exposure as an opportunity to put his fear to the test and completed the exposure without any reassurance-seeking. At the end of the session, he was allowed to call his grandparents to see whether they had been harmed by the exposure. Indeed, when they answered the phone and were happy to speak with Alex, Alex was surprised that (a) they were OK and that (b) he was able to manage the distress provoked by the exposure. Between sessions, Alex and his parents practiced similar exposures to the number 13 while thinking of other people having bad luck.

At the subsequent session, Alex reported that after three exposures to writing the number 13, he was much less concerned about bad luck and had even stopped calling people to make sure they were unharmed. Accordingly, Alex had no difficulties with the next exposure, which was to write the number 13 on the back of pictures of his loved ones (which Alex had brought from home). Alex wrote this number on 10 pictures of different friends and relatives, including a picture of his therapist! Alex's parents proved to be excellent coaches.

Reading information online about car accidents was the focus of session 3. Alex and his therapist reviewed statistics about the likelihood of getting into different types of accidents, the likelihood of survival, etc. The aim of the session was not to provide reassurance, but instead to help Alex think about these topics without needing to ritualize. Ordinarily, such material would provoke tapping rituals, which Alex was helped to resist. For practice between sessions, he was instructed to read additional information about accidents. Session 4 was similar to session 3 and included watching news stories about car accidents while refraining from tapping rituals and asking for reassurance.

The fifth and sixth sessions were similar to sessions 3 and 4, except the topic was household accidents and mishaps. Alex read information about such misfortunes, watch videos, and repeated similar exercises each day between sessions. At the end of session 6, Alex's therapist readministered the CY-BOCS and checked in with Alex's mother about how things were progressing at home. Alex scored 14 on the CY-BOCS, indicating a clinically noticeable reduction in obsessions and compulsions. This was corroborated by his mother's report that things were generally going smoothly at home.

The plan for session 7 was for Alex to write a story about his grandfather falling down the stairs and suffering life-threatening injuries. Alex became very anxious and was unable to complete this exposure during the session, saying that the risk of writing such a story was simply too high. Accordingly, this session focused on a discussion about risk, with the therapist encouraging Alex to consider the outcomes of the previous exposures he had completed (e.g., despite his worries, no bad luck had ever occurred).

The day before session 8 was scheduled, Alex's mother called the therapist to cancel the appointment because Alex was still refusing to do the planned exposure. After rescheduling for the following week, Alex's mother again called to cancel at the last minute because Alex was refusing to get into the car to go to the clinic. At that point, the therapist asked Alex's mother if she and Alex's father would attend a session together without Alex present. They agreed, and this was scheduled for the following week.

At the subsequent meeting, an assessment of family functioning revealed that not only was Alex refusing to do any exposures at home, but he had also gone back to demanding that his parents provide reassurance about his family's safety. This reassurance-seeking had begun to interfere with family routines, and Alex's parents had once again started helping Alex avoid any distressing stimuli. Even though they knew this was not helping Alex in the long term, his parents had a difficult time not giving in to OCD just to get through the day. Alex would become very angry and tantrum if he became anxious or did not receive "proper" reassurance. Attempts to talk Alex back into going to therapy had not worked either. Accordingly, the therapist used methods derived from SPACE to intervene with Alex's parents. Three additional sessions were scheduled (without Alex present).

The therapist and Alex's parents agreed that providing reassurance would be the first target. Accordingly, Alex's parents were instructed to draft a letter to Alex in which they explain that they love Alex, know he is capable, and that they will begin reducing their reassurance to help Alex face his fears. This letter was edited by the therapist and Alex's parents into the following format:

> Alex, it makes us sad to see you become anxious and upset because of your OCD. We are always happy to give you advice and comfort, but we also know that answering your questions about safety over and over doesn't help you feel better for very long. As your parents, it's our job to help you get better at facing your OCD so you can grow stronger and more independent. That's why, after discussing it together, we've decided that we will only answer your OCD questions one time. This is not a punishment. We are here to help you. We love you and we know that you're stronger than the OCD.

Shortly afterward, Alex's parents found a time when Alex seemed receptive and read Alex their letter. Afterward, when they handed the letter to Alex, he became angry, crumpled up the paper, and demonstratively threw it in his trash can. As instructed by the therapist, Alex's parents did not argue with Alex, but instead calmly left the room. Before long, Alex picked up the letter and read it again on his own.

For the next 5 weeks, Alex's parents met weekly with the therapist to discuss the gradual reduction of accommodation and how to best support Alex when he became anxious or angry. This process generally went smoothly, and Alex's parents reported that over time, Alex was asking less and less for reassurance. This gave them confidence to expand their focus to address additional targets that had interfered with the family routine. Although Alex occasionally became irritable when his parents refused to avoid or provide reassurance for him, his parents continued to use the supportive strategies introduced within the SPACE model. After five check-ins, Alex's parents felt they had the tools to continue supporting Alex on their own and discontinued appointments with the therapist.

At a 6-month check-in, Alex's parents reported that Alex was still refusing to attend individual therapy for his OCD. However, because Alex's parents continued to emphasize that Alex could face his fears and manage OCD-related discomfort, the degree of OCD's interference in the household had become relatively minimal and Alex's parents were satisfied.

8 Conclusion

Cognitive-behavioral therapy, particularly ERP, remains the most effective treatment for pediatric OCD, with substantial empirical support. By systematically confronting feared stimuli and eliminating compulsive rituals, children can learn to tolerate distress and break the cycle of obsessional fear. However, treatment success depends not only on the child's participation but also on the role of caregivers, who must reduce symptom accommodation and reinforce therapeutic gains at home. While most children experience significant improvement with ERP, individual factors such as comorbidities and insight levels can present challenges. Clinicians can optimize outcomes by tailoring exposures to developmental levels and leveraging parent-based interventions. With continued research and refinement of treatment approaches, the field will further enhance the effectiveness and accessibility of CBT for pediatric OCD, ultimately improving long-term outcomes and quality of life for affected youth and their families.

Conflicts of Interest

The authors have no conflicts of interest to declare that are relevant to the content of this chapter.

References

1. Scahill L, Riddle MA, McSwiggin-Hardin MA, Ort SI, King RA, Goodman WK, Leckman JF (1997) Children's Yale-Brown obsessive compulsive scale: reliability and validity. J Am Acad Child Adolesc Psychiatry 36(6):844–852
2. American Psychiatric Association (2013) Diagnostic and statistical manual of mental disorders, 5th edn. Author, Washington DC
3. March JS, Leonard HL (1996) Obsessive-compulsive disorder in children and adolescents: a review of the past 10 years. J Am Acad Child Adolesc Psychiatry 35:1265–1273
4. Storch EA, Lewin AB, Geffken GR, Morgan JR, Murphy TK (2010) The role of comorbid psychopathology in the functional impairment of clinically referred youth with obsessive-compulsive disorder. Child Psychiatry Hum Dev 41:543–554
5. Lebowitz ER, Panza KE, Bloch MH (2016) Family accommodation in obsessive-compulsive and anxiety disorders: a five-year update. Expert Rev Neurother 16(1):45–53
6. Heyman I, Fombonne E, Simmons H, Ford T, Meltzer H, Goodman R (2003) Prevalence of obsessive-compulsive disorder in the British nationwide survey of child mental health. Br J Psychiatry 182(5):437–442
7. James SC, Farrell LJ, Zimmer-Gembeck MJ (2017) Description and prevalence of OCD in children and adolescents. In: Abramowitz JS, McKay D, Storch EA (eds) The Wiley handbook of obsessive compulsive disorders, vol 1. Wiley, Hoboken, pp 5–23
8. Geller DA, Biederman J, Griffin S, Jones J, Lefkowitz TR (1998) Comorbidity of juvenile obsessive-compulsive disorder with disruptive behavior disorders. J Am Acad Child Adolesc Psychiatry 37(4):420–426
9. Piacentini J, Bergman RL, Keller M, McCracken J (2003) Functional impairment in children and adolescents with obsessive-compulsive disorder. J Child Adol Psychopharm 13(2):61–69
10. Rachman S, de Silva P (1978) Abnormal and normal obsessions. Behav Res Ther 16(4): 233–248
11. Beck AT (1979) Cognitive therapy and the emotional disorders. Penguin, New York
12. Cervin M, McNeel MM, Wilhelm S, McGuire JF, Murphy TK, Small BJ, Storch EA (2022) Cognitive beliefs across the symptom dimensions of pediatric obsessive-compulsive disorder: type of symptom matters. Behav Ther 53(2):240–254
13. Cervin M, McGuire JF, D'Souza JM, De Nadai AS, Aspvall K, Goodman WK, Andrén P, Schneider SC, Geller DA, Mataix-Cols D, Storch EA (2024) Efficacy and acceptability of cognitive-behavioral therapy and serotonin reuptake inhibitors for pediatric obsessive-compulsive disorder: a network meta-analysis. J Child Psychol Psychiatry 65(5):594–609
14. Calvocoressi L, Lewis B, Harris M, Trufan SJ, Goodman WK, McDougle CJ, Price LH (1995) Family accommodation in obsessive-compulsive disorder. Am J Psychiatry 152(3): 441–443
15. Storch EA, Lehmkuhl H, Pence SL, Geffken GR, Ricketts E, Storch JF, Murphy TK (2009) Parental experiences of having a child with obsessive-compulsive disorder: associations with clinical characteristics and caregiver adjustment. J Child Fam Stud 18(3):249–258
16. Lebowitz ER, Scharfstein LA, Jones J (2014) Comparing family accommodation in pediatric obsessive-compulsive disorder, anxiety disorders, and nonanxious children. Depress Anxiety 31(12):1018–1025
17. Piacentini J, Bergman RL, Chang S, Langley A, Peris T, Wood JJ, McCracken J (2011) Controlled comparison of family cognitive behavioral therapy and psychoeducation/relaxation training for child obsessive-compulsive disorder. J Am Acad Child Adolesc Psychiatry 50(11):1149–1161
18. Peris TS, Storch EA, McGuire JF (2020) Exposure therapy for children with anxiety and OCD: clinician's guide to integrated treatment. Academic Press
19. Craske MG, Treanor M, Conway CC, Zbozinek T, Vervliet B (2014) Maximizing

exposure therapy: an inhibitory learning approach. Behav Res Ther 58(1):10–23

20. Armstrong AB (2012) Acceptance and commitment therapy for adolescent obsessive compulsive disorder. Dissertation, Utah State University
21. Petersen J, Ona PZ, Blythe M, Möller CM, Twohig M (2022) Intensive outpatient acceptance and commitment therapy with exposure and response prevention for adolescents. J Context Behav Sci 23(1):75–84
22. Barney JY, Field CE, Morrison KL, Twohig MP (2017) Treatment of pediatric obsessive-compulsive disorder utilizing parent-facilitated acceptance and commitment therapy. Psychol Sch 54(1):88–100
23. Lebowitz ER, Omer H, Hermes H, Scahill L (2014) Parent training for childhood anxiety disorders: the SPACE program. Cogn Behav Pract 21(4):456–469
24. Maye CE, Pinciotti CM, Storch EA (2022) Intensive treatments for youth with obsessive-compulsive disorder and anxiety disorders. Adv Psychiatry Behav Health 2(1):133–140
25. Uhre CF, Uhre VF, Lønfeldt NN, Pretzmann L, Vangkilde S, Plessen KJ, Gluud C, Jakobsen JC, Pagsberg AK (2020) Systematic review and meta-analysis: cognitive-behavioral therapy for obsessive-compulsive disorder in children and adolescents. J Am Acad Child Adol Psychiat 59(1):64–77
26. Uhre CF, Uhre VF, Lønfeldt NN, Pretzmann L, Vangkilde S, Plessen KJ, Gluud C, Jakobsen JC, Pagsberg AK (2016) Systematic review and meta-analysis: cognitive-behavioral therapy for obsessive-compulsive disorder in children and adolescents. J Am Acad Child Adol Psychiat 59(1):64–77
27. Lewin AB, Wu MS, McGuire JF, Storch EA (2014) Cognitive behavior therapy for obsessive-compulsive and related disorders. Psychiat Clin 37(3):415–445
28. Garcia AM, Sapyta JJ, Moore PS, Freeman JB, Franklin ME, March JS, Foa EB (2010) Predictors and moderators of treatment outcome in the pediatric obsessive compulsive treatment study (POTS I). J Am Acad Child Adolesc Psychiatry 49(10):1024–1033
29. Rech M, Weinzimmer S, Geller D, McGuire JF, Schneider SC, Patyk KC, De Nadai AS, Cepeda SC, Small BJ, Murphy TK, Wilhelm S (2020) Symptom trajectories of early responders and remitters among youth with OCD. J Obsessive-Compul Related Disorders 27: 100580
30. Turner C, O'Gorman B, Nair A, O'Kearney R (2018) Moderators and predictors of response to cognitive behaviour therapy for pediatric obsessive-compulsive disorder: a systematic review. Psychiatry Res 261:50–60
31. Krebs G, Heyman I (2018) Obsessive-compulsive disorder in children and adolescents: epidemiology, diagnosis and management. Paediatr Drugs 20(6):515–523
32. Caporino NE, Kendall PC (2012) Comorbidity and childhood anxiety disorders: clinical issues and implications. J Clin Child Adolesc Psychol 41(5):603–615
33. Masi G, Pfanner C, Lenzi F (2020) Clinical presentation and treatment of pediatric obsessive-compulsive disorder. Curr Treat Options Psychiatry 7(4):457–474

Chapter 8

Trauma-Focused Cognitive Behavioral Therapy with the School-Age Child

Yahaira I. Márquez, Stephanie Cruthirds, Aimee Sirois-Geddie, and Esther Deblinger

Abstract

Trauma-focused cognitive behavioral therapy (TF-CBT) is an evidence-based treatment designed to address the therapeutic needs of children and their caregivers in the aftermath of trauma. The efficacy of this treatment model has been well established in 25 randomized trials, and as a result it has become the standard of care for childhood PTSD and other trauma-related emotional and behavioral difficulties. TF-CBT is a short-term model implemented in a structured manner following the PRACTICE acronym that represents the treatment components including psychoeducation/parenting, relaxation, affect expression/regulation, cognitive coping, trauma narration and processing, in vivo mastery, conjoint sessions, and enhancing safety and future development.

An overview of the application of the treatment is provided, highlighting the role of the child, the non-offending caregiver, and the therapist in this intervention. The implementation of the PRACTICE skills with school-age children is highlighted. Parenting guidance to help caregivers optimally support their children's trauma recovery is also discussed. In addition, strategies for helping caregivers use coping skills to overcome their own trauma-related distress and serve as effective coping role models for their children are offered. Cultural and clinical considerations are discussed and guidance regarding the value and implementation of pre- and posttreatment assessments is offered. Clinical examples, trauma-specific resources, and recommendations for accessing TF-CBT training and consultation are provided.

Key words TF-CBT, Childhood trauma, Evidence-based treatment, PTSD, Non-offending caregiver, Childhood adversity, School-age child, Traumatic stress symptoms, Cultural and clinical considerations, Treatment

1 Overview of the Model

Greater levels of trauma and adversity in childhood have been found to be associated with significant challenges with both medical and mental health difficulties across the life span [1]. However, there are evidence-based interventions that have the potential to disrupt such problematic life trajectories and help youngsters overcome trauma-related psychosocial difficulties. One such evidence-

Robert D. Friedberg and Micaela A. Thordarson (eds.), *Cognitive Behavioral Therapy with Youth: Tradition and Innovation*, Neuromethods, vol. 231, https://doi.org/10.1007/978-1-0716-5134-6_8,

based treatment model is trauma-focused cognitive behavioral therapy (TF-CBT) for children, adolescents, and their non-offending caregivers [2, 3]. TF-CBT was originally developed to be used to treat children impacted by child sexual abuse. Randomized controlled trials documented its efficacy in helping these children overcome PTSD, depression, and behavior problems with improvements being maintained over 1- and 2-year follow-up periods [4–9]. The efficacy of TF-CBT has since been replicated in over 25 randomized trials including studies in which the children have been impacted by diverse traumas including exposure to domestic violence, war-related violence, traumatic loss of loved ones, and natural and man-made disasters impacting children from around the world (e.g., Norway, Canada, Germany, Japan, Tanzania, Kenya, the Democratic Republic of the Congo). TF-CBT has been reported to have the strongest level of empirical support for addressing the therapeutic needs of children and adolescents who have experienced diverse traumas [10] with widespread dissemination efforts resulting in TF-CBT becoming the standard of care for trauma-exposed children, not only in the United States but in other countries as well. Although TF-CBT has been utilized and evaluated for its efficacy with children from 3 to 18 years of age and more recently with young adults [11], the focus of this chapter will be limited to the implementation of TF-CBT with school-age children and their non-offending caregivers.

TF-CBT is a short-term treatment model typically delivered in 12 to 20 sessions. The components of TF-CBT are summarized by the acronym PRACTICE: **P**sychoeducation and **P**arenting, **R**elaxation, **A**ffective expression and modulation, **C**ognitive coping, **T**rauma narration and processing, **I**n vivo mastery, **C**onjoint sessions, and **E**nhancing safety and future development. It is worth noting that with exception of in vivo mastery, which is utilized on an as-needed basis, each of the TF-CBT components are implemented with most clients.

TF-CBT treatment initially involves parallel individual sessions with the child and caregiver meeting separately with the therapist with increasing time devoted to conjoint caregiver-child sessions over the course of treatment. Therapists engage clients in gradual exposure to trauma reminders by discussing the trauma(s) the child experienced in general terms beginning in the first session with increasing focus and details discussed over the course of the three phases of TF-CBT treatment. The first phase focuses on stabilization and skill-building during which the therapist provides psychoeducation about trauma in general terms and encourages the use of coping skills in managing distressing reactions to trauma reminders. The second phase of treatment consists of trauma narration and processing with the youth, while the caregiver continues to develop their parenting and coping skills and prepares for conjoint sessions when the child's trauma narration is, when clinically appropriate, shared with the caregiver. The final phase of treatment is devoted to mastery and integration of skills with a focus on encouraging open caregiver-child communication as well as enhancing safety and future development.

Although cognitive behavioral principles underlie the structure and components of TF-CBT, it is important to acknowledge that concepts and ideas from other theoretical orientations and theories including humanistic, family systems, empowerment, as well as attachment theories have also been incorporated [2, 3]. For example, the emphasis on engaging caregivers is drawn from the attachment literature that emphasizes the powerful influence of the caregiver-child relationship on children's development, well-being, and healing. Other important developmental, cultural, and clinical considerations that are central to effective TF-CBT implementation are outlined below.

2 Cultural and Clinical/Structure Considerations

TF-CBT with school-age children thrives on the participation of the child and their non-offending caregivers. TF-CBT affirms the values of each particular family with this stance. In the weekly sessions, the goal of the therapist is to work in a parallel manner with the dyad. In this instance, parallel refers to teaching the same skills to the child and the caregiver during their individual sessions. Even when this structure is followed there is flexibility to take into consideration the individual and family needs. Although the identified client is the child, the school-age child needs the support of their caregiver in their healing journey. The caregiver may be a biological relative, a foster parent, or other adult in a primary caregiving role. The work with the caregiver is a long-term investment in the child's well-being. This work increases the likelihood that the therapeutic process may continue long after therapy ends as the caregiver models effective coping skills, praises the child's healthy coping, and maintains open and effective caregiver-child communication. In this way, a healthier and effective caregiver multiplies the opportunities for the child's treatment benefits to be maintained over time.

The language of the home, the recognition of the family's cultural and racial background, family composition, and the family's religious values are some of the many factors that should be considered when working with families and actively integrated in the implementation of the TF-CBT components. Thus, it is valuable for TF-CBT therapists to inquire about family traditions and/or cultural practices that the child and/or caregiver may utilize when coping with stress and/or adversity.

3 TF-CBT Appropriateness

TF-CBT was developed for children between the ages of 3 and 18 years where there is some evidence of childhood trauma including single, multiple, and complex traumas. It is imperative that disclosures of child abuse or neglect be reported and documented in some manner as per laws of that state in which the therapist is working (e.g., sexual violence reported to the child protection agency). The child must have memory of the trauma. There should also be a link between the child's trauma history and their current difficulties functioning including (but not limited to) symptoms of posttraumatic stress disorder (PTSD), symptoms of depression, or behavioral problems. Children do not need to meet diagnostic criteria for any disorder to receive TF-CBT.

It is also important to consider when TF-CBT may not be the treatment of choice at the moment. Issues related to safety should be the priority when considering the appropriateness of TF-CBT. For instance, if the child is actively suicidal, this should be carefully assessed, and the client should be provided with a higher level of care as necessary. This child should be reconsidered for TF-CBT after active suicidality is addressed, and weekly outpatient therapy is deemed appropriate. TF-CBT should also not be initiated when the child is residing with an adult who is being investigated for allegations of child abuse. Instead, the child's physical and psychological safety should be the priority, and trauma treatment should be considered after the investigation has been completed and safety parameters for the child have been established.

4 Treatment Modality

TF-CBT has been shown to be effective when implemented in outpatient mental health clinics as well as in schools, residential treatment programs, and in-home environment (e.g., mobile therapy) [12]. It can be successfully implemented in individual format as well as in group modalities [13] and has been shown to be a treatment easily applied via telehealth with clear benefits [14, 15]. Telehealth treatment not only assists clients with overcoming concrete barriers to treatment such as transportation, sudden illness, and childcare, but also may be invaluable for families in rural communities that are not able to travel long distances to access the mental health services that they require.

5 Assessment (Pre-post Treatment Assessment)

A thorough review of records and the referral information will begin the process of learning more about the child's and family's context. Clinical assessment prior to the first therapy session is essential in TF-CBT. The use of standardized assessment measures serves the practitioners well when determining the current emotional and behavioral functioning of the child. The type of instrument may vary but should include inventories that assess a history of traumatic experiences, the symptoms associated with these experiences, behavioral issues, and emotional functioning. Whenever possible, measures that gather information from both the child and caregiver's perspectives should be used. The assessment measures at this point have two main goals. The first goal is to provide information about current functioning and symptoms present and to assist with diagnosis and treatment planning. The second goal is to begin the gradual exposure process for the school-age child and their non-offending caregiver. This is accomplished by naming the reason for referral (e.g., sexual abuse) and answering questions about current difficulties. Community practitioners may worry about the cost of instruments, or the skills needed to score and interpret such measures. There are free-of-charge measures available that assess common symptomatology associated with trauma that are quick and easy to administer as well as score. Examples of no-cost assessment measures include the Child and Adolescent Trauma Screen 2 (CATS-2) [16] and the Pediatric Symptom Checklist, 35-Item Version (PSC-35) [17]. The CATS-2 is a great example of an assessment measure easy to score with a child version and a caregiver version. The two versions allow the therapist to obtain information from different reporters which may provide different perspectives. The CATS-2 is also available in English and in Spanish.

The process of using standardized assessment measures helps the TF-CBT therapist provide early psychoeducation about trauma in general as well as the reason for referral. Providing general information about childhood adversity helps the child and caregiver(s) feel less alone as they learn about the prevalence of the trauma(s) experienced and the common reactions experienced by children and their families. Such information will help in the meaning-making process that occurs later in the course of treatment following trauma narration. The posttreatment assessment, which typically takes the form of repeating the same measures when all TF-CBT components are completed, provides information not only about the progress of the child but also for any recommendations for services that may be needed at the end of treatment.

6 Engagement with Caregivers

As noted earlier, the caregiver's role is central to the healing of the child. Thus, effective engagement of the caregiver from the start of treatment is critical. Engagement should address both potential concrete obstacles to treatment such as transportation, as well as attitudinal obstacles such as ambivalence about the value of therapy. Thus, directly inquiring about caregivers' general beliefs about therapy as well as prior therapy experiences is recommended. When negative therapy experiences are shared, it can be helpful to highlight ways in which TF-CBT might differ from their prior experiences. In addition, the initial treatment session with the caregiver can highlight a review of the assessment findings, while linking the child's symptoms to the treatment plan proposed. Sharing the structure of treatment, the approximate time frame for completing therapy, and the collaborative nature of the treatment approach can also be useful as an engagement strategy. Perhaps most importantly, emphasizing the critical role of the caregiver in the child's healing, while also encouraging optimism regarding the impact of treatment, should be conveyed to the caregiver. This optimism is based on the child's presentation as well as the research that has established the efficacy of TF-CBT for treating the trauma-related difficulties reported by the child and caregiver on the assessment measures.

7 Psychoeducation

One of the goals of the psychoeducation component is to provide general information about the trauma(s) experienced to normalize the experience of the child and their caregiver. This is to say that a wide range of reactions are expected after experiencing trauma. This information may include a definition of trauma, children's typical reactions to traumatic experiences, survival responses, and general information about the traumas experienced. School-age children typically respond well to a variety of activities including reading books, watching educational videos, and playing educational games. Once such game is the *What Do You Know?* game that was designed to help TF-CBT therapists engage children in a general discussion about the trauma(s) experienced through a simple question and answer card game [18]. Alternatively, the therapist can create open-ended questions about trauma (e.g., bullying) and write them on index cards, then take turns with the child answering the questions (e.g., *What is bullying? How do children feel when they have been bullied? What could a child do if they witness someone else being bullied?*). The same information is provided to the caregiver during their individual session. The information related to the trauma(s) experienced can validate the experience of the family and dispel myths and concerns held by the participating caregiver. It may also aid in strengthening the therapeutic alliance with the caregiver as trauma-related concerns are addressed. Depending on the caregiver's emotional readiness, conjoint caregiver-child psychoeducational sessions may be as simple as reading a book together and exchanging praises, which establishes positive therapeutic routines. Preparation for the conjoint session occurs during the individual time with the child and with the caregiver. This is key to ensure that the conjoint session functions as a behavioral rehearsal time (e.g., practice the skill learned during the individual session). This structure will continue throughout treatment.

8 Parenting

The parenting component within TF-CBT helps to refocus the caregiver's attention on the child's strengths while simultaneously addressing children's traumatic reactions and/or challenging behaviors. Caregivers sometimes receive little support as they face the traumatic events experienced by their children. This component provides both support and guidance to the caregiver that will strengthen protective factors that may counteract the effects of adversity. Examples of these factors include enhancing caregiver-child connectedness and communication (e.g., using praise to increase positive interactions and strengthening their bond) and promoting feelings of safety and security (e.g., by encouraging the initiating or reestablishing of family routines). All adults in primary caregiving roles (e.g., biological parents, foster parents, grandparents) can benefit from this component.

The therapist utilizes this component to explore the current concerns of the caregiver and designs a plan to collaboratively address behavioral challenges. The behavioral difficulties of school-age children vary but often include issues with defiance (e.g., refusing to complete tasks or follow directions given), disrespectful behavior (e.g., talking back or hitting others), and tantrums (e.g., verbal, emotional, or physical displays of discomfort when what is wanted is not received). These behaviors typically interfere with the child's functioning in one or more settings (e.g., home, school, community). It is important to note that the parenting component is initiated during the first therapy session with the caregiver and continues throughout treatment.

During the initial sessions, the therapist and the caregiver explore the connection between the challenging behaviors and the lived traumatic experience. This connection should not be overlooked as some behavioral difficulties may be a trauma response. For instance, a caregiver may complain that their child is frequently coming home with soiled underpants and that the teacher reports that they are waiting until the last minute to ask to go to the restroom. After discussing the behavior further, the therapist and the caregiver realized that there is a high likelihood that the child is avoiding going to the bathroom in school as they have a history of being bullied by other students in the bathroom. This understanding provides insight into the function of the behavior and allows the therapist and the caregiver to use this information to find appropriate ways to address it. During session with the caregiver, the therapist can explore a plethora of behavioral management skills from global and specific praise (see PRAISE acronym) to reflective listening to encourage positive replacement behaviors such as sharing feelings rather than engaging in avoidant behaviors. The use of tools such as handouts, parenting books, and homework assignments are highly desirable to support the learning experience of the caregiver. Conjoint sessions with individual preparation in advance provide opportunities to practice in real time the use of parenting skills such as effective specific and global praise [19]. The therapist is able to model the use of the various skills and provide feedback to the caregiver as they practice the skills with the child during the session.

9 Relaxation

The goal of the relaxation component is to teach children about strategies they can use to manage feelings and trauma reminders. Relaxation is helpful in reducing the physical symptoms of PTSD or anxiety, such as increased heart rate, rapid breathing, and upset stomach. There are several creative ways relaxation can be introduced to children and their caregivers. Focused breathing (i.e., conscious slow breaths taken in through the nose and out through the mouth) is a great place to start as taking three focused breaths has been shown to reduce heart rate and blood pressure [20]. When working with school-age children, it can be helpful to call this skill "belly breathing."

Therapist	*Today, we are going to learn how to do belly breathing. Sit comfortably in the chair and place your hands on your belly. Now, take a slow breath in through your nose and then slowly let it out through your mouth. Like this.* [Therapist demonstrates how to do belly breathing.] *Now you try it.*
Child	practices one slow breath with hands on their belly
Therapist	*Great job! Our bellies get bigger when we breathe in, and they get smaller when we breathe out. Let's take two more slow belly breaths.*
Child	places hands on their belly and takes two more slow breaths
Therapist	*How do you feel now after those three belly breaths?*

Child *Good, I feel calmer.*
Therapist *That's great. Belly breathing can help us feel calmer especially if we are feeling nervous or worried.*

Progressive muscle relaxation (PMR) is another technique therapists can teach children during the relaxation component. PMR helps relax the body by first tensing specific muscles and then relaxing those muscles. PMR can be introduced by asking the child to imagine they are holding a lemon in each hand and asking them to pretend to squeeze the lemon very tightly for a few moments. Then the child is asked to "drop the lemons" and stretch out their fingers to relax their hands.

Mindfulness (i.e., to be fully aware in the present moment) is another important relaxation technique that can be introduced to children and their caregivers to help relax their mind and stay present in the moment. The "five senses exercise" is one way to teach mindfulness. During this exercise, the therapist asks the child to focus on their present surroundings and name five things they see, four things they can touch, three things they hear, two things they smell, and one thing they taste. Helping the child to make the connection between the use of the above-described skills to manage trauma reminders is essential to gradual exposure. The therapist can ask the child at the end of the session which technique they can use when they have trauma-related thoughts. For example, *You have learned a few different ways to relax your body today. Which relaxation skill do you think might be helpful when you have thoughts about the domestic violence this week?*

Caregivers can also benefit from learning to utilize relaxation skills to manage everyday stresses, distress related to their children's traumatic experiences, and the challenges of managing their children's behavior. In addition, by learning to use these skills, caregivers can then model the use of these skills in the home. Conjoint session(s) focused on relaxation may include children teaching their caregivers one of the skills they learned during that therapy session. This, in turn, provides caregivers an opportunity to praise their children's use of these newfound skills. As part of homework, the therapist may also encourage caregivers and children to practice during the upcoming week the skills that they learned. Homework is a tool for the therapeutic intervention to continue during the week.

10 Affective Expression and Modulation

Children who are exposed to trauma may have many emotions related to their experience. One benefit of helping children learn how to label and identify their feelings is that it provides them with not only an awareness of their emotions but also the language to communicate these experiences. Increasing children's emotional vocabulary in the first phase of treatment will serve them well later during trauma narration when feelings and thoughts related to the trauma(s) endured are elicited. Creating a list of feelings can be the start of this component. This list of emotions can then be expanded by using a variety of strategies, including games and books. Identifying emotions in self and in others, accepting and tolerating a wide range of emotions, as well as learning about the intensity of feelings are important skills for the school-age child. It is helpful for the therapist to keep a list of the feelings identified. Using the list of feelings to identify emotions felt during the experienced traumas is a simple way of initiating gradual exposure.

In conjunction with affective expression, the therapist teaches the child ways to regulate emotions. This is an opportunity to build on previously learned skills. The therapist may begin by having the child identify healthy ways they already manage emotions (e.g., listen to music, talk to a friend). The therapist then assists the child in exploring additional healthy modulation strategies to expand the child's repertoire (e.g., focused breathing, physical activity, journaling).

Working with caregivers on affective expression and modulation is important as well. An initial discussion with the caregiver exploring how emotions are expressed within their family and/or their culture can be helpful. Inviting caregivers to label different emotions and discussing their feelings related to their children's trauma exposure is an important factor in helping caregivers provide support. Identifying emotions and discussions around ways to regulate emotions can give caregivers the opportunity to reinforce children's healthy expression of emotions as well as the opportunity to model the healthy expression and regulation of emotions.

The therapist may have caregivers and children meet at the end of the session to review and reinforce what was learned during this component. For example, they may play feeling charades together identifying different facial expressions related to emotions and regulation strategies. Additionally, they may have a friendly competition to see who can list the most feelings in a certain amount of time.

11 Cognitive Coping

In this component, the child and the caregiver are taught about the relationship between thoughts, feelings, and behaviors. A triangle is an image often used to teach about this connection (e.g., cognitive triangle). The work begins by demonstrating the differences between thoughts, feelings, and behaviors. One way to accomplish this with school-age children is by presenting a list of words and phrases and having the child identify which are thoughts (e.g., "I can do it"), which are feelings (e.g., sad), and which are behaviors (e.g., running). After differentiating thoughts, feelings, and behaviors, the therapist gives the child a hypothetical, non-trauma-related situation and has them identify what thoughts would come into their mind, what feelings are connected to the thoughts, and what behaviors are connected to that thought and to that feeling. Teaching children that thoughts are not facts, and all thoughts are not helpful, is also important. When thoughts are connected to intense, negative emotions, the therapist can teach the child to ask themselves if the thought is accurate and/or helpful. After teaching the cognitive triangle, the therapist can guide the child in utilizing cognitive coping related to their personal **non-trauma**-related experiences. Please note, the therapist **is not** eliciting trauma-related thoughts at this point in treatment. Inquiring about the child's thoughts related to the trauma occurs during trauma narration. If a child spontaneously brings up thoughts related to their traumatic experiences, the therapist should utilize reflective listening and validation while letting the client know that they will have the opportunity later to discuss these thoughts further. It is helpful for the therapist to record any trauma-related thoughts the child expresses during this component to explore further during trauma narration and processing. Some children, particularly those with somatic symptoms, may benefit from using a cognitive diamond that includes body sensations. The image of a diamond is used to connect thoughts ("I hate taking tests"), feelings (nervous), body sensations (restless legs), and behaviors (bites nails).

When working with younger school-age children, therapists may need to adjust their expectations around teaching cognitive coping because of younger children's more concrete way of thinking. This may be due to their current developmental stage which is preparing them for later achieving other cognitive milestones that facilitate the understanding of more abstract thinking and cognitive flexibility. However, younger children may be able to understand the connection between thoughts and feelings, and the connection between thoughts and behaviors. The image of arrows can be used to represent these simpler connections. Therapists are encouraged to utilize games and creative strategies to teach this skill to younger children.

Cognitive coping is also very important for caregivers of children who have experienced trauma. Caregivers may feel a variety of emotions about their child's trauma, including, but not limited to, sadness, anger, and guilt. Cognitive coping can be a beneficial skill to aid caregivers in reducing overwhelming negative emotions. The therapist may teach caregivers the connection between thoughts, feelings, and behaviors using a hypothetical situation. For example, the therapist may present the following scenario to the caregiver: *I want you to imagine you are walking in a downtown area and you see someone you know across the street, and you wave at them. They don't wave back. What is the first thought that comes into your head?* The therapist then may make note of the thoughts on a white board or piece of paper and then elicit feelings (including the intensity of the feelings) and behaviors. If the caregiver's automatic thought is a helpful thought, like "maybe they didn't see me," for learning purposes, the therapist may offer an unhelpful thought, such as "they must be mad at me." Next, the therapist may provide some education around thoughts including how many thoughts people have in a day, the automatic nature of thoughts, and how thoughts are not always true and/or helpful. The therapist will then go through the cognitive coping process, asking the caregiver if the thought is accurate and/or helpful, encouraging the caregiver to identify a more accurate and/or helpful thought, and then explore the replacement thought's impact on feelings and behavior. Please note that the replacement thought may not eliminate the presence of a feeling; however, even a reduction in the intensity of the feeling is progress. Once the therapist has taught cognitive coping to the caregiver, the therapist then begins eliciting the caregiver's thoughts related to their child's trauma. This is a major difference between teaching cognitive coping to the child and teaching cognitive coping to the caregiver. As a reminder, children will identify their trauma-related thoughts during trauma narration. Since caregivers do not participate in their own trauma-related narration, the therapist elicits the caregiver's thoughts related to their child's trauma(s) and processes the thoughts with the caregiver during the cognitive coping component of treatment. For some caregivers, cognitive coping may span over a few sessions continuing during the period of time when the child is engaged in trauma narration and processing (see below). Additionally, if the therapist shares the child's trauma narration with the caregiver, the therapist may elicit thoughts related to the child's narration and encourage the use of cognitive coping with the caregiver at that time in the final phase of treatment.

The therapist may conduct a conjoint session with the caregiver and child around cognitive coping if it is determined that doing so will be beneficial to them. The child, for example, can present what they learned about the cognitive triangle and/or the child and caregiver can play a game like "thought Jenga" where each block has a general thought on it and when pulled the person indicates if the thought is helpful or hurtful and how it can be changed to a more helpful one.

12 Trauma Narration and Processing

When gradual exposure is incorporated into the initial treatment components as described earlier, the trauma narration and processing component often feels like a natural next step for the school-age child and the therapist to further explore children's thoughts and feelings related to their traumatic experiences. Moreover, the PRAC components have been laying an important foundation of skills that will be utilized during this component. Typically, the narration is initiated with a psychoeducational activity related to the reason for referral. The school-age child often enjoys picture books and stories that provide relatable information about a specific trauma. For example, in the case of sexual abuse, the therapist can select a book about this topic and after discussing the experience of the school-age child in the story, this can serve as a bridge to suggesting an activity where the child gets to write their own book about their traumatic experience. The therapist introduces the narration exercise and begins it in the same session. It is important to note that children may have had experiences reporting the trauma/abuse experiences to various professionals in the past in the context of child protection or other forensic investigations. The narration process, however, is different from those previous experiences in that it is not forensic or evaluative, but therapeutic in nature. The therapeutic process is directed by the therapist, and it is a collaborative effort between the child and the therapist not only to discover the details of the traumatic event but to also explore the thoughts and feelings related to the lived experiences. A book is not the only possible format for this therapeutic exercise as some children may prefer to share their experiences in the form of poetry, lyrics, or artwork. Regardless of the format, it is important for therapists to elicit from the child the thoughts and feelings they experienced during the traumatic experiences as well as the thoughts and feelings they have when confronted with trauma reminders and during related experiences (e.g., medical exam, court appearance, return to school).

Trauma narration is an opportunity for the therapist to bear witness to the children's traumatic memories. In cases of multiple, complex, and/or chronic trauma(s), a subset of children's experiences will be explored through this process. A book format allows for multiple traumas and/or related experiences to be represented in different chapters. It is not uncommon for school-age children to also include pictures in their book as the drawing process can not only be soothing, but it can also enhance the narration of their experiences. Examples of trauma narration chapters include an introductory chapter in which children can share facts about themselves including the things they enjoy, the school they attend, and other important details of their current life. This is also a wonderful opportunity to assist children in writing something, even just a single sentence, that describes the purpose of the book. The therapist gently reminds children that this book is about the reason for referral, often using the description that children have been using throughout treatment (e.g., "this book is going to be about the time daddy touched my privates"; "this book is going to be about the bad bully in my school"; "this book is going to be about the time my auntie was killed"). Trauma-specific chapters focus on the memory of the traumatic event, the related thoughts, and related feelings (e.g., "the first time I was sexually abused"; "when I went to see the doctor"; "when the police came to my house"). As the child and the therapist gradually work on writing about these memories, the chapters capture the feelings, thoughts, and bodily sensations experienced at the time of these experiences. Some thoughts revealed may be accurate and helpful, whereas others may reflect inaccurate interpretations of others' behaviors, misinformation, and/or accurate but hurtful/unhelpful thoughts. Through this process, therapists and children have the opportunity to process these thoughts by engaging in guided discovery, Socratic questioning, evidence gathering, and other strategies that may help restructure and address problematic thoughts, while replacing them with more accurate, helpful thoughts. The following example aims to give the reader a feel for narration work. It captures the memory of the event, as well as the feelings and the thoughts experienced by the child (*see* Fig. 1).

Fig. 1 A school-age child's drawing depicting a person standing under a dark rain cloud with a sad, tearful expression. This type of artwork often accompanies a trauma narration chapter, integrating a high-interest activity that supports the child in narrating their experience and illustrating the associated emotions and related thoughts

> *The time I found out my little brother was killed*
>
> *I was in school. My mom came to pick me up, so they called me to the office. I was confused. I was thinking, I don't know why my mom is here. Her eyes and her face was red. I thought, oh, no, something is up. We got in the car and she told me something happened at home. I didn't know what to think. She said "little Carl is in heaven now" and started crying. I said, wait what? I started crying too. I was confused. I saw him this morning. I didn't understand. She told me we were going to pick up my sister. I cried the rest of the way and she cried too. It was like a dark cloud was on top of me.*

As mentioned above, the caregiver does not write a narration of their own. During the time that the therapist and the child are immersed in trauma narration and processing, the therapist continues to engage the caregiver in cognitive processing through the exploration of trauma-related thoughts. For example, the therapist elicits the caregiver's thoughts when they first learned about the trauma(s) their children experienced. It is a critical time to provide validation and support to caregivers through the processing of these thoughts which often are as problematic as the beliefs found in the narration of their children. Caregivers are also engaged in strategies that encourage cognitive flexibility and the exploration of their internal dialogue. Often thoughts of their roles as caregivers, the management of the entire disclosure process, and the interaction with the system are discussed. These sessions are valuable in addressing the caregivers' distress related to their children's experiences. For some caregivers, the completion of an automatic thought record in session as well as for homework between sessions can help them practice catching and challenging dysfunctional thoughts related to their children's traumatic experiences. Moreover, this important work with the caregiver can be helpful in preparing them for the sharing of the trauma narration later in conjoint sessions (when deemed clinically appropriate).

If it is determined clinically appropriate, the child will be prepared to share their trauma narration project with the non-offending participating caregiver. The caregiver is also carefully prepared to engage in this conjoint session with the child. The result is often sessions filled with pride for the work done, family tenderness, and spontaneous demonstration of love and support.

13 In Vivo Mastery

The in vivo mastery component assists therapists in decreasing specific trauma-related behavior that interferes with everyday functioning. For example, the fear of riding cars after a car accident. A child who was involved in a car accident with their caregivers that resulted in the death of a grandparent may begin to fear riding in cars. This fear may evolve not only into the refusal of getting into a car but also into sudden tantrums that seem irrational. Riding in a car, typically an innocuous stimulus, has now turned into a trauma reminder and has been connected with extreme emotions such as fear and terror (a trauma response). This connection is problematic for school-age children who are often transported by caregivers (often in cars or other vehicles) everywhere they need to go. This component allows the therapist and the family to work collaboratively to design a detailed plan that will expose the child to the feared stimulus supported by people and skills that aid in the management of the physical, emotional, and cognitive reactions that come with exposure. Once the plan is developed, it is imperative that all parties agree to the implementation of the plan. The therapist explains the plan to the adults involved in implementing such a plan in the home or the community. It is important that all the adults involved in implementing the plan are confident that the plan will work.

In vivo mastery plans are developed by identifying the problematic avoidant behavior as well as the thoughts and feelings that are driving the response. The input of the child and caregiver helps understand the appearance, frequency, and intensity of the problematic behavior. This information is used to create a hierarchy of the feared stimuli and a plan to use a variety of skills to manage the distress that exposure creates. For example, a family came home to find an intruder in their home. The intruder hid in the house and was found by the police to be hiding in the child's bedroom. The child now refuses to go in or near their room. This was identified as a problematic behavior by the caregiver. The child reported that some of the thoughts that are present are that "an intruder may be hiding in the room" and "the intruder in the room will hurt the child and the family." The child is also able to identify the feelings related to these thoughts. The primary feeling is fear, but the child may also feel anxious, uncertain, and alone. The next step should be to agree on the goal of this exercise. The caregiver may want the child to not be fearful of the room as there is no intruder there. The child may not want to feel the way that they do as they miss being in their room with their toys and video games. The hierarchy of fear (and/or other feelings) is created. Small steps are designed that will lead the child to move closer to the goal of being able to enter the room and tolerate being inside the room for a short period of time. As each step is taken, the plan should state the skills (e.g., relaxation skills, mindfulness, distractions) that will be used to increase the tolerance to the feelings that come with each movement forward. The plan should also state the person responsible for assisting the child with each step as the therapist may not be physically present to assist with the implementation of the plan. The use of praise and a variety of reinforcers may be particularly helpful to recognize the efforts and progress of the child.

14 Conjoint Sessions

Throughout treatment, the conjoint sessions have reinforced the sentiment that caregiver participation is integral in TF-CBT. Generally, brief conjoint time with the child and the caregiver begins early in treatment and increases in time over the course of treatment. As a review, during the skill-building phase of treatment, the therapist prepares the dyad during their individual sessions and conducts a conjoint caregiver-child meeting at the end of sessions for the mutual exchange of praise as well as conjoint activities to reinforce the coping skills taught during each component.

One of the goals in the final phase of treatment is working with the caregiver and child to bring them together for a conjoint session where the child's narration is shared with the caregiver. The therapist begins assessing caregiver readiness and appropriateness for sharing narration when treatment begins and continues to assess throughout treatment to ensure that the caregiver can be supportive of the child during the sharing of trauma narration during the conjoint session. Observing interactions between the child and caregiver as well as observing the caregiver's response to trauma-related discussions and material can help the therapist determine if they will share the narration with the caregiver in individual session as well as determine if the child will share their narration with the caregiver in conjoint session. It is important to note that most caregivers are able to hear the trauma narration, and the few that may not be able to hear all of the narration can usually hear the introduction chapter and the final chapter. Once the therapist has decided to proceed with having a conjoint session around sharing the narration, they begin preparing the caregiver and child independent of one another for the upcoming session as the caregiver should not hear trauma narration for the first time in front of the child. Sharing the narration with the caregiver initially outside the presence of the child gives the caregiver the opportunity to have their initial, authentic, and oftentimes emotional reaction, without worrying about the child's response. During the individual session with the caregiver, the therapist elicits the caregiver's thoughts and feelings related to the child's narration. If the caregiver shares inaccurate or unhelpful cognitions, the therapist may utilize cognitive coping to help process the thoughts with the caregiver. Also, during this session, the therapist helps the caregiver practice engaging in reflective listening and offering positive feedback and praise in response to the child's narration.

At the beginning of treatment, there are discussions about confidentiality, privacy, but also the importance of open communication. Reducing secrecy is key as this has been a mechanism used in maltreatment to keep the violence occurring. The therapist should use their clinical judgment in determining what is shared and whether the narration should be shared with the caregiver in the final phase of treatment. This practice of open communication and the use of age-appropriate rationales are helpful to the therapist in guiding the conversation of sharing the narration with the caregiver. When preparing the child for the conjoint session around trauma narration, it is not recommended that therapists ask a closed-ended question about whether the child would like to share their narration as such a question often triggers a child's avoidance and desire to protect their caregiver from the painful details. The suggested approach is for the therapist to enthusiastically explain to the child that the next step in therapy is sharing their narration with the caregiver and several reasons to support such a step. For example, the therapist may begin the discussion by explaining to the child the benefits of sharing with the caregiver such as giving the caregiver an opportunity to see how much work the child has done as well as giving the caregiver and child the space to openly discuss the child's traumatic experience(s). Openly discussing the child's narration with a supportive caregiver can be beneficial, as the caregiver will continue to provide support to the child once therapy is completed. If the child is hesitant to share, the therapist can have a discussion with them regarding their thoughts and feelings related to their caregiver hearing the trauma narration. The child's concerns about sharing the narration may be related to guilt, embarrassment, shame, or distorted cognitions. In this case, the therapist can help the child process their thoughts and emotions. There are times when the child has a valid reason for not sharing the complete narration. In these instances, the therapist and child collaborate on deciding what will be shared. Next, the therapist and child will discuss and practice who will read the narration or how the narration will be presented to the caregiver. Once the caregiver and child have been independently prepared for the session, the therapist brings them together to share the trauma narration. During the sharing, the caregiver is encouraged to provide the positive feedback they have prepared. The session may end with a positive ritual of the child's choosing.

15 Enhancing Safety and Final Elements of Treatment

A goal of the enhancing safety component is to help the child feel more confident and prepared in the event that they experience adversity in the future. Often children who have experienced trauma feel worried or anxious that they may be revictimized or experience additional traumas in the future. While a therapist should not assure a child that trauma will not happen again, therapists can teach children what they can do to respond to threats in the future while also reminding them of their strength in overcoming the trauma(s) they endured.

There are many ways therapists can introduce safety skills to children. Therapists may choose to teach children assertiveness skills, including "I statements." This may be particularly helpful in bullying situations. Role-plays are tools for children to practice recognizing a potentially harmful situation, utilize appropriate language, and learn how to best respond including how to get help. A caregiver may also be brought in for a conjoint session to practice these role-plays with the child. For older school-age children, enhancing safety may include discussing healthy relationships and the signs of an unhealthy relationship. Therapists may decide to make a relationship timeline with the child and discuss what they would like in a future relationship versus what they would not want in a relationship. For children who have experienced sexual abuse, it may be important to introduce information about puberty and age-appropriate sex education; however, it is important to note that it necessary to obtain the consent of the caregiver before introducing this topic. Most importantly, the therapist wants to teach children the differences between sexual abuse and healthy sexuality. The therapist can also teach children about respect and consent. Conjoint sessions are typically fun and full of practical information for families.

During this final stage of treatment, it is important that therapists help children to prepare for the end of treatment. One way this can be accomplished is to complete post-assessment measures with children and their caregivers. It may also be helpful to share these assessment results with the caregiver and, when appropriate, with the child to help them see the progress that has been made. Another way to prepare for the end of treatment could be to discuss upcoming dates and/or anniversaries that may be difficult for the child and their family. This may be especially important for a child who has experienced traumatic grief.

The final session of treatment is a time to celebrate the child, the family, and all the progress that has been made with a graduation. There are many ways that a therapist could choose to celebrate graduation with the family. The therapist may decorate the space that is being used for this session and make the client a graduation certificate. The client could make an item to take with them, such as a graduation hat or a picture that they drew. The client may choose to play a game with their caregiver and the therapist, including bringing a game from home. Graduation is a great time to reflect on all the hard work and achievements that the client and caregiver (s) have made in therapy. A jubilant graduation celebration also may help a child create a positive memory of their success in therapy that is associated with feelings of mastery, relatedness, and confidence in their ability to face and overcome stressful life experiences in the future.

16 Conclusion

The efficacy of TF-CBT in addressing the therapeutic needs of children and their caregivers in the aftermath of trauma has been well documented over the last several decades. However, as an evidence-based practice, the model continues to evolve based on ongoing research, client feedback, clinical applications, and dissemination evaluations. More specifically, research has (1) replicated TF-CBT outcomes with diverse populations [21, 22], (2) identified therapy mechanisms that have been proposed to underline the efficacy of TF-CBT [22–25], (3) examined differential treatment responses among youth exposed to multiple and complex traumas [26, 27], and (4) evaluated training formats, organizational support, and the incorporation of self-care strategies [28], as well as adaptations for training in low-resource settings to optimize training experiences and client outcomes [29–31].

The current chapter attempts to reflect the state of the art in implementing TF-CBT with school-age children and their caregivers. However, given the significant potential impact of childhood traumatic experiences, the complexity of child and caregiver reactions, and TF-CBT implementation challenges, face-to-face training and supervision/consultation are highly recommended for therapists new to TF-CBT to optimize outcomes. To facilitate such training, an introductory online course was developed that can be accessed through https://tfcbt2.musc.edu/. Following the Web-based TF-CBT course, participation in a multiday in-person and/or virtual training and consultation/supervision by expert TF-CBT trainers are strongly encouraged as therapists learn to implement the model over the course of 6 to 12 months. Finally, more information about TF-CBT, training opportunities, and TF-CBT implementation manuals tailored to address the needs of youth and caregivers presenting in the aftermath of different traumas, with distinct clinical presentations and with families from diverse cultures and backgrounds, are accessible through https://tfcbt.org/. Additional valuable resources for learning more about TF-CBT and for implementing the model successfully with clients are listed below in Box 1.

Box 1: Helpful Resources
Books

1. *Treating trauma and traumatic grief in children and adolescents, second edition*
2. *Child sexual abuse: A primer for treating children, adolescents, and their nonoffending caregivers, second edition*
3. *Trauma-focused CBT for children and adolescents: Treatment applications*

(continued)

Box 1 (continued)
Websites
Official TF-CBT page—tfcbt.org
TF-CBT web 2.0—tfcbt2.Musc.Edu (available in English and Spanish)
nctsn.org

Conflicts of Interest

Yahaira Marquez, Stephanie Cruthirds, and Aimee Sirois have no conflicts of interest to declare that are relevant to the content of this chapter. Esther Deblinger is codeveloper of trauma-focused cognitive behavioral therapy (TF-CBT) and receives royalties, honoraria, and grant funding for TF-CBT-related activities.

References

1. Felitti VJ, Anda RF, Nordenberg D et al (1998) Relationship of childhood abuse and household dysfunction to many of the leading causes of death in adults: the adverse childhood experiences (ACE) study. Am J Prev Med 14(4):245–258. https://doi.org/10.1016/S0749-3797(98)00017-8
2. Cohen JA, Mannarino AP, Deblinger E (2017) Treating trauma and traumatic grief in children and adolescents, 2nd edn. The Guilford Press, New York
3. Deblinger E, Mannarino AP, Cohen JA et al (2015) Child sexual abuse: a primer for treating children, adolescents, and their nonoffending parent, 2nd edn. Oxford University Press, New York
4. Cohen J, Deblinger E, Mannarino A et al (2004) A multisite, randomized controlled trial for children with sexual abuse-related PTSD symptoms. J Am Acad Child Adolesc Psychiatry 43(4):393–402. https://doi.org/10.1097/00004583-200404000-00005
5. Cohen JA, Mannarino AP (1997) A treatment study for sexually abused preschool children: outcome during a one-year follow-up. J Am Acad Child Adolesc Psychiatry 36(9):1228–1235. https://doi.org/10.1097/00004583-199709000-00015
6. Cohen JA, Mannarino AP (1998) Interventions for sexually abused children: initial treatment outcome findings. Child Maltreat 3(1):17–26. https://doi.org/10.1177/1077559598003001002
7. Deblinger E, Lippmann J, Steer R (1996) Sexually abused children suffering posttraumatic stress symptoms: initial treatment outcome findings. Child Maltreat 1(4):310–321. https://doi.org/10.1177/1077559596001004003
8. Deblinger E, Mannarino AP, Cohen JA et al (2006) Follow-up study of a multisite, randomized, controlled trial for children with sexual abuse-related PTSD symptoms: examining predictors of treatment response. J Am Acad Child Adolesc Psychiatry 43(4):393–402. https://doi.org/10.1097/00004583-200404000-00005
9. Deblinger E, Steer RA, Lippmann J (1999) Two-year follow-up study of cognitive behavioral therapy for sexually abused children suffering post-traumatic stress symptoms. Child Abuse Negl 23(12):1371–1378. https://doi.org/10.1016/s0145-2134(99)00091-5
10. Dorsey S, McLaughlin KA, Kerns SEU et al (2017) Evidence base update for psychosocial treatments for children and adolescents exposed to traumatic events. J Clin Child Adolesc Psychol 46(3):303–330. https://doi.org/10.1080/15374416.2016.1220309
11. Peters W, Rice S, Cohen J et al (2021) Trauma-focused cognitive-behavioral therapy (TF-CBT) for interpersonal trauma in transitional-aged youth. Psychol Trauma 13(3):313–321. https://doi.org/10.1037/tra0001016
12. Cohen JA, Mannarino AP, Deblinger E (2012) Trauma-focused CBT for children and adolescents: treatment applications. The Guilford Press, New York

13. Deblinger E, Pollio E, Dorsey S (2015) Applying trauma-focused cognitive behavioral therapy in group format. Child Maltreat 21(1): 59–73. https://doi.org/10.1177/1077559515620668
14. Stewart RW, Orengo-Aguayo RE, Cohen JA et al (2017) A pilot study of trauma-focused cognitive-behavioral therapy delivered via telehealth technology. Child Maltreat 22(4): 324–333. https://doi.org/10.1177/1077559517725403
15. Stewart RW, Orengo-Aguayo R, Young J et al (2020) Feasibility and effectiveness of a telehealth service delivery model for treating childhood posttraumatic stress: a community-based, open pilot trial of trauma-focused cognitive–behavioral therapy. J Psychother Integr 30(2): 274–289. https://doi.org/10.1037/int0000225
16. Sachser C, Berliner L, Risch E et al (2022) The child and adolescent trauma screen 2 (CATS-2) – validation of an instrument to measure DSM-5 and ICD-11 PTSD and complex PTSD in children and adolescents. Eur J Psychotraumatol. https://doi.org/10.1080/20008066.2022.2105580
17. Jellinek MS, Murphy JM, Robinson J et al (1988) The pediatric symptom checklist: screening school-age children for psychosocial dysfunction. J Pediatr 112(2):201–209
18. Deblinger E, Neubauer F, Runyon M et al (2019) What do you know? A therapeutic card game about childhood trauma, sex education, and personal safety. CARES Institute, Rowan Medicine, Stratford
19. Deblinger E, Mannarino AP, Cohen JA et al (2015) Child sexual abuse: a primer for treating children, adolescents, and their nonoffending parents, 2nd edn. Oxford University Press, New York
20. Altman D (2010) The mindfulness code. New World Library, Novato
21. Last BS, Johnson C, Dallard N et al (2023) Implementing trauma-focused cognitive behavioral therapy in Philadelphia: a 10-year evaluation. Implement Res Pract. https://doi.org/10.1177/26334895231199467
22. Lee P, Lang JM (2023) Comparing trauma-focused cognitive-behavioral therapy to commonly used treatments in usual care for children with posttraumatic stress disorder. Psychol Trauma. https://doi.org/10.1037/tra0001555
23. Hayes AM, Yasinski C, Grasso D et al (2017) Constructive and unproductive processing of traumatic experiences in trauma-focused cognitive-behavioral therapy for youth. Behav Ther 48(2):166–181. https://doi.org/10.1016/j.beth.2016.06.004
24. Jensen TK, Holt T, Mørup Ormhaug S et al (2018) Change in post-traumatic cognitions mediates treatment effects for traumatized youth – a randomized controlled trial. J Counsel Psychol 65(2):166–177. https://doi.org/10.1037/cou0000258
25. Pfeiffer E, Sachser C, de Haan A et al (2017) Dysfunctional posttraumatic cognitions as a mediator of symptom reduction in trauma-focused cognitive behavioral therapy with children and adolescents: results of a randomized controlled trial. Behav Res Ther 97:178–182. https://doi.org/10.1016/j.brat.2017.08.001
26. Jensen TK, Braathu N, Birkeland MS et al (2022) Complex PTSD and treatment outcomes in TF-CBT for youth: a naturalistic study. Eur J Psychotraumatol. https://doi.org/10.1080/20008066.2022.2114630
27. Sachser C, Keller F, Goldbeck L (2017) Complex PTSD as proposed for ICD-11: validation of a new disorder in children and adolescents and their response to trauma-focused cognitive behavioral therapy. J Child Psychol Psychiatry 58(2):160–168. https://doi.org/10.1111/jcpp.12640
28. Deblinger E, Pollio E, Cooper B et al (2020) Disseminating trauma-focused cognitive behavioral therapy with a systematic self-care approach to addressing secondary traumatic stress: PRACTICE what you preach. Community Ment Health J 56(8):1531–1543. https://doi.org/10.1007/s10597-020-00602-x
29. Ascienzo S, Sprang G, Eslinger J (2020) Disseminating TF-CBT: a mixed methods investigation of clinician perspectives and the impact of training format and formalized problem-solving approaches on implementation outcomes. J Eval Clin Pract 26(6):1657–1668. https://doi.org/10.1111/jep.13351
30. Helseth SA, Peer SO, Are F et al (2020) Sustainment of trauma-focused and evidence-based practices following learning collaborative implementation. Adm Policy Ment Health 47: 569–580. https://doi.org/10.1007/s10488-020-01024-328
31. Orengo-Aguayo R, Dueweke AR, Nicasio A et al (2022) Trauma-focused cognitive behavioral therapy with Puerto Rican youth in a post-disaster context: tailoring, implementation, and program evaluation outcomes. Child Abuse Negl. https://doi.org/10.1016/j.chiabu.2022.105671

Chapter 9

Adapting Cognitive Behavioral Therapy for Children with Autism Spectrum Disorder and Co-occurring Anxiety and Obsessive-Compulsive Disorders

Kelly N. Banneyer and Rachel Fein

Abstract

This chapter explores the adaptation of cognitive behavioral therapy (CBT) for children with autism spectrum disorder (ASD), with a focus on the theoretical foundations that support its application, the unique treatment adaptations required, and evidence-based programs specifically developed for this population. Traditional CBT, a widely recognized approach for addressing anxiety and other emotional challenges, must be tailored for children with ASD to account for differences in communication, social understanding, and cognitive flexibility. We examine specific adaptations to the core components of exposure-based CBT, including assessment, psychoeducation, goal setting, exposure and response prevention, treatment termination, and telehealth adaptations, to better align with the needs of children with ASD. Additionally, we highlight several modular programs accessible to clinicians to address anxiety symptoms in children with ASD.

Key words Cognitive behavioral therapy (CBT), Autism spectrum disorder (ASD), Anxiety, Obsessive compulsive disorder (OCD), Children

1 Introduction

Autism spectrum disorder (ASD) is a neurodevelopmental disorder characterized by social communication deficits as well as atypical behaviors including restricted and repetitive behaviors, interests, and/or activities [1]. When making a diagnosis of ASD, the individual must display symptoms early in development, and they must currently exhibit impairment across settings and/or areas of functioning. Notably, a diagnosis of ASD may or may not be accompanied by a diagnosis of language disorder and/or intellectual disability. Based on information from the Centers for Disease Control and Prevention (CDCP), ASD occurs in 1 out of 36 children and is more common in males (~4% of boys) when compared to

Robert D. Friedberg and Micaela A. Thordarson (eds.), *Cognitive Behavioral Therapy with Youth: Tradition and Innovation*, Neuromethods, vol. 231, https://doi.org/10.1007/978-1-0716-5134-6_9,

females (~1% of girls) [2]. Importantly, ASD occurs along a spectrum of severity along multiple domains.

While the research community has shifted away from looking for a cure for ASD, there are several established treatments that can be implemented with the ASD population. One of the most effective treatments for children with ASD is applied behavior analysis (ABA). Specifically, ABA involves identifying behaviors in need of remediation, defining them, and finding objectives ways to measure them. Behaviors may include those that are maladaptive (e.g., self-injurious behavior, motor stereotypy, restricted interests) and are in need of reduction, whereas other behaviors may include those that are prosocial (e.g., functional communication, social skills, appropriate play skills) and are in need of augmentation. ABA uses a number of strategies including, but not limited to, prompting, shaping, chaining, task analysis, positive reinforcement, and extinction to increase prosocial behaviors while decreasing maladaptive behaviors. Based on findings from the National Standards Project [3], ABA has years of sound research to support its efficacy in treating children with ASD [4–6].

In addition to the core deficits associated with ASD, many children with ASD display symptoms of anxiety. In fact, some researchers have estimated that ~40% of children with ASD meet full diagnostic criteria for an anxiety disorder [7], while between 22% and 84% of children with ASD display impairing symptoms of anxiety, though not necessarily meeting the diagnostic cutoff for a disorder [8]. In terms of co-occurring anxiety disorders and obsessive-compulsive and related disorders, specific phobia (29.8%), obsessive-compulsive disorder (OCD) (17.4%), and social anxiety disorder (16.6%) were the most common types [7, 9]. Similarly, a study of 108 children with ASD, who were enrolled in anxiety treatment trials, found that almost all children met criteria for two or more anxiety disorders, with greatest prevalence of social phobia (41.7%) and generalized anxiety disorder (25.9%) [10].

Although anxiety frequently co-occurs in children with ASD, with equal prevalence among boys and girls [11], the presentation of anxiety symptoms varies across individuals. With regard to specific phobia, children with ASD may display unusual phobias compared to neurotypical peers. For instance, phobias related to shots/needles and crowds are most commonly found within the ASD population; however, phobias more commonly found in neurotypical peers (e.g., tunnels, flying, bridges) were less common [9, 12]. Similarly, although the fear of loud noises is less frequent among neurotypical children, approximately 10% of children with ASD report this specific phobia [9, 12]. Additionally, the phobias of children with ASD are often associated with medical visits, animals, or highly specific situations related to the individual [13].

In terms of social anxiety disorder, Gillott and colleagues [14] found children with high-functioning ASD (i.e., children with functional communication and broadly average cognitive abilities) were less likely to display fears of negative social evaluation. In contrast, others have observed similar rates of parent-reported self-consciousness and avoidance among neurotypical teenagers compared to those with high-functioning ASD [12, 15].

When examining symptoms of OCD seen in children with ASD, Leyfer and colleagues [9] found the most commonly observed compulsions involved repetitive comments and question-asking as well as verbal and behavioral rituals involving another individual. Other researchers have found children with ASD displayed more compulsions related to repeating, touching, tapping, and hoarding while exhibiting less compulsions involving cleaning, checking, and counting [16]. Additionally, some children with ASD exhibit distress or anxiety in response to changes in their routine, engage in repetitive behaviors to alleviate symptoms of anxiety, and/or respond to disruption of repetitive behaviors with distress [17, 18]. Furthermore, distress caused by disruptions in routines, such as transitioning from home to school or from caregiver to caregiver, may often mirror symptoms of separation anxiety disorder [19].

Interestingly, symptoms of anxiety occur across the full range of intellectual functioning of children with ASD [19]; however, the presentation and rate of anxiety may differ depending on the child's intellectual functioning. Specifically, parents of children with an IQ below 70 typically provide lower ratings on measures of anxiety compared to parents of children with an IQ of 70 or greater. Given how common functional communication deficits are within the ASD population, especially with children with lower IQ, it is not unusual that parents are not endorsing symptoms that require language (e.g., "complains" or "worries") [19, 20]. Although children with lower IQ may not verbally express symptoms of anxiety, this does not mean they do not experience symptoms of anxiety. Notably, higher parent ratings of symptoms of anxiety are correlated with functional verbal communication, an IQ of 70 or above, as well as higher rates of inappropriate speech (e.g., stereotyped language), irritability, and behavioral difficulties (e.g., hyperactivity) [19]. Additionally, children rated as having the most impairing forms of anxiety were also rated as having higher levels of ASD symptom severity compared to those who were rated as less anxious. These findings suggest that anxiety may potentially exacerbate other behavioral difficulties, or rather, symptoms of anxiety may manifest as core symptoms of ASD or as other behavioral difficulties [19].

2 Theoretical and Empirical Foundations

Exposure-based cognitive behavioral therapy (CBT) is the primary treatment for anxiety and obsessive-compulsive disorders in children and adolescents [21]. Behavioral intervention research has focused on adapting CBT to address the co-occurrence of autism spectrum disorder (ASD) and anxiety [22]. A 2013 meta-analysis found that CBT has a large effect size in treating anxiety in youth with ASD, showing significantly greater improvements in anxiety symptoms than waitlist control or standard treatment options [23]. Additionally, children with ASD and anxiety show more substantial treatment gains in group CBT [24] and family-based group CBT [25, 26] compared to those in waitlist control or standard treatments.

3 Intervention Adaptations

This section will serve as a guide and give information about adaptations to exposure-based CBT programs for use with children with co-occurring ASD and anxiety or obsessive-compulsive disorders. While CBT programs are more effective at addressing anxiety symptoms in children with ASD in comparison to treatment as usual, CBT programs that have been specifically adapted for youth with ASD are more efficacious than CBT programs designed for neurotypical children, as indicated in a randomized clinical trial with children 7–13 years old [27]. Primary areas or modules to be incorporated include assessment, psychoeducation, goal setting, exposure and response prevention (ERP), and treatment termination. Adaptations will have a focus on incorporating a behavioral approach, significant parental involvement, and the use of reinforcement systems. Additionally, recent researchers have also worked to adapt these interventions to be delivered in an online or telemedicine format.

4 Assessment

As with intervention in any domain, evidence-based treatment will involve measurement-based care. Assessment typically begins with a diagnostic interview, at which time diagnoses are established. For children with ASD, it will be important to gather information related to cognitive functioning and pragmatic language and insight, as treatment will have to be modified depending on these areas. An anxiety or obsessive-compulsive disorder diagnosis could be made using a semi-structured diagnostic interview and diagnostic criteria for the disorders. When conducting a diagnostic interview, the clinician should be sure to gather information related to history of presenting problem(s), previous assessment or intervention, family background, family mental health history, prenatal and developmental history, medical history, previous and current medication usage, behavioral symptoms, emotional functioning, social functioning, and academic history. A clinician may also choose to use a more structured interview, such as the Anxiety and Related Disorders Interview Schedule for DSM-5 (ADIS-5 L) [28] or Children's Yale-Brown Obsessive Compulsive Scale, second edition (CY-BOCS-II) [29] to diagnose an anxiety or obsessive-compulsive disorder. These interviews schedules will also include questions related to the domains listed above, so the clinician can be informed about symptoms, strengths, and challenges to provide an accurate diagnosis and form a comprehensive biopsychosocial case conceptualization.

After the initial assessment, the clinician should be able to create a running list of anxiety-provoking/distressing triggers for the child. While the child's anxiety or obsessive-compulsive ritual does not need to be fully mapped out at this point, the clinician should be able to form a preliminary list that will then be incorporated into active treatment.

Anxiety in children with ASD may look similar to anxiety presentations in typically developing children or there may be distinct anxiety presentations. For example, children with ASD can have "traditional" fears, such as anxiety about separating from parents or fear of dogs. However, there are also "distinct" anxiety presentations that are more common in the ASD population, such as anxiety around change in routines and anxiety related to sensory sensitivities. Additionally, some of the anxiety symptomatology may also present as "distinct." For example, anxiety might present as an increase in disruptive behavior or stereotypies [22]. As discussed above, clinicians should be aware of both "traditional" and "distinct" trends when completing assessment.

During the assessment process with this population, it can sometimes be difficult to differentiate between obsessions and restricted interests and between compulsions and inflexibility around routines. Generally speaking, individuals with OCD engage in compulsions or avoidance to alleviate a distressing or anxious thought or impulse. Individuals with ASD may have restricted, special interests, but they focus on the activity or topic because it is enjoyable to them. A compulsion brings about relief from distress, but it is not pleasurable.

In addition to a thorough diagnostic intake, measurement-based care will incorporate regular progress monitoring. Because children with ASD may have limited insight into their anxious or obsessive-compulsive symptoms, regular progress monitoring should incorporate parent-report questionnaires related to these domains. The DSM-5 [1] offers open-access symptom tracking forms that can be suited to this purpose. Clinicians may also use parent rating forms, such as the Screen for Child Anxiety Related Emotional Disorders (SCARED) [30] or the Spence Children's Anxiety Scale (SCAS) [31], which were originally developed to assess anxiety in neurotypical children. The SCARED has shown acceptable psychometric properties in high-functioning children with ASD compared to neurotypical children [9]. Similarly, the SCAS also demonstrated acceptable psychometric properties in children with ASD [32–34]. Clinicians should use results from behavioral questions to guide treatment, which means they should be completed on a regular basis throughout treatment. If possible, clinicians may try to incorporate a short symptom and interference checklist at each session, in addition to regularly giving clinician-based ratings of symptom improvement.

5 Psychoeducation

The initial treatment session should focus on psychoeducation, with both the child and caregivers present, so that the entire family has a good understanding of anxiety/OCD and of exposure-based CBT. The initial part of psychoeducation should provide the basis for a CBT framework to address anxiety/OCD by discussing the interaction between thoughts, feelings, and behaviors and the cycle that negatively reinforces anxiety and avoidance. While traditional CBT protocols make use of metaphors to understand the interaction between thoughts, feelings, and behaviors, terms that are more direct should be used to help children with ASD to have an accurate understanding. The use of visuals and/or the child's interests will certain aid in explaining the anxiety/OCD as well as the treatment to address it. While explaining this cycle, the feelings that occur may take the form of anger or irritation on the part of the child when accommodation does not occur. Disruptive behavior may be the primary avoidance behavior (e.g., a child has a tantrum or yells at the parents, so they do not have to engage in a feared activity). Since children with ASD have difficulties with emotion regulation, disruptive behaviors often occur and should be explained as an anxiety reaction within this cycle. Further, normalizing this reaction and providing the family with preparation and appropriate expectations related to the potential reactions of their child will be critical to treatment.

Some CBT protocols make use of cognitive techniques to change maladaptive/anxious thinking. However, the primary aim of exposure-based CBT, especially when working with children with ASD, is to rely on a more behavioral approach. Trying to change feelings or thoughts completely with cognitive restructuring will likely be ineffective. Instead, the clinician should discuss a change in behavior, both in the part of the parent and the child, as the primary mechanism of change. Once the child is engaging in more adaptive behaviors (e.g., confronting feared stimuli), the child's anxious thoughts and feelings will subsequently decrease. Use the following dialogue as an example:

Clinician: *In any situation, people's thoughts, feelings, and actions all work together to impact their reactions and future thoughts, feelings, and actions in that situation. For example, when seeing a dog on the sidewalk, one child may think, "I really like dogs. I want to pet it." After thinking this, the child may feel happy, and their action may be to ask the owner to pet it. If the child is allowed to pet the dog, the child may think, "What a soft dog. I really like it!" and continue to feel happy and excited. The next time this child sees a dog, the child will be likely to engage in similar thoughts, feelings, and actions. However, another child may see a dog and think, "Dogs are so scary! The last time I saw a dog, it chased me!" and then feel afraid. This child may act by running away or asking a parent to remove the dog. When the dog is removed or the child runs away, he or she will likely feel some relief. This feeling of relief strengthens the previous anxious thoughts, feelings, and actions, and the next time the child sees a dog, the same cycle is likely to occur.*

[The clinician may use this opportunity to apply the cycle of thoughts, feelings, and actions to a stimulus that the child enjoys and to a trigger of the child's anxiety. As noted previously, incorporating visuals will likely be beneficial.]

Clinician: [Speaking to child] *If you are feeling afraid, and your mom tells you to calm down or to not be scared, does that help? Do you not feel afraid anymore?*

Child: *No, I still feel scared.*

Clinician: [Speaking to parent] *I'm sure there are many times you have tried to talk through and change your child's feelings. Is this usually effective at making him feel better?*

Parent: *Usually not. I've told him to calm down a million times and nothing changes.*

Clinician: *Talking about feelings usually does not change feelings. The same is true for thoughts.* [Speaking to child] *Have there been times when your Mom has told you there is no need to worry about something because it is perfectly safe?*

Child: *Yes.*

Clinician: *And are you still worried?*

Child: *Yes.*

Clinician: *Trying to change thoughts and feelings by talking about them is not effective. However, what we can do is learn different actions to take to help change this cycle. Once we start to act bravely and change the cycle of anxiety by no longer avoiding the things that scare us, our thoughts and feelings start to be braver as well.*

After this discussion, the family should understand that the purpose of exposure practice is to change the child's behaviors from that of avoidance or other compulsive behaviors to brave or "approach" behaviors.

Either during the initial or subsequent treatment session, the clinician should introduce the family to the concept of differential attention as a behavioral parental tool to help shape behavior. The clinician should explain that attention is one of the parent's most powerful tools. When parents pay attention to anxious or disruptive behaviors, such as consistently answering reassuring-seeking questions, the behavior is reinforced or maintained. Parental attention to these behaviors is thought of as one way anxiety is accommodated. One aim of treatment will be to help the family pay attention to brave and on-task behaviors and ignore anxious behaviors. However, this should be done in a systematic manner, so it is not overwhelming for the family. Fading accommodation using differential attention strategies is one concept that will be discussed during ERP.

6 Goal Setting

When working with families of children with ASD, it will be important for all family members to agree on appropriate goals. However, there are certain goals which ERP will not be able to target in this population. For example, a goal such as "I want my child to be less awkward when interacting with peers," is not something to be targeted within ERP. This goal may be better addressed with social skills training. Other inappropriate goals involve a desire to decrease perceived negative emotions. For example, "To not be anxious anymore" is inappropriate for several reasons. First, anxiety is an innate emotion that is important for all individuals, and it would be unrealistic for an individual never to experience the emotion again. Anxiety helps maintain an individual's safety and ability to perform when needed (i.e., someone who did not feel any worries about taking tests would never study). Second, while decreasing interference from anxiety is a primary goal of treatment, a goal should focus on what the child would like to be able to do or achieve.

Consistent with how goals are developed within ABA programs, goals or target behaviors identified within the context of CBT should be observable and measurable (i.e., operationally defined). Additionally, goals should be stated in a positive, strengths-based manner. For example, for a child that has anxiety about falling asleep and has a 2-h compulsive ritual before bedtime, an appropriate goal may be: to go to sleep independently before

9 PM. Another example of an appropriate goal for a child with contamination OCD might be: to wash hands for 20 s or less with only one pump of soap after using the bathroom.

7 Exposure and Response Prevention

7.1 Map Out Symptoms

After delivering psychoeducation and helping the family to establish treatment goals, the clinician will start to form a plan for ERP. This is the point when the child's OCD or anxiety symptoms should be more thoroughly mapped out. If the clinician previously developed a list of anxiety-provoking or distressing triggers during the assessment process, these should be reviewed and discussed in terms of associated anxious thoughts (or obsessions) and anxious behaviors (or compulsions). It is important to understand what is driving avoidance or fear to plan ERP. For a child who is scared at night and co-sleeping with parents, this fear could be driven by separation anxiety, a specific phobia (e.g., fear of the dark), generalized worries (e.g., child says different worries keep him/her from falling asleep), an obsessive ritual (certain characteristics have to be in place so something "bad" does not happen"), or in the case of a child with ASD, a rigid behavior (i.e., this is the way it has to be). Treatment will differ based on which of these thoughts is driving the fear. For example, if a child is afraid of the dark, the practice should focus on increasing exposure to the dark.

7.2 Form Hierarchy

A practice hierarchy should be created based on the family's goal (s) and list of anxiety-provoking situations or triggers. Some children may present with one difficulty, but others may present with a vast list of anxiety-provoking triggers. Each target for ERP will need to have a hierarchy formed. At the beginning of treatment, collaborative hierarchy formation will need to occur in great detail, so the family learns how to create and establish a hierarchy. Depending on the insight and the functioning of the child with ASD, an initial target for ERP should be chosen collaboratively with the child and the family. For some families, relying on the parent's perceptions and observations of the child's behavior may be more informative and appropriate. Ideally, the initial target is a frequently occurring situation everyone is motivated to change that is moderately distressing. Practicing a target that causes minimal distress is not necessary and choosing a target that will cause overwhelming distress will not be productive. The child will not be able to achieve new learning if they are overwhelmed, and if the child and parent become overwhelmed, they may decide to leave treatment.

7.3 Create Subjective Measurement Scale

Once a target has been identified, a practice hierarchy should be formed collaboratively with the family. A practice hierarchy is not simply an ordered list of triggers. A practice hierarchy incorporates different steps of increasing exposure to the feared or distressing stimulus. In order to form a meaningful hierarchy, it can be helpful for the child to identify a way to measure their distress in different situations. This may take the form of a 0–10 subjective units of distress scale (SUDS). Because individuals with ASD can sometimes interpret things very literally, a simpler rating scale, like the 0–5 hierarchy used in the Incredible 5 Point Scale, may be more suitable [34]. If the child has difficulty applying numerical ratings to distress, labels such as "easy/medium/hard," colors such as red/-yellow/green, or other visuals could be used. Additionally, if the child has a restricted interest of some kind, using visuals related to the child's interest could serve as a meaningful depiction of the hierarchy. For instance, if a child is interested in weather, a tornado could serve to represent the most distressing situations, and a cloudless sky could represent the least distressing situations. As noted previously, the parent may have more insight into how distressing different situations will make the child as well as which type of hierarchy would be the most appropriate.

7.4 Establish Reinforcement System/Token Economy

Depending on the insight and cognitive abilities of the child with ASD, he or she may benefit from external motivation to engage in ERP. When ERP practice is introduced, a token economy system should also be discussed in detail. Specifically, target behaviors should be identified and assigned a point value, and children should be rewarded each time they engage in ERP activities. For instance, the target behavior for a child with anxiety linked to loud sounds may be for the child to remain in the kitchen, while the blender is on for 5 s in order to receive one point. If the child engages in this behavior, he or she should immediately receive the designated point value. In order to make the reinforcement more salient for the child, the child can earn points that are visually displayed on a chart or the child could be given tokens or other tangible indicators of points, such as tickets, coins, or marbles. This system will be most effective if the child is able to exchange tokens for rewards he or she is motivated to earn. Some children enjoy creating a "rewards menu," to which parents can then assign point values. Rewards options may include screen time (e.g., television, tablet, video games, phone), ability to choose a family activity, or even physical rewards of toy or games. Each reward option can be assigned a different point value depending on the size of the reward, the ease of accessing the reward, and how motivated the child is to earn it. Most importantly, parents must be in agreement for items placed on the reward menu. Families should be informed that token economies are only effective when target behaviors are clearly identified, tokens are provided immediately, point values correctly match the rewards, there is a clear and regular time to "cash-in" rewards, and families follow through consistently.

7.5 In-Session Exposure Practice

Once the hierarchy has been formed and a reinforcement system is in place, practice should commence. The hierarchy should form a guideline for practice. To begin, the clinician should engage the child in a practice toward the bottom of the hierarchy that is not overly distressing. Many clinicians like to progress linearly from the bottom of the hierarchy to the top, although it is not necessary. What is more important is to have the family's engagement in the practice. Oftentimes, clinicians will offer children the opportunity to choose where they prefer to start on their hierarchy using open-ended questions. However, children with ASD may benefit from close-ended choices. For instance, in the context of symptoms of separation anxiety disorder, a clinician could ask, "Pick one goal. Would you rather begin with mom staying outside the room for 30 seconds or you staying outside the room for 30 seconds?"

For most exposure practices, length of time is a variable to consider and should be increased as practice continues. For example, for children with contamination-based OCD, touching a "contaminated" or "dirty" object for 1 s may be less distressing than doing so for 30 s. In this case, practices should work in increasing the amount of time the child touches the object. After and during each practice, the child should be instructed to be brave and endure the practice rather than engage in a safety behavior or compulsion. Many clinicians like to use the SUDS scale during exposure practice to monitor the distress of the child. This helps the clinician to have an idea about how difficult exposure practice is for the child and helps to keep the child engaged in the practice. Sometimes the clinician can monitor whether SUDS scores decrease within and between trials of ERP. Furthermore, depending on the insight of the child with ASD, he or she may not be able to accurately articulate or recognize varying levels of distress using this scale.

Most importantly, the majority of each session should focus on repeated exposure practice. Sometimes the same step may be repeated multiple times, but the clinician should work to make sure the child is challenging his or her anxiety/OCD by continuing to engage in more difficult practice, as clinically appropriate. The clinician should also take opportunities to model exposure practice for parents, so parents understand how to complete the practice and will be able to do so at home while modeling brave behavior and providing minimal reassurance. While ERP oftentimes goes "beyond" what the child will need to do in everyday life, ERP should never involve an activity that is dangerous or that is not aligned with the family's goals or values.

Decreasing or fading preexisting parental accommodation rituals or routines is another area to be addressed during exposure practice. Fading may be part of a child's active exposure hierarchy, but separate hierarchies may have to be developed when safety behaviors primarily involve the parents. In these situations, the bottom level of the hierarchy is the current level of

Table 1
Hierarchy for fading avoidance of reassurance-seeking statement

Challenge	Silence
Goal statement	"See you in the morning" (alternative/different statement each night)
Levels of the hierarchy	"Goodnight" "Goodnight. I love you." "Goodnight. I love you. I'll stay in my room all night and listen for any noises." "Goodnight. I love you. I'll stay in my room all night and listen for any noises. You will sleep well and I will protect you." Repeated one time "Goodnight. I love you. I'll stay in my room all night and listen for any noises. You will sleep well and I will protect you." Repeated two times
Current statement	"Goodnight. I love you. I'll stay in my room all night and listen for any noises. You will sleep well and I will protect you." Repeated three times

Note. The length of time each level is completed will depend on child's discomfort and parent's ability to tolerate discomfort. For example, a certain level may plan on occurring one night, but another level may need to be practiced seven nights before a child is ready to move on. A child may also be willing to skip and move ahead to more difficult levels

accommodation, and the top of the hierarchy may be the family's goal. For example, when a parent is attempting to separate from the child before bedtime, the bottom level of the hierarchy might be: Parent stays in child's bed for entirety of night. The goal might be: Parent completes 5-min bedtime routine and leaves room, checking on child only one time during night. Creating a hierarchy to decrease reassurance may also involve fading of accommodating statements. For example, before bed, the child may prefer for the parent to repeat the following statement three times: "Goodnight. I love you. I'll stay in my room all night and listen for any noises. You will sleep well and I will protect you." If it is too overwhelming for the child to completely remove this part of the routine, the parent may need to advance through a hierarchy of fading this accommodating statement. See Table 1 for an example. When forming these steps and assigning practice, it will be important to have parents and family members in agreement with the plan, because they will be the primary actors of change.

7.6 Between-Session Practice

Engaging in exposure practice and reducing accommodation between treatment sessions is crucial for treatment progress to occur. Each treatment session should conclude with an agreement of home-based practice. This practice should be concretely stated, and it is helpful to be written out and provided to the family on paper or electronically. Some children with ASD like having a chart where they can indicate when they engage in practice each day. Practice may consist of continuing exposures completed in session or may consist of completing exposure that were unable to be completed in the therapy environment. For example, fading accommodation before nighttime is a topic that would be discussed and planned in the lesson, and practice would occur primarily between sessions. Practice should also be very concrete. Vague instructions, such as "Practice exposures at nighttime," are less likely to be completed than instructions such as "Practice sitting by yourself in your room with the lights off for five minutes each day at 7 PM." Establishing a time when families review whether they have completed their homework each day as well as a reminder about the token economy (i.e., reward menu, point values, and "cash-in" procedures) will also be beneficial to treatment. Each follow-up ERP session should begin with a review of practice completed the previous week.

7.7 Challenges and Obstacles to ERP

Working with children who have co-occurring anxiety/OCD symptoms and ASD can bring about some unique challenges to ERP. Because of the nuances to ERP and modifications to standard protocols described above, therapists may need additional training and expertise to deliver the interventions effectively, as has been demonstrated in a transportability study [35]. Additionally, ERP intervention can be more intensive for these children, requiring additional sessions [36] to help generalize outcomes. The new learning that occurs during ERP can take more repetition for individuals with ASD, who as part of ASD characteristics, demonstrate more difficulties with cognitive flexibility. Therapists should also be aware that low insight is associated with poorer outcomes for OCD intervention in youth [37], so this finding can be especially relevant to the population of youth with co-occurring OCD and ASD.

8 Treatment Termination

The clinician should regularly monitor progress (via behavioral measures as discussed above) and communicate with the family about progress toward achieving goals. If progress has plateaued or goals have not been met, there are several factors to consider. First, medication management of anxiety/OCD symptom may be warranted. Second, the clinician should determine whether there are any other factors interfering with treatment. For example, some children with ASD who have some insight into deficits in social skills may experience anxiety due to actual differences and difficulties in comparison to peers. If social anxiety is based on an identifiable real-life trigger (e.g., teasing or bullying), another intervention is warranted, such as communication with the school about establishing a safe environment and the possibility of social skills training for the child. Other potential barriers to treatment progress may be related to families continuing to provide accommodation and reassurance to the child, inconsistent implementation of home-based practice, and/or lack of consistency in implementing the token economy.

If adequate progress is made toward achieving goals, treatment sessions can be spread out and eventually treatment termination will occur. Before terminating treatment, treatment maintenance strategies should be discussed. The family should be aware of how to respond if a new trigger emerges by reducing avoidance/accommodation and creating a hierarchy for exposure practice if needed. Many times, establishing intermittent booster sessions can be helpful for the family stay on track with strategies and for the child to maintain progress. Finally, consultation with other providers (e.g., school psychologists or child psychiatrists) may be necessary to ensure ongoing treatment maintenance.

9 Telehealth

Prior to the COVID-19 pandemic, pilot studies were starting to examine the efficacy of interventions with youth with autism spectrum disorder to be delivered in an online, or telemedicine, format. These pilot and feasibility studies showed promising results with interventions able to be adapted acceptably and resulting in improvements of anxiety symptoms [38, 39]. Since the pandemic, this area of research has achieved more focus with programs working to achieve fidelity with in person approaches while working to incorporate strategies for engagement with autism youth on screen [40].

10 Specific Treatment Programs

Research in this field has emphasized a modular approach to CBT. Significantly more children with both ASD and clinical anxiety show positive responses to modular CBT compared to waitlist control [41]. We have chosen to highlight several programs specifically designed or adapted for clinicians to use to address anxiety symptoms when working with children who have ASD.

10.1 Behavioral Interventions for Anxiety in Children with Autism (BIACA)

One such approach is the BIACA program [42] a family-based modular CBT program that includes both parent and child modules, focusing on developing coping skills and in vivo exposure practice. The program also offers optional social skills modules. When compared to an active treatment-as-usual condition, modular CBT using the BIACA program showed significantly greater treatment gains, with large effect sizes between groups, and these improvements were largely maintained at a 3-month follow-up [43]. Implementing BIACA in schools also led to significant gains for youth compared to standard treatment [44]. The program has been adapted for adolescents (ages 11–15) as well, with findings showing significant reductions in rated anxiety compared to waitlist and usual care controls, though not in full anxiety remission rates [45].

10.2 Learning to Understand and Navigate Anxiety (LUNA)

Guzick and colleagues [46] developed LUNA (Learning to Understand and Navigate Anxiety), which is an online parent-led cognitive behavioral therapy for anxiety in children. It includes both caregiver and child/teen web-based modules. The caregiver and teen modules include a specific module dedicated to anxiety related to core features of autism, such as insistence on sameness, need to stick to routines, and/or sensory sensitivities.

10.3 Supportive Parenting for Anxious Childhood Emotions (SPACE)

The creators of SPACE (Supportive Parenting for Anxious Childhood Emotions) have also begun to pilot their program with families of children with autism, with a focus on reducing family accommodation [47]. Results of the study indicated that 86.66% of participants experienced reliable change (reduction in anxiety symptom severity and family accommodation) posttreatment with continued maintenance of symptom reduction at 2-month post-treatment. This study offered preliminary evidence that SPACE produces improvement in symptoms of anxiety in children with autism.

11 Conclusion

As described above, there are several key aspects of adapting exposure-based CBT intervention for children with co-occurring ASD and anxiety or obsessive-compulsive disorders. These factors include an emphasis on behavioral aspect of CBT, high incorporation and reliance on parental involvement, and need to establish a reinforcement system. Similar to when conducting intervention with typically developing children, the primary treatment modality for the treatment of anxiety and OCD for children with co-occurring ASD is exposure-based CBT.

Conflicts of Interest

The authors have no conflicts of interest to declare that are relevant to the content of this chapter.

References

1. American Psychiatric Association (2013) Diagnostic and statistical manual of mental disorders (DSM-5®). American Psychiatric Publsihing
2. Maenner MJ, Warren Z, Williams AR et al (2023) Prevalence and characteristics of autism spectrum disorder among children aged 8 years – autism and developmental disabilities monitoring network, 11 sites, United States, 2020. MMWR Surveill Summ 72(SS-2):1–14. https://doi.org/10.15585/mmwr.ss7202a1
3. Manning-Courtney P, Murray D, Currans K et al (2013) Autism spectrum disorders. Curr Probl Pediatr Adolesc Health Care 43:2–11
4. Lovaas OI (1987) Behavioral treatment and normal educational and intellectual functioning in young autistic children. J Consult Clin Psychol 55:3–9

5. McEachin JJ, Smith T, Lovaas OI (1993) Long-term outcome for children with autism who received early intensive behavioral treatment. Am J Ment Retard 97:359–372. discussion 373–91
6. Sallows GO, Graupner TD (2005) Intensive behavioral treatment for children with autism: four-year outcome and predictors. Am J Ment Retard 110:417–438
7. van Steensel FJ, Bögels SM, Perrin S (2011) Anxiety disorders in children and adolescents with autistic spectrum disorders: a meta-analysis. Clin Child Fam Psychol Rev 14:302–317
8. White SW, Oswald D, Ollendick T et al (2009) Anxiety in children and adolescents with autism spectrum disorders. Clin Psychol Rev 29:216–229
9. van Steensel FJ, Deutschman AA, Bögels SM (2013) Examining the screen for child anxiety-related emotional disorder-71 as an assessment tool for anxiety in children with high-functioning autism spectrum disorders. Autism 17(6):681–692
10. Ung D, Wood JJ, Ehrenreich-May J, Arnold EB, Fuji C, Renno P, Murphy TK, Lewin AB, Mutch PJ, Storch EA (2013) Clinical characteristics of high-functioning youth with autism spectrum disorder and anxiety. Neuropsychiatry 3(2)
11. Worley JA, Matson JL, Sipes M et al (2011) Prevalence of autism spectrum disorders in toddlers receiving early intervention services. Res Autism Spectr Disord 5:920–925
12. Kerns CM, Kendall PC (2012) The presentation and classification of anxiety in autism spectrum disorder. Clin Psychol Sci Pract 19:323–347
13. Evans DW, Canavera K, Kleinpeter FL et al (2005) The fears, phobias and anxieties of children with autism spectrum disorders and down syndrome: comparisons with developmentally and chronologically age matched children. Child Psychiatry Hum Dev 36:3–26
14. Gillott A, Furniss F, Walter A (2001) Anxiety in high-functioning children with autism. Autism 5:277–286
15. Russell E, Sofronoff K (2005) Anxiety and social worries in children with Asperger syndrome. Aust N Z J Psychiatry 39:633–638
16. McDougle CJ, Kresch LE, Goodman WK et al (1995) A case-controlled study of repetitive thoughts and behavior in adults with autistic disorder and obsessive-compulsive disorder. Am J Psychiatry 152:772–777
17. Bodfish JW, Symons FJ, Parker DE et al (2000) Varieties of repetitive behavior in autism: comparisons to mental retardation. J Autism Dev Disord 30:237–243
18. Volkmar F, Cook EH, Pomeroy J et al (1999) Practice parameters for the assessment and treatment of children, adolescents, and adults with autism and other pervasive developmental disorders. J Am Acad Child Adolesc Psychiatry 38:32S–54S
19. Hallett V, Lecavalier L, Sukhodolsky DG et al (2013) Exploring the manifestations of anxiety in children with autism spectrum disorders. J Autism Dev Disord 43:2341–2352
20. Witwer AN, Lecavalier L (2010) Validity of comorbid psychiatric disorders in youngsters with autism spectrum disorders. J Dev Phys Disabil 22:367–380
21. James AC, James G, Cowdrey FA et al (2013) Cognitive behavioural therapy for anxiety disorders in children and adolescents. Cochrane Database Syst Rev:CD004690
22. Kerns CM, Renno P, Storch EA et al (2017) Anxiety in children and adolescents with autism spectrum disorder: evidence-based assessment and treatment. Academic Press
23. Sukhodolsky DG, Bloch MH, Panza KE et al (2013) Cognitive-behavioral therapy for anxiety in children with high-functioning autism: a meta-analysis. Pediatrics 132:e1341–e1350
24. Sofronoff K, Attwood T, Hinton S (2005) A randomised controlled trial of a CBT intervention for anxiety in children with Asperger syndrome. J Child Psychol Psychiatry 46:1152–1160
25. Chalfant AM, Rapee R, Carroll L (2007) Treating anxiety disorders in children with high functioning autism spectrum disorders: a controlled trial. J Autism Dev Disord 37:1842–1857
26. Reaven J, Blakeley-Smith A, Culhane-Shelburne K et al (2012) Group cognitive behavior therapy for children with high-functioning autism spectrum disorders and anxiety: a randomized trial. J Child Psychol Psychiatry 53:410–419
27. Wood JJ, Kendall PC, Wood KS, Kerns CM, Seltzer M, Small BJ, Lewin AB, Storch EA (2020) Cognitive behavioral treatments for anxiety in children with autism spectrum disorder: a randomized clinical trial. JAMA Psychiatry 77(5):474–483
28. Brown TA, Barlow DH (2014) Anxiety and related disorders interview schedule for DSM-5 (ADIS-5L)-lifetime version: client interview schedule 5-copy set (treatments that work)
29. Storch EA, McGuire JF, Wu MS, Hamblin R, McIngvale E, Cepeda SL, Schneider SC,

Rufino KA, Rasmussen SA, Price LH, Goodman WK (2019) Development and psychometric evaluation of the children's Yale-Brown obsessive-compulsive scale second edition. J Am Acad Child Adolesc Psychiatry 58(1): 92–98

30. Birmaher B, Khetarpal S, Brent D, Cully M, Balach L, Kaufman J, Neer SM (1997) The screen for child anxiety related emotional disorders (SCARED): scale construction and psychometric characteristics. J Am Acad Child Adolesc Psychiatry 36(4):545–553
31. Spence SH (1997) Structure of anxiety symptoms among children: a confirmatory factor-analytic study. J Abnorm Psychol 106(2):280
32. Magiati I, Chan JY, Tan WL, Poon KK (2014) Do non-referred young people with autism Spectrum disorders and their caregivers agree when reporting anxiety symptoms? A preliminary investigation using the Spence children's anxiety scale. Res Autism Spectr Disord 8(5): 546–558
33. Ozsivadjian A, Hibberd C, Hollocks MJ (2014) Brief report: the use of self-report measures in young people with autism spectrum disorder to access symptoms of anxiety, depression and negative thoughts. J Autism Dev Disord 44:969–974
34. Zainal H, Magiati I, Tan JW, Sung M, Fung DS, Howlin P (2014) A preliminary investigation of the Spence children's anxiety parent scale as a screening tool for anxiety in young people with autism spectrum disorders. J Autism Dev Disord 44:1982–1994
35. Reaven J, Blakeley-Smith A, Beattie TL, Sullivan A, Moody EJ, Stern JA, Hepburn SL, Smith IM (2015) Improving transportability of a cognitive-behavioral treatment intervention for anxiety in youth with autism spectrum disorders: results from a US–Canada collaboration. Autism 19(2):211–222
36. Boyd BA, Woodard CR, Bodfish JW (2011) Modified exposure and response prevention to treat the repetitive behaviors of a child with autism: a case report. Case Rep Psychiatry 2011(1):241095
37. Garcia AM, Sapyta JJ, Moore PS, Freeman JB, Franklin ME, March JS, Foa EB (2010) Predictors and moderators of treatment outcome in the Pediatric obsessive compulsive treatment study (POTS I). J Am Acad Child Adolesc Psychiatry 49(10):1024–1033
38. Conaughton RJ, Donovan CL, March S (2017) Efficacy of an internet-based CBT program for children with comorbid high functioning autism spectrum disorder and anxiety: a randomised controlled trial. J Affect Disord 218:260–268
39. Hepburn SL, Blakeley-Smith A, Wolff B, Reaven JA (2016) Telehealth delivery of cognitive-behavioral intervention to youth with autism spectrum disorder and anxiety: a pilot study. Autism 20(2):207–218
40. Kalvin CB, Jordan RP, Rowley SN, Weis A, Wood KS, Wood JJ, Ibrahim K, Sukhodolsky DG (2021) Conducting CBT for anxiety in children with autism spectrum disorder during COVID-19 pandemic. J Autism Dev Disord:1–9
41. Wood JJ, Drahota A, Sze K et al (2009) Cognitive behavioral therapy for anxiety in children with autism spectrum disorders: a randomized, controlled trial. J Child Psychol Psychiatry 50: 224–234
42. Wood JJ, Drahota A (2005) Behavioral interventions for anxiety in children with autism. UCLA Law Rev
43. Storch EA, Arnold EB, Lewin AB et al (2013) The effect of cognitive-behavioral therapy versus treatment as usual for anxiety in children with autism spectrum disorders: a randomized, controlled trial. J Am Acad Child Adolesc Psychiatry 52:132–142.e2
44. Fujii C, Renno P, McLeod BD et al (2013) Intensive cognitive behavioral therapy for anxiety disorders in school-aged children with autism: a preliminary comparison with treatment-as-usual. School Ment Health 5:25–37
45. Storch EA, Lewin AB, Collier AB et al (2015) A randomized controlled trial of cognitive-behavioral therapy versus treatment as usual for adolescents with autism spectrum disorders and comorbid anxiety. Dep Anx 32(3): 174–181
46. Guzick AG, Perozo AG, Kook M, McNeel M, Riddle D, Greenberg R, Lee M, Schneifer SC, Sotrch EA (2023) LUNA (Learning to Understand and Navigate Anxiety). https://www.bcm.edu/research/faculty-labs/luna-learning-to-understand-and-navigate-anxiety. Accessed 1 Dec 2024
47. Rozenblat S, Shimshoni Y, Lebowitz ER, Perez M, Koller J (2023) A pilot trial of SPACE (supportive parenting for anxious childhood emotions) in autism. Child Psychiatry Hum Dev 23:1–5

Chapter 10

Cognitive-Behavioral Therapy with Angry and Aggressive Youth: The Coping Power Program

John E. Lochman, Francesca Kassing, Devon E. Romero, and Sara L. Stromeyer

Abstract

This chapter will discuss cognitive-behavioral therapy for anger and aggression in youth, focusing on the Coping Power Program. The contextual social-cognitive model of anger and aggression as well as a variety of risk factors will be discussed, which present a conceptual framework for the Coping Power Program. An overview of other CBT programs for anger and aggression in youth will be presented, followed by a detailed description of the Coping Power intervention, including sample dialogue and handouts. The Coping Power Program was originally developed for late elementary to early middle school children and includes both a child and a parent component, addresses topics including children's emotion awareness, ability to cope with anger and frustration, and social problem-solving skills, as well as parenting skills related to building positive family relationships and managing children's behavior. Additionally, a number of adaptations have examined the Coping Power Program with individuals instead of groups, upwardly and downwardly extended age groups, altered intervention length, and different settings, indicating that the intervention can be flexibly applied. Research on Coping Power's efficacy and effectiveness will be discussed, followed by implications for dissemination and optimization of the intervention. Information about intervention manuals, workbooks, and training is available at www.copingpower.com.

Key words Anger, Aggression, Children, Cognitive-behavioral therapy, Intervention dissemination

1 Introduction

The Coping Power program is an effective cognitive-behavioral intervention for anger and aggression in youth. This chapter will first highlight the contextual social-cognitive model that the program is based on, as well as a variety of risk factors for anger and aggression. Key components of cognitive-behavioral interventions for anger and aggression will be discussed, followed by a detailed description of the Coping Power program, empirical findings, and considerations for dissemination and optimization.

Robert D. Friedberg and Micaela A. Thordarson (eds.), *Cognitive Behavioral Therapy with Youth: Tradition and Innovation*, Neuromethods, vol. 231, https://doi.org/10.1007/978-1-0716-5134-6_10,

2 Anger and Aggression in Youth

Anger and aggression are often considered within the broader category of externalizing problems and disruptive behavior disorders, which have been consistently linked with negative outcomes, such as delinquency, substance use, violence, academic difficulties, and other mental health problems [e.g., 1]. As a result, childhood anger and aggression have been a focus of preventive intervention and treatment in order to disrupt trajectories of further development of rule- and norm-violating behaviors later into adolescence and adulthood.

Some levels of anger and aggression may reflect appropriate responses to threatening environmental contexts that serve to protect an individual; however, it is necessary to distinguish between adaptive versus atypical and/or inappropriate expressions of anger and aggression. The developmental context is also important to consider, given that younger children are more likely to exhibit anger when a goal is blocked, whereas older children are more likely to exhibit anger, and even aggression, when they perceive threats to their self-esteem [2].

3 The Contextual Social-Cognitive Model

Childhood aggressive behavior is thought to result from a set of contextual and personal factors and potentially lead to adolescent and adult antisocial behavior [3]. The contextual social-cognitive model of prevention highlights two main mediators of adolescent antisocial behavior that may be targets for intervention: child-level factors (e.g., low social competence and social-cognitive skills) and social-level factors, especially within the family (e.g., poor caregiver involvement and discipline).

The contextual social-cognitive model of prevention was derived from a conceptual model of angry aggression as attributed to children's interrelated cognitive misinterpretations of events, related physiological hyperarousal and subsequent effects on deficient cognitive problem-solving [4], and from subsequent research on a more elaborated model of social information processing [5]. Children who exhibit higher levels of aggressive behaviors have been shown to demonstrate deficits and cognitive distortions during both phases of the social-cognitive model: ***appraisal*** and ***problem solution***. During ***appraisal***, children struggle to accurately encode social information and interpret both social events and intentions of others. Research has demonstrated that children with deficiencies during the appraisal stage recall fewer relevant event details, use fewer cues to form interpretations of others, and are more likely to attend to hostile versus neutral cues [6]. These children may also approach social interactions with a hostile attribution bias or assumptions that others act in hostile or provocative manners [6–8].

During *problem solution*, children may develop maladaptive solutions to potential problems and overestimate the success of aggressive solutions to their perceived problems. Research suggests that children who exhibit aggressive behavior are more likely to offer aggressive and direct-action solutions than verbal problem solutions (e.g., verbal assertion, compromise) [9]. It has been found that aggressive children are more likely to offer aggressive solutions in part due to their dominance and revenge-oriented social goals [10, 11] and to their expectations that aggression will lead to desired outcomes [6, 12]. These appraisal and problem-solving processes, which occur during children's interactions with their peers, are affected by their stored schematic expectations and beliefs [13–15]. When aggressive and normative children rate how angry and controlling they expect an unknown peer will be, and then make similar ratings of how they perceived the peer to actually behave after a 12-min discussion task, the normative children's perceptions of how aggressive their peer was were predicted by what observers coded rather than their prior expectation [13]. In contrast, the aggressive children's perceptions after the interaction were best predicted by what they had expected beforehand rather than what observers saw, indicating the powerful effect of aggressive children's impermeable schematic expectations for others will interact with them.

Children's social-cognitive processes and their aggressive behavior develop from a variety of contextual factors, including violent and disadvantaged neighborhoods [16–18], peer rejection and aggression and involvement with deviant peers in the school and community contexts [19–22], and contextual parental factors. Such parental factors may include harsh and/or inconsistent discipline, poor parental problem-solving, vague commands, and poor supervision or monitoring of children's behaviors [23, 24]. Research has also consistently linked children's externalizing behaviors with low levels of parental warmth and positive involvement and high levels of punitive discipline (e.g., spanking, verbal aggression) [25]. In addition to direct effects of the parenting context on children's aggressive behavior, parent and broader family factors may also affect children's social-cognitive processes. For example, many of the parental risk factors described above (i.e., harsh and inconsistent discipline, low parental involvement, vague commands, poor monitoring) have been linked with deficiencies in emotion regulation and social-cognitive processing, which may contribute to aggressive behavior [26]. Research has also demonstrated that the relationship between child and parent-related contextual risk factors is bidirectional, with ineffective parenting leading to children's externalizing behaviors, which in turn may also lead to the deterioration of parenting practices in response to difficult and problematic child behaviors [27].

Other Risk Factors

In addition to children's social-cognitive processes and parental context, there are other child-centered and contextual factors that have been shown to influence the development of anger and aggression. For example, several biological and temperamental factors, including infant fussiness, activity level, and positive affect, predict later displays of conduct problems [28]. Aggressive behavior in children has also been associated with certain patterns of physiological functioning (e.g., lower resting heart rate [29], atypical autonomic functioning [24]), and neurobiology (e.g., dysfunctions of the amygdala and prefrontal cortex, [30]). While previous research has highlighted associations between biological factors and aggressive behavior, these relationships are often moderated by environmental and personal factors. For example, research has demonstrated that children with certain genetic risk factors, including the monoamine oxidase –A (MAOA) gene, are more likely to develop conduct problems in adolescence *only* in the presence of environmental risk factors (e.g., parental neglect, interparental violence, inconsistent discipline) [31]. Similarly, when personal characteristics are considered, the low activity variant of the MAOA gene predicted children's conduct problems only for children who have poor affective decision-making [32]. Thus, it is clear that contextual and child factors can impact the relationships between biological risk factors and externalizing behavior.

In addition to the parental factors described above, other family contextual factors have been shown to influence aggressive behavior. For example, family structure (e.g., coming from a single-parent household), family poverty, marital conflict, and weak attachments to parents have been associated with conduct problems in children [33]. Other family factors, such as parental criminality, substance abuse, and depression, have also been associated with aggressive behavior [34]. These risk factors often have an influence on children's aggressive behavior by disrupting competent parenting behaviors. For example, marital conflict has proximal effects on parents' increased use of harsh parenting behaviors [35], and maternal depression can increase inconsistent discipline which in turn predicts increased aggressive behavior [36]. Given these multifactorial risk factors and processes for the development of anger and aggression and children, intervention often involves multiple targets and components.

4 Overview of Cognitive-Behavioral Therapy for Anger and Aggression

There are a number of cognitive-behavioral therapy programs that have been developed for anger and aggression in youth [37, 38]. Common elements include emotion awareness and regulation strategies, skills training and practice, and caregiver involvement in behavior management. Although this chapter will focus on the Coping Power Program, other effective programs deserve mention as well. Several programs initially evolved out of the Hanf [39] approach and include both parents and children. The Incredible Years program is designed for children aged 3–8 years and has a child component, as well as a parent component with videotape modeling and group discussion [40]. Helping the Noncompliant Child was developed to address noncompliance in children aged 3–8 and is typically employed with individual families and involves in-session modeling [41]. Parent-Child Interaction Therapy, for 2- to 8-year-olds, contains many of the same components as Helping the Noncompliant Child, with a particular focus on the parent-child relationship [42]. Some interventions focus on parents, such as Parent Management Training [43] and Triple P-Positive Parenting Program [44]. Parent Management Training is employed with parents of children aged 3–12 and involves weekly meetings plus phone contact, while Triple P is a multilevel intervention ranging from universal prevention to individual treatment for serious conduct problems, with extensions from birth to age 16. Other programs focus on children, such as Problem-Solving Skills Training [45], which implements cognitive-behavioral skills with youth ages 7–13. Additionally, the Anger Coping Program is a group, school-based cognitive-behavioral intervention for aggressive children from fourth to sixth grade [46]. The 18-session Anger Coping Program was eventually expanded to become the Coping Power Program, which will be discussed in detail below.

5 Coping Power Intervention

Coping Power is an evidence-based, multicomponent, cognitive-behavioral preventative intervention program based on the contextual social-cognitive model [47]. Coping Power evolved from an earlier cognitive-behavioral Anger Coping Program, which involved 12–18 child group sessions [4, 46, 48]. Coping Power was developed for preadolescent and early adolescent children (late elementary to early middle school) displaying significant anger and aggressive behavior, placing them at-risk for the development of later conduct problems and substance use. Support for the Coping Power program has been demonstrated through both efficacy and effectiveness studies [49, 50].

Coping Power includes a child component, which consists of a 34-session individual or group intervention, and a complementary 16-session parent group component. Both the child and parent components are designed to be delivered concurrently over a 16- to 18-month period. In its original format, children participate in weekly group sessions at school and meet with the therapist monthly on an individual basis. The child and parent program manuals are highly structured, providing detailed instructions and clear guidance for disseminating and completing session objectives and activities.

6 Child Component

Early sessions of the Coping Power child component [51] focus on long-term and short-term goal setting. Throughout the program, the child, with the therapist's assistance, identifies clearly observable and appropriate behavioral goals. These goals are monitored daily by the teacher or parent using goal sheets. Each session, the child can earn points based on their success. The use of goal-sheet tracking encourages generalization of treatment effects to the classroom or home. Other main areas of focus include organization and study skills, emotional awareness, anger management and self-regulation, perspective-taking to reduce hostile attributions, social problem-solving and consequence identification, entering prosocial peer groups and cooperating with peers, and managing peer pressure. Below is a description of specific Coping Power units and activities that address deficits in emotion awareness, self-regulation, perspective taking, and social problem-solving in angry and aggressive children.

6.1 Awareness of Feelings and Physiological Arousal Related to Anger

This unit of the intervention program focuses on teaching children to recognize the physiological cues associated with varying levels of anger and to utilize a range of coping skills to manage anger more effectively. This unit begins by helping children identify the behavioral (e.g., pushing, screaming), cognitive (e.g., "He did that on purpose," "I hate my sister"), and physiological (e.g., heart racing, face flushed) components of different emotions (e.g., sad, happy, afraid), with a particular emphasis on their experience of anger. Using a thermometer analogy, children learn to assess various levels of anger intensity they experience (e.g., annoyance, frustration, rage) and types of problems that trigger these different levels of anger. Children then practice noticing and describing situational triggers and anger intensities, such as "I felt medium anger when my classmate started to tease me about my shoes at lunch," "I felt a little bit annoyed when my baby sister kept crying when I was trying to sleep," "I was fuming hot when my cousin broke my brand new video game." During this part of the intervention, children receive

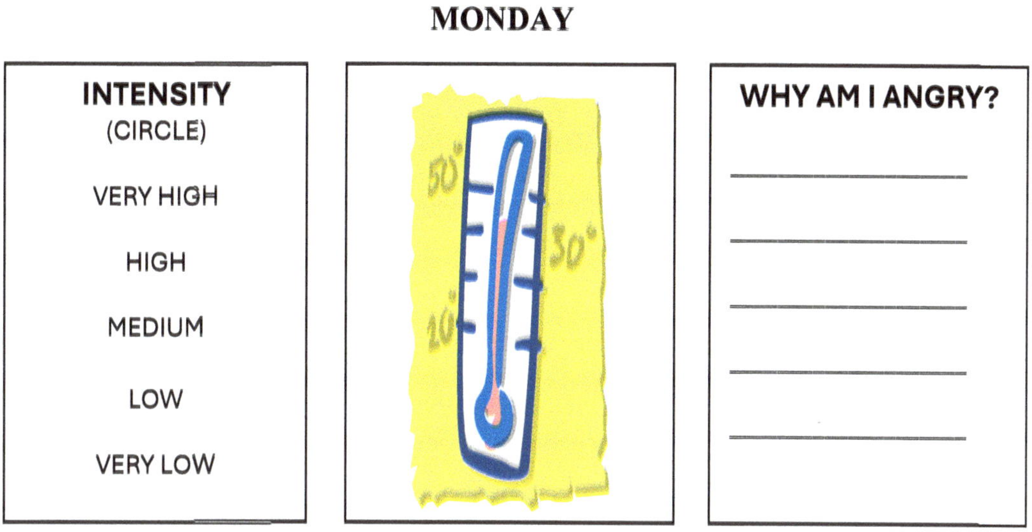

Fig. 1 Segment of anger thermometer worksheet

thermometer worksheets for homework (Fig. 1). This assignment is used to track feelings of anger each day, for 1 week. Children are asked to identify the anger intensity they experience each day and to identify the general trigger for this feeling.

6.2 Practice Anger Coping and Self-Control

The remaining sessions of the unit described in the previous section focus on teaching and practicing emotion regulation strategies such as distraction, relaxation, and coping self-statements. Distraction techniques are taught as a means to divert attention away from anger-provoking situations. Guided imagery, deep breathing, and progressive muscle relaxation techniques are used to promote the regulation of physiological responses and prevent anger escalation. Finally, coping self-statements (e.g., "It's not worth getting angry"; "I'm not going to let this get to me"; "I'll grow up, not blow up"; "I can keep my cool and handle this") are taught through a series of in vivo activities designed to be increasingly more realistic.

Children first use anger management strategies during practice exercises with peers to build their self-efficacy with the skills and mimic real-world situations. For example, children engage in puppet role-plays in which they use puppets to tease another child's puppet. The child holding the puppet being teased is instructed to state aloud the coping statements the puppet is using to avoid anger escalation. After each turn, the group leader can then ask one or more of the following questions: (1) "What was the puppet thinking or saying to himself/herself?" (2) "What level of anger did the puppet experience during the taunting?" (3) "What coping statements did the puppet use to cope with his/her anger?" (4) "Did the puppet use different coping strategies or coping statements for different levels of anger?" As a next step, children apply these skills in practice taunting tasks with peers.

Homework for these subsequent sessions uses a variation of the thermometer worksheet to reinforce new skills as they are learned. For example, during week 2 of this unit, in addition to the homework prompts described above, children are also instructed to identify what they did to cope. For week 3 of this unit, children are asked to identify more specifically the coping statements they used to cope.

6.3 Perspective Taking

Role-playing and discussion activities are used to demonstrate how difficult it often is to accurately determine and understand others' intentions. As an activity, the clinician can (a) have children role-play different characters in an ambiguous situation or (b) show an image of a scenario involving one or more individuals to facilitate discussion. During these activities, children are encouraged to consider the different viewpoints of each of the characters portrayed. Although the primary focus of these activities is on peer situations, aggressive children also often misperceive teachers' intentions regarding classroom management, expectations, and disciplinary procedures. As a homework assignment, children are asked to schedule and conduct an interview with a teacher of their choosing. This assignment serves to assist children in understanding that teachers also have different perspectives.

6.4 Social Problem-Solving Skills

After children learn to manage their anger and take time to view problem situations from others' perspectives, they are taught additional steps for solving problems. Children are introduced to a problem-solving model where problem-solving steps are summarized by the acronym PICC. Problem-solving steps of the PICC model include (1) Problem Identification, which entails noticing that there is a problem, picking the problem apart, and describing the problem as specifically as possible; (2) brainstorming a variety of Choices to solve the problem; and (3) identifying the possible Consequences of each choice and using that information to determine which choice to try first. A series of sessions addresses how the PICC model could be used in common social problem situations with teachers, peers, and family. This unit of the intervention involves the steps and the use of the PICC problem-solving model with children's own problems. Brainstorming discussions, as well as hands-on activities and PICC worksheets (Fig. 2), are used to introduce the range of choices, or possible solutions, that children have in most social problem situations, and then to introduce the range of positive and negative consequences that result from these various choices.

To illustrate a portion of this process in a group session, the clinician could reinforce the group for coming up with a lot of

The problem is: ____

____

Possible choices/solutions:	**Consequences of the choices/solutions:**
1.______________________	1.______________________
2.______________________	2.______________________
3.______________________	3.______________________

The choice I think will work out best is:

____

____

Reporting back: Try out your choice and describe how it actually went

____

____

Fig. 2 PICC model problem-solving worksheet

choices for a problem involving a child (Jermaine) whose friend, James, did not respond when he said he wanted to play with him. The clinician would then indicate "let's go through and evaluate the likely consequences, good and bad, of each choice." Antonio could say "well, he could find someone else to play with, but he might never know why Bob did not answer James did not respond. Charlie then said that if Jermaine had directly asked James why he did not respond, James might have said "I can't play because I have to get home for dinner" or James might have said "I don't want to be friends right now." The group acknowledges the potential for either good or bad consequences for Jermains being verbally assertive here, but the group then decides that this choice, while risky, is more likely to have positive consequences than their other choices.

7 Parent Component

The Coping Power parent component [52] focuses on academic support in the home, stress management, identifying positive and negative child behaviors, praising appropriate child behaviors, ignoring minor disruptive behaviors, giving clear and effective instructions, establishing age-appropriate rules and expectations, giving effective and appropriate discipline and punishment, and family cohesion and communication. In the program, parents learn skills to support and promote the social-cognitive and problem-solving skills that their children learn in the child component. For example, parents learn to apply the PICC model to problems in the home context, so that skills learned in the child component will be reinforced at home. Similar to the child component, parents also learn stress and mood management and the influence that thoughts have on how a person feels, acts, or responds to a situation. Descriptions of specific Coping Power activities that foster positive family relationships and manage children's behavior using basic principles of social learning are provided below.

In the child component, children learn a variety of skills based on social learning theory that they may use to control their behavior and manage their anger in their interactions with others. Relatedly, parent interactions with their children influence their child's behavior. As such, parents are taught the basic principles of social learning. Specifically, the A-B-C (antecedents-behavior-consequences) model is introduced and is used in all subsequent parent sessions. *Antecedents* are the events that happen just before the child's *behavior*, and *consequences* are the events that happen just after the child's *behavior*. As new parenting skills are introduced throughout the program, they are added to the appropriate column on an A-B-C chart. For example, rewarding and praising their child for positive behaviors are added to the C (consequences) section. Other examples of age-appropriate consequences discussed in the program include ignoring minor disruptive behaviors, the use of time-outs, work chores, and privilege removal.

Immediately following the introduction of the A-B-C model, parents are asked to consider which problem behaviors they are currently observing in their child and the positive behaviors they would like to see more of. To facilitate discussion, parents complete a checklist of positive (e.g., expresses anger appropriately, follows directions) and negative (e.g., curses, defies authority) child behaviors with reference to their child. For homework (Fig. 3), parents refer to this checklist and select negative behaviors and their positive opposites to track. Parents should indicate whether the behavior was observed and whether praise or some other form of positive consequence was used. For example, a parent may be concerned

BEHAVIOR	Monday	Tuesday	Wednesday	Thursday	Friday	Saturday	Sunday
Problem Behavior __________ —	**Happened** ? YES NO ________ —	**Happened** ? YES NO ________ —	**Happened** ? YES NO ________ —	**Happened** ? YES NO ________ —	**Happened** ? YES NO ________ —	**Happened** ? YES NO ________ —	**Happened** ? YES NO ________ —
Positive Behavior	**Happened** ? YES NO **Praised?** YES NO	**Happened** ? YES NO **Praised?** YES NO	**Happened** ? YES NO **Praised?** YES NO	**Happened** ? YES NO **Praised?** YES NO	**Happened** ? YES NO **Praised?** YES NO	**Happened** ? YES NO **Praised?** YES NO	**Happened** ? YES NO **Praised?** YES NO

Fig. 3 Positive and negative behavior tracking worksheet

about a son's pattern of shouting frequently at his brother, leading to escalating conflicts. A positive opposite was talking and playing positively with the sibling. When the parent saw that occurring, the parent said "I really appreciate how you have been playing together so well with your brother in the last hour after you came home."

Through discussion and role-plays, parents learn to modify their child's behavior by using positive consequences to reward good behavior. Parents also learn to "catch" their children being good with further emphasis on rewarding and using labeled and unlabeled praise as a positive consequence. Labeled praise is a praise statement that states the good behavior exactly and is particularly useful when teaching a new skill or reinforcing a desired behavior. Examples are (a) "You did a good job of taking out the garbage when I asked you to," (b) "Thank you for playing nicely with your little brother for 30 minutes," and (c) "I like the way you got your homework done on time tonight." Unlabeled praise, on the other hand, is a praise statement that acknowledges good behavior without specifying the exact behavior (e.g., "Good job," "Thank you," "I like that"). Both forms of praise are effective, though each may be more or less appropriate to any given situation.

Defiant behavior can lead to negative social interactions and create a negative family environment. Thus, the final component of the A-B-C model covers the use of praise and giving effective instructions as parenting skills that can help increase compliance. Instructions are the antecedents to child compliance or noncompliance. To further illustrate, leaders present examples of "good" instructions (i.e., those that elicit compliance) and "bad" instructions (i.e., those that elicit noncompliance from children). Some characteristics of bad instructions include repeating instructions over and over again and giving instructions in the form of a question rather than a declaration. A few qualities of good instructions are giving no more than one or two instructions at a time and following instructions with a 10-second period of silence, so that the child has an opportunity to comply. Consistent with other sessions, parent's discuss, role-play, and complete homework assignments to develop and improve their parenting skills.

8 Empirical Findings

8.1 Efficacy and Effectiveness

To date, research has demonstrated a wide range of support for the Coping Power Program. In an initial study examining efficacy, 183 aggressive boys were randomly assigned to receive the full Coping Power program (child and parent components), the child component only, or no treatment [47]. At the 1-year follow-up, participants (particularly in the full Coping Power program) demonstrated lower rates of covert delinquent behavior, lower parent-reported substance abuse, and improved teacher-rated school behavioral functioning compared to the control group [49]. Providing support for the contextual social cognitive model underlying Coping Power, boys in both Coping Power groups displayed significantly fewer hostile attributions of others' behavior and expectations that aggression would lead to a successful outcome, as well as a higher internal locus of control. These child factors, along with more consistent parenting, led to decreased levels of delinquency, substance use, and school behavior problems at the 1-year follow-up [47]. In a subsequent effectiveness study, 245 aggressive boys and girls participated in the full Coping Power program. Following the intervention, participants demonstrated lower ratings of parent- and teacher-rated proactive aggression (i.e., aggression driven by motivation toward a reward), as well as improvements in social competence, problem-solving, and anger coping skills [53]. Furthermore, at the 1-year follow-up, Coping Power resulted in lower rates of substance use and self-reported delinquency compared to the control group [26]. The positive effects of the intervention continued to be evident through a 3-year follow-up [50].

8.2 Cross Cultural Adaptations

Further effectiveness studies have examined the Coping Power Program's effects in a variety of populations and international samples. For example, the Coping Power intervention has demonstrated positive effects with aggressive deaf children [54] and preventative effects on early adolescent substance use and delinquent behavior in Dutch children with disruptive behavior disorders [55]. In another Dutch sample of children 9- to 16-year-olds with mild to borderline intellectual disabilities, Coping Power led to reductions in children's externalizing behavior and improvements in parenting behavior and parent–child relationships [56]. In Pakistan, where the sample was at higher risk for violence exposure, Coping Power demonstrated reduced aggressive behavior and improvements in behavior, social skills, and social cognitive processes [57]. In Sweden, researchers combined Coping Power with Parent Management Training, noting reductions in behavior problems, improvements in prosocial behavior, and enhanced parenting skills [58]. Finally, Italian researchers found Coping Power effective in reducing problem behaviors with clinical samples of children [59, 60] and creatively developed an adapted Universal Coping Power, which has proven effective in randomized clinical trials as a universal preventive intervention for full classrooms of preschool children [61, 62], elementary school children [63, 64], and middle school youth [65].

Given these findings, Coping Power has demonstrated the ability to be effectively applied with different samples and has been placed on peer-reviewed lists of evidence-based treatments, including What Works Clearinghouse, Blueprints for Healthy Youth Development, Model Programs for the Office of Juvenile Justice and Delinquency Prevention, and the California Evidence-Based Clearinghouse for Child Welfare.

9 Dissemination and Optimization

9.1 Dissemination

When assessing the effectiveness of interventions, it is especially important to consider an intervention's ability to be disseminated to the real world or widely circulated to a broader audience. Characteristics of the training process, the clinician, and organizational characteristics can influence the implementation of programs and should be examined from an implementation science perspective. In a prevention-oriented dissemination trial, counselors from 57 elementary schools in Alabama were randomly assigned to 1 of 3 conditions: Coping Power-Intensive Training, Coping Power-Basic Training, or Care-as-Usual [66]. Counselors in both training conditions received 3 days of training before the intervention as well as monthly additional trainings throughout the intervention. The Intensive Training group also had access to individualized problem solving via e-mail or a telephone hotline and received feedback based on audio recordings of individual sessions. Training intensity was found to have a significant impact on outcomes, with the Coping Power-Intensive Training group showing significantly greater reductions in teacher-, parent-, and self-reported externalizing behaviors and greater improvements in social and academic behaviors than the other groups [67]. Significant improvements in children's behavior only occurred in the Intensive Training groups, emphasizing the importance of intensive training and the availability of feedback to counselors throughout the intervention. A follow-up study of this dissemination trial found that children with counselors who received the Coping Power-Intensive Training had better outcomes in language arts grades than children with counselors in the other groups after 2 years [68]. Clinician and organizational characteristics also affected implementation, as clinicians who had personality traits of being more conscientious and agreeable, and who came from schools with less rigid managerial control and more principal and teacher support, implemented the program with greater quality and sustained their use of the program in later years more completely [67, 69].

9.2 Optimization

A critical next step in intervention research, after the evidence-base of an intervention has been established, is to develop methods for optimizing the intervention to enhance its ability to be successful with more subgroups of clients and then to rigorously test those adaptations. A variety of adaptations have been, and are currently being, explored for Coping Power. These include the integration of Coping Power with other interventions and adaptations to the delivery format and setting.

Program Integration Contemplative practices (mindfulness, yoga, and compassion-building practices) have been integrated with the cognitive behavioral practices from traditional Coping Power in Mindful Coping Power (MCP) [70]. A randomized comparison study has examined whether MCP further enhanced program effects on child and parent emotional reactivity relative to standard Coping Power with a sample of 96 child/parent dyads. MCP has been found to be feasible and has been well-received by child and parent participants. In comparison to traditional Coping Power, MCP children have more strongly improved their emotional, cognitive, and behavioral regulation, both on self-report and physiological measures, and improved teacher-rated social skills [71, 72].

Clinicians and researchers recognize the difficulty in engaging parents in preventive interventions and maintaining their involvement across time. One proven approach to enhancing parent engagement is the Family Check-Up [73]. In the Family Check-Up, families are assessed and then receive feedback on strengths and areas of concern that can be targeted for change. The Coping Power parent component has been adapted for tailored, modular use, to address only the specific areas of need identified for a family, using the Family Check-Up [74]. Seven modules have been created from the Coping Power parent component, which can be used alone or in combination. This form of tailored, or adaptive, intervention is designed to offer parents only the training relevant to their concerns and may also reduce the overall length of intervention.

Adaptations to the Delivery Format and Setting A central concern for parents and clinicians in schools and mental health settings has been the length of time required to implement evidence-based prevention programs, reducing schools' likelihood of adopting the programs and reducing parents' attendance and engagement in the program. Interventions often need to be modified to meet the realities of treating children and families in applied settings [75]. One area of difficulty may be delivering lengthy evidenced-based programs efficiently, yet effectively. Two randomized controlled trials have examined the outcomes of an abbreviated version of the Coping Power Program that can be implemented in 1 school year (24 child sessions and 10 parent sessions). Three years after the end of intervention, abbreviated Coping Power, in comparison to the control condition, had more generalized long-term effects on reducing children's externalizing behavior problems regardless of the degree of parental attendance and also showed reductions in impulsivity traits and callous-unemotional traits [76].

To make the intervention even briefer, while incorporating technology that can be engaging to children, a randomized prevention trial with 96 child/parent dyads tested the feasibility and preliminary effects of a hybrid web and in-person delivery format for the adapted Coping Power—Internet Enhanced program (CP-IE) [77]. This more efficient format allows the full content to be delivered in a smaller number of in-person sessions, while children and parents have access to the content of the full Coping Power program through interactive web and video content, including the *Adventures in Captain Judgment* animated video series. This much briefer, hybrid version of Coping Power was found to have beneficial preventive effects by eliminating the significant increase in children's conduct problems in the school setting displayed by control children. The size of these effects was similar to the effects of the longer version of Coping Power. Additionally, the website materials appear to successfully engage children and parents' use of the website predicted children's changes in conduct problems.

A subsequent internet-based version of Coping Power (Coping Power—Predominantly Internet; CP-PI) was adapted to have only one in-person session and then eight internet modules for children and four modules for the parents. The webpage content for CP-PI is similar to that provided for CP-IE but also includes a new set of brief YouTube-like videos of an adolescent encountering problem situations and using the Coping Power skills. CP-PI was found to be acceptable and feasible in a small, randomized pilot study with 40 at-risk students, showing significantly lower rates of child-reported proactive aggression (i.e., aggression drive by motivation toward a reward) and child- and parent-reported reactive aggression (i.e., aggression drive by responses to threat) by the post-intervention assessment [78]. These initial findings suggest that a version of Coping Power that could be delivered almost completely through the internet is practical and promising.

Coping Power has also been adapted for use in middle school settings, addressing the social issues of early adolescents by including content and activities more relevant for this developmental period [79]. In a large randomized trial conducted across 40 middle schools with 709 youth in Maryland and Alabama, those who received the adapted version of the Early Adolescent Coping Power (EACP) program exhibited lower rates of teacher-rated externalizing behavior problems than did youth in the school-as-usual condition. This indicates that the Coping Power program and its contextual social cognitive model can be effectively adapted for students with conduct problems in this often difficult-to-treat age period.

In an effort to adapt a briefer version of Coping Power that could be used in rural schools with limited resources, a Coping Power—Rural (CP-R) has been developed from the EACP curriculum by condensing the content to 12 weekly small group sessions, complemented by 12 parallel lessons for the full classrooms [80]. The multiyear effort to adapt the program, in partnership with the local schools, for rural elementary and middle schools, involved an innovative effort to integrate intervention delivery across classrooms and the small groups for at-risk children (Tiers 1 and 2) and to use a fully remote training and coaching model. Teachers in rural schools were found to be receptive to the CP-R program, and the program is now ready for larger-scale testing.

Another area that may pose a challenge is the delivery format. There is often a concern about iatrogenic effects when adapting an intervention targeting disruptive behavior for a group delivery format. Due to Coping Power typically being delivered in small groups, a large-scale study was conducted to investigate how children fare if they received Coping Power in a group versus individual format [81]. Because of the primary research question, this study only used the child component of Coping Power and not the parent component. Results indicated both intervention delivery methods led to similar significant reductions in parent-rated externalizing problems through a 1-year follow-up period. However, although teacher-rated externalizing problems declined significantly for both intervention conditions, the reductions were significantly greater for children receiving Coping Power in an individual format. The main effect was moderated by children's baseline levels of inhibitory control. Children with fewer inhibitory control problems responded in similar positive ways to either the group or individual format. Children who were more oriented to social rewards (indexed by an oxytocin receptor gene) and more emotionally dysregulated (indexed by the autonomic nervous system indicators) also had weaker teacher-rated outcomes through a 1-year follow-up [82, 83]. These results are encouraging in terms of effectively adapting an evidence-based program to the specific needs of an applied setting, whether it be working individually or in small groups with children.

Crossover Effects Although Coping Power was developed to target mechanisms associated with children's aggressive behavior, the mechanisms related to emotion dysregulation and poor problem-solving can also impact other negative outcomes. Although not the original focus of the program, Coping Power has been found to reduce children's internalizing problems as well as their externalizing problems, according to adult reports [81]. In another instance of crossover effects, the individual delivery format of Coping Power has been found to produce a greater reduction in caretaker depression than did the group delivery version of Coping Power. Reduced caretaker depression, in turn, subsequently predicted greater decreases in children's hyperactive behaviors [84]. These findings indicated the potential for crossover effects of a child-focused intervention on the parents' affective adjustment. In a larger investigation of crossover effects for a key focus of mental health prevention efforts, a study using integrative data analysis with 11 Coping Power datasets found that 2 variants of Coping Power: MCP, and CP-IE, have significantly reduced youths' suicidal behaviors over time, as rated by teachers and parents, in comparison to untreated children [85]. These findings all point to the importance of the multifinality effects of certain key intervention mechanisms and indicate how an intervention such as Coping Power can have broad effects on multiple externalizing and internalizing outcomes.

The Role of the Therapist and the Therapeutic Alliance Another focus of intervention optimization can be how aspects of the therapy process, particularly the therapist and the therapeutic alliance, can affect intervention implementation and outcomes even for evidence-based interventions like Coping Power. The dissemination study described earlier indicated how clinician personality traits can predict their ability to implement and sustain the program well [67, 81]. The therapists' traits that affect their emotion regulation and flexible response to stress may contribute to clinicians' actual behavior in sessions. In a study of the child and therapist in-session behaviors that predicted children's outcomes years after the group-based Coping Power intervention, therapists' clinical skills played a major role. Children with therapists who interacted with greater warmth and flexibility and less irritability were likely to have lower externalizing behavior rates up to 4 years after intervention [86, 87]. In a potentially related manner, therapists' reports of greater therapeutic alliance with their clients in Coping Power also have been found to predict greater reductions in externalizing and internalizing behaviors through a 1-year follow-up period [88]. These findings indicate that careful training and supervision of Coping Power therapists should include a focus on how they can monitor and use effective clinical skills and develop stronger therapeutic alliances.

10 Conclusion

In conclusion, the Coping Power program has been demonstrated as an effective intervention for anger and aggression in youth, with outcomes ranging from reducing existing aggression and disruptive behavior to improving deficits and distortions in social-cognitive processes and to preventative effects on later delinquent behavior and substance abuse. The intervention is highly structured, provides detailed instructions and a selection of activities for completing session objectives, and addresses multiple risk factors associated with aggressive behavior through targeted session content, using a contextual social-cognitive framework. As demonstrated, Coping Power can be flexibly applied with different age groups, populations, and settings, as well as with individuals instead of groups, and with a shorter number of sessions. More recent studies have examined integrating the intervention with a Mindfulness component to boost outcomes, with the Family Check-Up to enhance parent engagement, or with interactive web and video content to increase efficiency. An important consideration for dissemination and optimization is intensive training and regular feedback for clinicians implementing the program, as well as some clinician and organizational characteristics. Information about intervention manuals, workbooks, and training is available at www.copingpower.com.

Conflicts of Interest

Research described in this chapter was funded by grants from the National Institute of Drug Abuse (R34 DA035946, R34 DA034295, R01 DA023156, R41 DA022184, R01 DA016135, R01 DA08453), the National Institute of Mental Health (R01 124,438, R01 MH39989), the National Institute of Child Health and Human Development (R01 HD079273), the Institute for Education Sciences (R305 A140070), the Centers for Disease Control and Prevention (R49/CCR418569), the Substance Abuse and Mental Health Administration (KD1 SP08633, UR6 5,907,956), the Office of Juvenile Justice and Delinquency Prevention (2006JLFX0232), and US Department of Justice (COPS CKWX0091).Dr. Lochman receives royalties from Oxford University Press for the Coping Power Implementation Guides and Workbooks. Drs. Kassing, Romero, and Stromeyer have no conflicts of interest to declare.

References

1. Matthys W, Lochman JE (2017) Oppositional defiant disorder and conduct disorder in childhood, 2nd edn. Wiley-Blackwell, Chichester
2. Lochman JE, Powell N, Clanton N, McElroy H (2006) Anger and aggression. In: Bear G, Minke K (eds) Children's needs III: development, prevention, and intervention. National Association of School Psychologists, Washington, DC
3. Conduct Problems Prevention Research Group (1992) A developmental and clinical model for the prevention of conduct disorder: the FAST track program. Dev Psychopathol 4: 509–527. https://doi.org/10.1017/S0954579400004855
4. Lochman JE, Nelson WM III, Sims JP (1981) A cognitive behavioral program for use with aggressive children. J Clin Child Psychol 10: 146–148
5. Crick NR, Dodge KA (1994) A review and reformulation of social information-processing mechanisms in children's social adjustment. Psychol Bull 115(1):74–101. https://doi.org/10.1037/0033-2909.115.1.74
6. Lochman JE, Dodge KA (1994) Social-cognitive processes of severely violent, moderately aggressive and nonaggressive boys. J Consult Clin Psychol 62:366–374. https://doi.org/10.1037/0022-006X.62.2.366
7. De Castro BO, Veerman JW, Koops W, Bosch JD, Monshouwer HJ (2002) Hostile attribution of intent and aggressive behavior: a meta-analysis. Child Dev 73:916–934
8. Yaros A, Lochman JE, Rosenbaum J, Jimenez-Camargo LA (2014) Real-time hostile attribution measurement and aggression in children. Aggress Behav 40:409–420
9. Dunn SE, Lochman JE, Colder CR (1997) Social problem-solving skills in boys with conduct and oppositional defiant disorders. Aggress Behav 23:457–469. https://doi.org/10.1002/(SICI)1098-2337(1997)23:6<457::AID-AB5>3.0.CO;2-D
10. Lochman JE, Wayland KK, White KJ (1993) Social goals: relationship to adolescent adjustment and to social problem solving. J Abnorm Child Psychol 21:135–151
11. McDonald KL, Lochman JE (2012) Predictors and outcomes associated with trajectories of revenge goals from fourth grade through seventh grade. J Abnorm Child Psychol 40:225–236
12. Pardini DA, Lochman JE, Frick PJ (2003) Callous/unemotional traits and social cognitive processes in adjudicated youth. J Am Acad Child Adolesc Psychiatry 42:364–371
13. Lochman JE, Dodge KA (1998) Distorted perceptions in dyadic interactions of aggressive and nonaggressive boys: effects of prior expectations, context, and boys' age. Dev Psychopathol 10:495–512
14. Lochman JE, Lenhart (1995) Cognitive behavioral therapy of aggressive children: effects of schemas. In: Van Bilsen HPJG, Kendall PC, Slavenburg JH (eds) Cognitive behavioral approaches for children and adolescents:

challenges for the next century. Plenum, New York, pp 145–166
15. Matthys W, Schutter D (in press) Involving parents in cognitive behavioral thrapy for children and adolescents with conduct problems: goals, outcome expectations, and normative beliefs about aggression are targeted in sessions with parents and their child. Clin Child Fam Psychol Rev
16. Burt SA, Klump KL, Gorman-Smith D, Neiderhiser JM (2016) Neighborhood disadvantage alters the origins of children's nonaggressive conduct problems. Clin Psychol Sci 4:511–526
17. Fite PJ, Winn P, Lochman JE, Wells KC (2009) The effect of neighborhood disadvantage on proactive and reactive aggression. J Community Psychol 37:542–546
18. Kassing F, Lochman JE, Vernberg E, Hudnall M (2022) Using crime data to assess longitudinal relationships between community violence and aggressive behavior among at-risk youth. J Early Adolesc 42(3):431–448. https://doi.org/10.1177/02724316211042941
19. Barth JM, Dunlap ST, Dane H, Lochman JE, Wells KC (2004) Classroom environment influences on aggression, peer relations, and academic focus. J School Psychol 42:115–133
20. Dishion TJ, Tipsord JM (2011) Peer contagion in child and adolescent social and emotional development. Annu Rev Psychol 62:189–214. https://doi.org/10.1146/annurev.psych.093008.100412
21. Powers CJ, Bierman KL, Conduct Problems Prevention Research Group (2013) The multifaceted impact of peer relations on aggressive-disruptive behavior in early elementary school. Dev Psychol 49:1174–1186
22. Thomas DE, Bierman KL, Thompson C, Powers CJ, Conduct Problems Prevention Research Group (2009) Double jeopardy: child and school characteristics that predict aggressive-disruptive behavior in the first grade. School Psychol Rev 37:516–532
23. Pardini DA, Waller R, Hawes SW (2015) Familial influences on the development of serious conduct problems and delinquency. In: Morizot J, Kazemian L (eds) The development of criminal and antisocial behavior: theory, research and practical applications. Springer, Cham
24. Kassing F, Lochman JE, Glenn A (2018) Autonomic functioning in reactive versus proactive aggression: the influential role of inconsistent parenting. Aggress Behav 44:524–536
25. Stormshak EA, Bierman KL, McMahon RJ, Lengua LJ, Conduct Problems Prevention Research Group (2000) Parenting practices and child disruptive behavior problems in early elementary school. J Clin Child Psychol 29(1):17–29
26. Lochman JE, Wells KC (2003) Effectiveness of the coping power program and of classroom intervention with aggressive children: outcomes at a 1-year follow-up. Behav Ther 34: 493–515
27. Pardini DA (2008) Novel insights into longstanding theories of bidirectional parent–child influences: Introduction to the special section. J Abnorm Child Psychol 36:627–631
28. Lahey BB, Van Hulle CA, Keenan K et al (2008) Temperament and parenting during the first year of life predict future child conduct problems. J Abnorm Child Psychol 36:1139–1158
29. Lorber MF (2004) Psychophysiology of aggression, psychopathy, and conduct problems: a meta-analysis. Psychol Bull 130:531–552
30. Huebner T, Vloet TD, Marx IVO, Konrad K, Fink GR, Herpertz SC, Herpertz-Dahlmann B (2008) Morphometric brain abnormalities in boys with conduct disorder. J Am Acad Child Adolesc Psychiatry 47:540–547
31. Foley DL, Eaves LJ, Wormley B et al (2004) Childhood adversity, monoamine oxidase A genotype, and risk for conduct disorder. Arch Gen Psychiatry 61:738–744
32. Levantini V, Powe CE, Lochman JE, Glenn A (2022) Monoamine Oxidase-A and conduct problems in children: the role of affective decision-making. J Psychopathol Behav Assess 44:227–234. https://doi.org/10.1007/s10862-021-09917-9
33. Moffitt TE, Scott S (2008) Conduct disorders of childhood and adolescence. In: Rutter M, Bishop DVM, Pine PS, Scott, Stevenson SJ, Taylor E, Thapar A (eds) Rutter's child and adolescent psychiatry, 5th edn. Blackwell, Oxford, pp 543–564. https://doi.org/10.1002/9781444300895.ch35
34. Barry TD, Dunlap ST, Cotton SJ, Lochman JE, Wells KC (2005) The influence of maternal stress and distress on disruptive behavior problems in children. J Am Acad Child Adolesc Psychiatry 44:265–273
35. Sherrill RB, Lochman JE, DeCoster J, Stromeyer SL (2017) Spillover between interparental and parent-child conflict within and across days. J Fam Psychol 31:900–909
36. Barry TD, Dunlap S, Lochman JE, Wells KC (2009) Inconsistent discipline as a mediator between maternal distress and aggression in boys. Child Fam Behav Ther 31:1–19

37. Kaminski JW, Claussen AH (2017) Evidence base update for psychosocial treatments for disruptive behaviors in children. J Clin Child Adolesc Psychol 46(4):477–499
38. Sukhodolsky DG, Kassinove H, Gorman BS (2004) Cognitive-behavioral therapy for anger in children and adolescents: a meta-analysis. Aggress Violent Behav 9(3):247–269
39. Hanf C (1969) A two-stage program for modifying maternal controlling during mother-child (M-C) interaction. Paper presented at the meeting of the Western Psychological Association, Vancouver
40. Webster-Stratton C, Reid M (2003) The incredible years parents, teachers, and children training series: a multifaceted treatment approach for young children with conduct problems. In: Kazdin AE, Weisz JR (eds) Evidenced-based psychotherapies for children and adolescents. Guilford, New York, pp 224–240
41. McMahon RJ, Forehand RL (2003) Helping the noncompliant child: Family-based treatment for oppositional behavior, 2nd edn. Guilford, New York
42. Brinkmeyer MY, Eyberg SM (2003) Parent-child interaction therapy for oppositional children. In: Kazdin AE, Weisz JR (eds) Evidenced-based psychotherapies for children and adolescents. Guilford, New York, pp 204–223
43. Patterson GR, Reid JB, Jones RR, Conger RE (1975) A social learning approach to family intervention: families with aggressive children, vol 1. Castalia, Eugene
44. Sanders MR (1999) Triple p-positive parenting program: towards an empirically validated multilevel parenting and family support strategy for the prevention of behavior and emotional problems in children. Clin Child Fam Psychol Rev 2:71–90
45. Kazdin AE (2003) Problem-solving skills training and parent management training for conduct disorder. In: Kazdin AE, Weisz JR (eds) Evidence-based psychotherapies for children and adolescents. Guilford, New York, pp 241–262
46. Larson J, Lochman JE (2011) Helping schoolchildren cope with anger: a cognitive-behavioral intervention, 2nd edn. Guilford, New York
47. Lochman JE, Wells KC (2002) Contextual social-cognitive mediators and child outcome: a test of the theoretical model in the Coping Power Program. Dev Psychopathol 14(4): 945–967
48. Lochman JE, Burch PR, Curry JF, Lampron LB (1984) Treatment and generalization effects of cognitive behavioral and goal setting interventions with aggressive boys. J Consult Clin Psychol 52:915–916
49. Lochman JE, Wells KC (2004) The Coping Power Program for preadolescent boys and their parents: outcome effects at the 1-year follow-up. J Consult Clin Psychol 72(4): 571–578
50. Lochman JE, Wells KC, Qu L, Chen (2013) Three year follow-up of coping power intervention effects: evidence of neighborhood moderation? Prev Sci 14:364–376
51. Lochman JE, Wells KC, Lenhart LA (2008) Coping power child group program: facilitator guide. Oxford University Press, New York
52. Wells KC, Lochman JE, Lenhart L (2008) Coping power parent group program: facilitator guide. Oxford University Press, New York
53. Lochman JE, Wells KC (2002b) The Coping Power Program at the middle school transition: universal and indicated prevention effects. Psychol Addict Behav 16(4S):S40–S54
54. Lochman JE, FitzGerald DP, Gage SM, Kannaly MK, Whidby JM, Barry TD, Pardini DA, McElroy H (2001) Effects of social-cognitive intervention for aggressive deaf children: the Coping Power Program. J Am Deafness Rehabil Assoc 35:39–61
55. Zonnevylle-Bender MJS, Matthys W, van de Wiel NMH, Lochman J (2007) Preventive effects of treatment of DBD in middle childhood on substance use and delinquent behavior. J Am Acad Child Adolesc Psychiatry 46: 33–39
56. Schuiringa H, van Nieuwenhuijzen M, Orobio de Castro B, Lochman JE, Matthys W (2017) Effectiveness of an intervention for children with externalizing behavior and mild to borderline intellectual disabilities: a randomized trial. Cogn Ther Res 41:237–251
57. Mushtaq A, Lochman JE, Tariq PN, Sabih F (2017) Preliminary effectiveness study of Coping Power program for aggressive children in Pakistan. Prev Sci 18:762–771
58. Helander M, Lochman J, Högström J, Ljótsson B, Hellner C, Enebrink P (2018) The effect of adding Coping Power Program-Sweden to parent management training-effects and moderators in a randomized controlled trial. Behav Res Ther 103:43–52
59. Muratori P, Milone A, Manfredi A, Polidori L, Ruglioni L, Lambruschi F, Masi G, Lochman JE (2017) Evaluation of improvement in

externalizing behaviors and callous-unemotional traits in children with disruptive behavior disorder: a 1-year follow up clinic-based study. Admin Policy Ment Health 44: 452–462. https://doi.org/10.1007/s10488-015-0660-y

60. Muratori P, Milone A, Levantini V, Ruglioni R, Lambruschi F, Pisano S, Masi G, Lochman JE (2019) Six-year outcome for children with ODD or CD treated with Coping Power Program. Psychiatry Res 271:454–458

61. Muratori P, Giofre D, Bertacchi I, Darini A, Giuli C, Lai E, Moderna A, Lochman JE, Mammarella I (2022) Testing the efficacy of coping power universal on behavioral problems and pre-academic skills in preschoolers. Early Child Educ J 50:613–625. https://doi.org/10.1007/s10643-021-01179-0

62. Muratori P, Lochman J, Bertacchi, Giuli C, Guarguagli E, Pisano S, Gallani A, Mammarella IC (2019) Universal Coping Power for pre-schoolers: effects on children's behavioral difficulties and pre-academic skills. Sch Psychol Int 40(2):128–144

63. Muratori P, Bertacchi I, Giuli C, Nocentini A, Lochman JE (2017) Implementing Coping Power adapted as a universal prevention program in Italian primary schools: a randomized control trial. Prev Sci 18:754–761. https://doi.org/10.1007/s11121-016-0715-7

64. Muratori P, Bertacchi I, Masi G, Milone A, Nocentini A, Powell NP, Lochman JE, Jones S, Kassing F, Romero D (2019) Effects of a universal prevention program on externalizing behaviors: exploring the generalizability of findings across school and home settings. J School Psychol 77:13–23

65. Muratori P, Bertacchi I, Catone G, Mannucci F, Nocentini A, Pisano S, Lochman JE (2020) Coping Power Universal for middle school students: the first efficacy study. J Adolesc 79:49–58

66. Lochman JE, Boxmeyer CL, Powell N, Qu L, Wells K, Windle M (2009) Dissemination of the Coping Power program: importance of intensity of counselor training. J Consult Clin Psychol 77:397–409

67. Lochman JE, Powell NP, Boxmeyer CL, Qu L, Wells KC, Windle M (2009) Implementation of a school-based prevention program: effects of counselor and school characteristics. Prof Psychol Res Pract 40:476–482

68. Lochman JE, Boxmeyer CL, Powell NP, Qu L, Wells K, Windle M (2012) Coping power dissemination study: intervention and special education effects on academic outcomes. Behav Disord 37:192–205

69. Lochman JE, Powell NP, Boxmeyer CL, Qu L, Sallee M, Wells KC, Windle M (2015) Counselor-level predictors of sustained use of an indicated preventive intervention for aggressive children. Prev Sci 16:1075–1085

70. Miller S, Boxmeyer CL, Romero D, Powell NP, Jones S, Lochman JE (2020) Theoretical model of mindful coping power: optimizing a cognitive behavioral program for high risk children and their parents by integrating mindfulness. Clin Child Fam Psychol Rev 23:393–406

71. Boxmeyer CL, Miller S, Romero DE, Powell NP, Jones SB, Qu L, Tueller S, Lochman JE (2021) Mindful coping power: comparative effects on children's reactive aggression and self-regulation. Brain Sci 11:1119. https://doi.org/10.3390/brainsci11091119

72. Boxmeyer CL, Stager CG, Miller S, Lochman JE, Romero DE, Powell NP, Bui C, Qu (2023) Mindful coping power effects on children's autonomic nervous system functioning and long-term behavioral outcomes. J Clin Med. https://doi.org/10.3390/jcm12113621

73. Dishion TJ, Kavanagh K (2005) Intervening in adolescent problem behavior: a family-centered approach. Guilford Press, New York

74. Herman KC, Reinke WM, Bradshaw CP, Lochman JE, Boxmeyer CL, Powell NP et al (2012) Integrating the family check-up and the parent coping power program. Adv Sch Men Health Promot 5:208–219

75. Stirman SW, Crits-Christoph P, DeRubeis RJ (2004) Achieving successful dissemination of empirically supported psychotherapies: a synthesis of dissemination theory. Clin Psychol Sci Pract 11:343–359

76. Lochman JE, Baden RE, Boxmeyer CL, Powell NP, Qu L, Salekin KL, Windle M (2014) Does a booster intervention augment the preventive effects of an abbreviated version of the coping power program for aggressive children? J Abnorm Child Psychol 42:367–381

77. Lochman JE, Boxmeyer CL, Jones S, Qu L, Ewoldsen D, Nelson WM III (2017) Testing the feasibility of a briefer school-based preventive intervention with aggressive children: a hybrid intervention with face-to-face and internet components. J School Psychol 62:33–50

78. Heilman M, Lochman J, Bui C, White BA (under review) Preliminary investigation of a novel internet-based aggression prevention program for children. Unpublished manuscript

79. Bradshaw CP, McDaniel HL, Pas ET, Debnam KJ, Bottiani JH, Powell NP, Ialongo NS, Lochman JE (under review) Randomized controlled trial of the early adolescent coping power program effects on emotional and behavioral

problems in middle schoolers. Unpublished manuscript

80. Nguyen AJ, Hersh J, Beahm L, Smith LH, Newman C, Birchfield K, Michael K, Bradshaw CP (2024) Coping power- riral: iterative adaptation of an evidence-based preventive intervention for rural upper elementary and middle schools. Sch Ment Health. https://doi.org/10.1007/s12310-024-09632-2
81. Lochman JE, Dishion TJ, Powell NP, Boxmeyer CL, Qu L, Sallee M (2015) Evidence-based preventive intervention for preadolescent aggressive children: one-year outcomes following randomization to group versus individual delivery. J Consult Clin Psychol 83:728–735
82. Glenn AL, Lochman JE, Dishion T, Powell NP, Boxmeyer C, Qu L (2018) Oxytocin receptor gene variant interacts with intervention delivery format in predicting intervention outcomes for youth with conduct problems. Prev Sci 19:38–48
83. Glenn AL, Lochman JE, Dishion T, Powell NP, Boxmeyer C, Kassing F, Qu L, Romero D (2019) Toward tailored interventions: sympathetic and parasympathetic functioning predicts responses to an intervention for conduct problems delivered in two formats. Prev Sci 20:30–40
84. Saavedra LM, Lochman JE, Morgan-Lopez AA, McDaniel HL, Bradshaw C, Powell NP, Qu L, Budavari A, Yaros AY (2024) Collateral effects of coping power on caregiver symptoms of depression and long-term changes in behavior. Dev Psychopathol. https://doi.org/10.1017/S095457942300144X
85. McDaniel HL, Budavari AC, Tonigan AT, Michael AE, West SG, O'Shea NG, Powell NP, Qu, Saavedra LM, Yaros AC, Bradshaw CP, Lochman JE, Morgan-Lopez AA (in press) Preventing suicidal thoughts and behaviors among youth: integrative data analysis of crossover impacts of the Coping Power preventive intervention. JAACAP Open
86. Lochman JE, Dishion TJ, Boxmeyer CL, Powell N, Qu L (2017) Variations in response to evidence-based group preventive intervention for disruptive behavior problems: a view from 938 Coping Power sessions. J Abnorm Child Psychol 45:1271–1284. https://doi.org/10.1007/s10802-016-0252-7
87. Lochman JE, Glenn AL, Powell NP, Boxmeyer CL, Bui C, Kassing F, Qu L, Romero D, Dishion T (2019) Group versus individual format of intervention for aggressive children: moderators and predictors of outcomes through four years after intervention. Dev Psychopathol 31(5):1757–1177
88. Lochman JE, Laird RD, McDaniel HL, Boxmeyer CL, Braun SS, Powell NP, Qu L (2024) Intervention format and therapist-child agreement associated with therapeutic alliance and outcomes. J Consult Clin Psychol 92(1):26–43. https://doi.org/10.1037/ccp0000841

Chapter 11

Exposure-Based Cognitive Behavioral Therapy Plus Parent Management Training for the Treatment of Youth with Severe Irritability

Ramaris E. German, Jamell D. White, Lisa Cullins, Lisa Felber, and Melissa A. Brotman

Abstract

Irritability, an increased proneness to anger relative to same-aged peers, is a significant public health issue. Along with the burden experienced by youth and parents across development, severe and impairing irritability is associated with an increased likelihood of depression, anxiety disorders, and suicidality in adulthood. Although there have been several treatments for disruptive behavior disorders that have been proven to be effective, limited interventions specifically target severe irritability. This chapter is focused on a treatment that has been developed to specifically treat severe and impairing irritability. The treatment consists of 12 sessions, including both child and parent components. The child component involves both psychoeducation about irritability and in vivo exposures to anger-inducing stimuli. The parent component covers topics often taught in parent behavior management training with the goal of teaching concrete skills for managing youth's irritability. Efficacy data suggests that this novel treatment may decrease severe irritability and positively impact overall level of functioning. Clinical trials of this treatment are ongoing with active recruitment of participants. Overall, this treatment demonstrates promise as an effective treatment for severe irritability in youth.

Key words Child adolescent irritability, Exposure therapy, Parent management training, Irritability, Skills training, CBT for anger in youth, Youth irritability treatments, DMDD

1 Introduction

Clinical irritability, or an increased proneness to anger, is a significant public health issue in youth. In epidemiological studies, the prevalence of severe and impairing irritability is between 0.8% and 3.3% [1]. Coupled with the burden and challenges experienced by youth and their family members, chronic irritability has been shown to increase the risk of developing depression and/or anxiety disorders in adulthood [2–5]. This potential disease progression has been the impetus in the development and scientific investigation of

Robert D. Friedberg and Micaela A. Thordarson (eds.), *Cognitive Behavioral Therapy with Youth: Tradition and Innovation*, Neuromethods, vol. 231, https://doi.org/10.1007/978-1-0716-5134-6_11,

a novel exposure-based cognitive behavioral therapy (CBT) with parent management training (PMT) approach for the treatment of severe irritability in youth at the National Institute of Mental Health (NIMH). Of note, the development and scientific investigation of this novel treatment approach was supported by the Intramural Research Program (IRP) of the National Institute of Mental Health, National Institutes of Health (ZIAMH002786-15, ZIAMH002778-17), and conducted under NIH Clinical Study Protocols 02-M-0021 and 15-M-0182 [ClinicalTrials.gov identifier: NCT02531893 (Brotman)].

2 Theoretical and Empirical Foundations

Irritability presents as a continuum—from a healthy adaptive emotion to harmfully severe moods and behavior. Most recently, DSM-5 codified disruptive mood dysregulation disorder (DMDD) characterized by severe and impairing irritability in multiple domains [6]. Particularly in youth, severe irritability may be persistent, pervasive, and maladaptive, reaching a clinical level that may benefit from therapeutic intervention. To determine if irritability meets clinical level criterion, three features are examined: *frequency*, *duration*, and *threshold*. As compared to typically developing youth, youth with severe irritability show a propensity toward anger multiple times a week; their anger outbursts last longer; and they have a lower frustration tolerance. In addition, these youth's severe irritability presents and demonstrates impairment in multiple domains: home, school, and/or with peers. Youth who have clinically severe irritability have difficulty engaging effective emotion regulation skills [7], remain angry for longer periods of time, and tend to have a lower threshold for what makes them irritable, revealing lower distress tolerance [8]. At home, these children experience problems with caregivers and other family members due to becoming easily upset and angry over "little things" (e.g., being asked to brush their teeth). At school, these youth may be frequently sent to the principal's office or have been asked to leave school with either a suspension or expulsion due to their disruptive behavior. Youth with clinically severe irritability also often have challenges with peer interactions; they may have lost friends or have decreased invitations to social events (e.g., play dates) due to their anger outbursts and irritability. In sum, youth with severe, clinical levels of irritability may experience significant impairment in interpersonal relationships as well as with navigating the school environment, which underscores the importance of effective therapeutic interventions that specifically target this symptomatology.

2.1 Diagnostic Framework

Irritability is a criterion in multiple mood and anxiety disorders and an associated feature in others. Disruptive mood dysregulation disorder (DMDD), introduced into the Diagnostic and Statistical Manual of Mental Disorders, Fifth Edition (DSM-5) in 2013 [6], is characterized by severe irritability. DMDD consists of a (phasic) behavioral component (recurrent severe temper outbursts with increased motor activity and verbal/physical aggression) and a (tonic) mood component (persistently irritable, grumpy, or angry mood on most days between outbursts [9]). Temper outbursts can be mild (e.g., mild arguing, name-calling, or snapping or talking back), moderate (e.g., yelling or screaming, throwing things without the intention to break them, verbal threats, clenching fist or raising arm to hit), or severe (e.g., shoving, slapping, pushing, kicking, using object to harm, and intentional destruction of property). For a child to be diagnosed with DMDD, they must meet the following criteria [6]:

- Severe, recurrent temper outbursts occur on average, three or more times per week in the form of verbal rages or physical aggression toward people or property that are grossly out of proportion in intensity or duration to the situation or provocation and inconsistent with developmental level.
- The mood between outbursts is persistently irritable or angry most of the day nearly every day and is observable by others.
- The temper outbursts and the irritable mood between outbursts lead to impairment that is present in at least two of three domains (home, school, peers), where the impairment is severe in at least one domain, and at least mild in another domain.
- These symptoms must be present for at least 12 months, without a 3-month break.
- Onset prior to age 10 and diagnosis cannot be made until after age 6.
- These symptoms cannot be better explained by another mental health disorder or physical or neurological condition, and the child must not have had mania or hypomania symptoms.

The prevalence of DMDD among youth is approximately 3% of the population [3]. At the same time, it is important to note that irritability as a construct is transdiagnostic. "Irritability" or "anger" appears as a symptom in 20 disorders in the DSM-5 [10]. Therefore, comprehensive assessment and evaluation are critical to (1) understand the etiology of the irritability, (2) design an effective treatment plan, and (3) determine whether conducting anger-inducing exposures (the novel and active ingredient of the treatment) is clinically indicated.

2.2 Rationale for Exposure-Based CBT + PMT for the Treatment of Severe Irritability in Youth

In this chapter, we present a novel approach to the treatment of severe irritability. There are several evidence-based treatments for disruptive behavior disorders, such as cognitive behavioral therapy (CBT) for anger and aggression [11] and parent management training (PMT) [12–18]. These treatments have been shown to be effective, but there are no interventions yet developed that specifically target the tonic and phasic irritability as that present in DMDD.

2.2.1 The Translational Model of Irritability

To better understand irritability, a translational model was developed to operationalize the neural/brain-based and behavioral deficits associated with irritability. Critically, this model was created to identify targets for new treatments and interventions [9]. The model was developed from evolutionarily conserved principles present in humans, as well as other animals.

The model focuses on the interplay between the child's environmental factors and related brain-mechanisms that lead to the mood (increased anger and frustration) and behavioral responses (increased motor activity and aggression) observed in irritability. One environmental factor relates to instrumental learning—"a type of learning where behaviors are strengthened or weakened by their consequences" [19]. For example, parents reward irritable behavior by "giving in" to child's outbursts (e.g., by allowing more screen time or no homework) which results in strengthening the likelihood of the child engaging in irritable behavior in the future, or where parents' inconsistent responses to the child's behaviors make it hard to know when to expect a positive or negative consequence, so that the child learns that irritable behavior causes parents to respond.

These instrumental learning deficits relate to the coercive pattern [20, 21] explanation for the development and maintenance of conduct behaviors in youth. Coercion theory [20, 21] describes a process of reciprocal reinforcement of aggressive behaviors between parents and child such that children learn that if they act out, they will likely get what they want (i.e., no chores or more attention from parent). For example, (1) a parent makes a request to the child; (2) the child refuses to comply with the request and/or responds in an aggressive manner; (3) the parent engages in a negative response such as raising their voice or threatening the child; (4) the child becomes more noncompliant, defiant, and/or aggressive; and (5) the negative feedback cycle of coercion escalates between both parent and child until one "wins" the exchange [22] in response to the other's coercive responses. In this cycle, both the parent and child learn that coercion works, and the parent's and child's maladaptive behaviors are reinforced [23–25]. The result of this negative cycle is that it reinforces negative, maladaptive behaviors. Consequently, the child learns to motivate others to act by using coercion, which can ultimately pervade other areas of the child's life, such as with teachers and peers.

2.2.2 Child Component: Theoretical Foundation

Youth with severe irritability tend to have challenges with reward and threat processing, which is mediated by widespread cortical and subcortical brain areas. Reward processing refers to how the brain reacts to stimuli that is rewarding, such as praise, earning a desired toy, accomplishing a goal, or a favorite food. Threat processing refers to how the brain reacts to stimuli that seem dangerous or scary, such as being yelled at, seeing an angry/hostile face, or confronting an unfamiliar, stressful situation. Youth with impairing irritability tend to overreact, underreact, or respond in unexpected ways to rewarding and threatening stimuli. They tend to have a lower threshold for experiencing a negative valence emotion, such as frustration/anger/irritability when an expected reward is not provided. Their anger occurs with greater intensity and for longer duration than typically developing youth.

Studies have shown that youth with irritability also tend to have a lower threshold for perceiving stimuli as threatening which causes them to respond with aggression (having a temper outburst) [26]. Thus, when confronted with what they perceive to be a threatening stimulus (e.g., a parent expressing anger to the child either with an angry face or using a strong tone of voice), they respond by approaching the threat with an angry outburst, as opposed to avoiding and/or moving away or retreating [9]. Youth with severe irritability are more sensitive to rewards and tend to have stronger behavioral responses, like temper outbursts, when they are prevented from getting what they want (e.g., a parent removing electronics), or when they do not receive a reward that they expect (e.g., parent says "no dessert" after dinner) [27].

Because irritability and anxiety have been found to overlap in their brain-related mechanisms and both involve responding to threatening stimuli in atypical ways, the translational model uses what is known about atypical threat processes in anxiety and applies it to irritability. Youth with anxiety exhibit atypical avoid responses, while youth with irritability exhibit atypical approach responses.

In both anxiety and irritability, stimuli or cues that are perceived as threatening trigger a sudden, momentary state of high levels of physiological arousal. In youth with anxiety, the high arousal state may manifest as worry, rumination, and avoidance. In youth with irritability, the high-arousal state may manifest as anger and/or frustration and increased motor activity such as behavioral aggression.

Exposure therapy has been exceedingly effective in the treatment of anxiety [28, 29]. Similarly, the novel treatment described in this chapter focuses on exposing the child to anger-inducing stimuli with the goal of increasing toleration. Through exposures, this treatment attempts to normalize the threat and reward processing deficits by increasing the child's threshold for anger and aggressive responses.

2.2.3 Exposure-Based Cognitive Behavioral Therapy: Theoretical Foundation

Exposure therapy consists of therapist and client working together to generate a "fear hierarchy" from least to most fear-inducing stimuli. The client is then tasked to confront or be "exposed" to fear-inducing stimuli. For example, a client with a fear of spiders may start treatment by looking at pictures of spiders while in session with the therapist. As clients make progress in tolerating their anxiety with the pictures of the spider in later sessions, the therapist assists clients working their way up the fear hierarchy, ultimately to tolerating the presence of spiders. With repeated exposure, the client learns not to avoid the fear-inducing stimuli (e.g., spiders) and learns new ways of responding to the stimuli (being present with a spider).

Exposure therapy has been mainly used to treat anxiety-related disorders. This novel treatment uses the same exposure therapy principles with youth with severe irritability. In this exposure-based treatment, youth with irritability are tasked to generate an "anger hierarchy" of anger-inducing stimuli and to work their way up the hierarchy to expose themselves to the range of anger-inducing stimuli. The child learns to acclimate to anger-inducing stimuli or triggers. Often, with repeated exposures, anger levels decrease when children with severe irritability are presented with the anger-inducing stimuli. For example, they no longer become angry when asked to stop a video game to do homework.

Prior to introducing exposures to anger-inducing stimuli, the therapist provides psychoeducation and discusses helpful and unhelpful ways to express anger. The aim is to normalize the feeling of anger for the child such that the child learns that feeling angry is not problematic but rather that maladaptive responses to anger, such as temper outbursts, are problematic. During the exposure, the therapist also works on increasing children's physiological awareness of their body cues or sensations when angry by asking them to identify where in the body they feel their anger. Having the child recognize their somatic signals for varying levels of anger helps them recognize their anger sooner and/or in the moment when it is occurring, so they can proactively deploy the distress tolerance and emotion regulation resources practiced during the exposures. Over time and with practice, children can learn positive adaptive responses to anger-inducing stimuli in "real life" settings.

2.2.4 Parent Component: Theoretical Foundation

In this novel treatment, deficits observed in instrumental learning in youth with irritability are further addressed by drawing on principles from parent behavior management training (PMT), which has been shown to be efficacious for disruptive behavior disorders [13, 15]. PMT teaches parents strategies to manage children's behaviors and targets the rules or consequences that parents use to influence their child's behaviors. In the development of this novel treatment, it was hypothesized that integrating both a child component using exposure to anger-inducing stimuli to address threat and reward processing deficits, and a parent component using PMT to address instrumental learning deficits, would lead to less severe and/or shorter responses to the anger-inducing stimuli and fewer temper outbursts.

3 Interventions

3.1 Case Example

To help with understanding and contextualizing the type of child who may benefit from this novel treatment, below is a case example of a child named: "Junior." This example is fictionalized and is based on the typical cases treated with this treatment.

Junior is a 9-year-old boy with a lifelong history of chronic irritability and temper outbursts. At age 4, Junior was asked to leave preschool because of disruptive behaviors and was diagnosed with attention deficit hyperactivity disorder (ADHD). Despite starting treatment for ADHD, he continued to experience irritability and temper outbursts when frustrated. He frequently became frustrated when told to do something he didn't want to do, such as chores. Transitions were hard for Junior, especially, when he needed to stop doing something enjoyable, like playing outside, to start homework or to pick up his room.

His temper outbursts tended to be mostly verbal—yelling, screaming, name-calling, and sometimes cursing. He had at least one physical outburst a week, typically involving his younger brother, whom he described as "annoying." When the outbursts became physical, he would slam doors, throw things in anger and kick, or hit his brother and sometimes, his parents. He also has a history of yelling at and sometimes hitting peers and received very few invitations to parties or playdates. At school, he often yelled at teachers when feeling he was unfairly asked to stop talking or when he perceived the schoolwork as "unnecessary or unfair." Parents and teachers described feeling like they were "walking on eggshells" to avoid him having an outburst. To have peaceful morning and evening routines, parents were making many accommodations, such as decreasing expectations for him to help with chores or complete schoolwork. During the day, Junior was mostly grumpy. On many days, it seemed like he "just wakes up on the wrong side of the bed." Junior was diagnosed with disruptive mood dysregulation disorder (DMDD) because of his chronic irritability and the frequency, severity, and duration of his temper outbursts along with significant impairment due to irritability at home, school, and with peers.

3.2 Child Component: Exposure-Based CBT

3.2.1 Phase 1

The exposure-based CBT treatment consists of a total of 12 sessions with the child and 12 sessions with the parents which are done during the same time period. This treatment is primarily focused on behavioral interventions, with exposures to anger-inducing stimuli as the core ingredient.

Table 1
Phases of child sessions and therapist tasks

Phase	Sessions	Therapist tasks
Phase 1	Sessions 1–3	Psychoeducation, goal setting, motivational interviewing, generating anger hierarchy, operationalizing ok ways and not ok ways to show anger
Phase 2	Sessions 4–10	Engaging in exposure-based behavioral interventions *As needed, motivational interviewing and assessing buy-in
Phase 3	Sessions 11–12	Processing termination, focus on maintenance of gains and encapsulating learning, discussion of aftercare plan

The child component of the treatment, the exposure-based CBT, can be divided into three phases (see Table 1). The first phase involves acclimatizing the child to treatment (including the structure and process for the sessions) and discussing the child's goals in terms of their anger, irritability, and frustration. This primer is particularly important for children who have not participated in CBT treatment previously and may not be accustomed to the active and behaviorally oriented nature of this treatment. As previously mentioned, during this initial phase of treatment, the child receives psychoeducation on irritability; the connection between thoughts, feelings, and behaviors; and normalizing the experience and prevalence of irritability in youth. The child is also oriented to the brain-body connections, with emphasis on the fight or flight response in relation to anger-inducing stimuli, and body cues/signals for anger/irritability at different levels of anger/irritability ranging from *1–mildly angry or irritable* to *10–very angry or irritable*. During this initial phase of treatment, the therapist also uses motivational interviewing techniques to assess the child's motivation and evoke their level of commitment to the exposure treatment. The therapist engages the child in a discussion about the positives/pros of their anger/irritability and the negatives/cons and utilizes the "motivation ruler" to assess readiness for change from 1–*not at all ready to change my anger* to 10–*really working on changing my anger* to quantify how motivated or ready the child feels to engage in this treatment. If indicated, the therapist will guide the child with exploring factors that could possibly move their rating closer to feeling fully ready at a "10" in order to identify and resolve their ambivalence, so they can move forward with the exposure treatment. This motivation assessment tool is revisited throughout the child's 12 treatment sessions as motivation for change may ebb and flow depending on the content and tasks in the treatment course as well as therapeutic factors that may arise. Along with motivational interviewing, relaxation techniques such as deep breathing are also reviewed with the child during this initial phase of treatment as a tool that a child can activate when they begin to feel dysregulated.

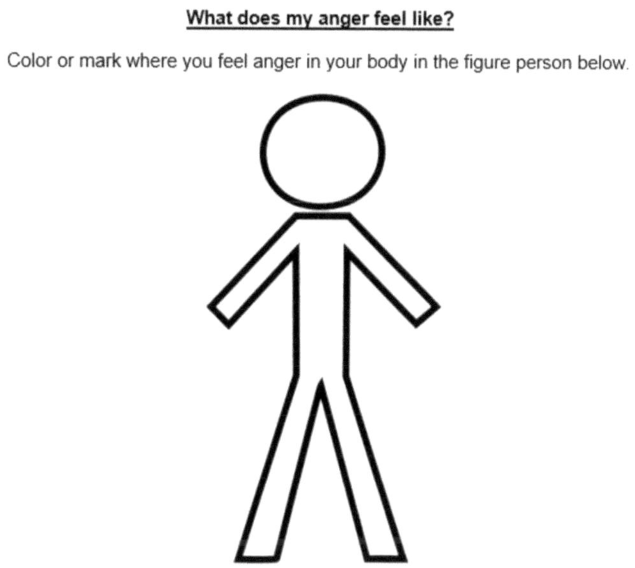

Fig. 1 *What does my anger feel like?* Worksheet

To help children better understand the mind/body connection, a primary worksheet used in session is the "*What does my anger feel like?*" worksheet (Fig. 1). In this worksheet, the child identifies body areas where they feel anger. Some examples of responses include arms and legs or mouth feeling active, sometimes feeling "heat" in their gut or stomach or chest, or head feeling foggy and full of angry thoughts. For example, when using this worksheet, Junior identified feeling anger mostly in his hands, since he tended to ball his hand into a fist when angry; on his feet and legs, since he often wanted to kick or stomp; and in his mouth, since he tended to yell or name call. This worksheet helps children to become more aware of their body's sensations and, potentially, to track changes in sensations activated by their anger.

3.2.2 Phase 2

At the start of treatment, as part of the orientation to anger, the child is guided in developing an anger hierarchy of the child's triggers, ranging from 1–*mildly angry or irritable* to 10–*very angry or irritable* (Fig. 2). This anger hierarchy provides the framework for the middle phase of the treatment, which is focused on anger-inducing exposures tailored to the child's specific perceived triggers. While the anger triggers/stimuli can be wide-ranging and abundant, common examples include parents asking child to stop using an electronic device, complete a chore (e.g., doing dishes, sorting laundry), practice a musical instrument, complete an activity of daily living (i.e., brushing teeth, taking a shower, getting hair done), and start homework; parents or siblings engaging in "annoying" quirks (e.g., humming, singing, making faces, or repeating an innocuous phrase), losing a game (e.g., video game,

Fig. 2 Anger hierarchy

basketball, soccer, board game), being rushed, not getting what the child wants (e.g., no ice cream, no screens, or parents refusal to purchase a specific toy or game), not being passed the ball at a team sport, and having to discuss a difficult or contentious topic with a parent; etc.

During exposure sessions, the therapist slowly introduces anger-inducing situations from the child's hierarchy, starting from those on the lowest end (e.g., for Junior, having to repeat himself was labeled at a level 2 on the anger hierarchy) and moving up to the higher/highest levels (e.g., having to get off electronics to start homework for Junior was labeled at a 9) on the hierarchy as the child demonstrates progress in their level of tolerance of the triggers. During each session, the thermometer is used for the anger hierarchy to assist the child in creating language to describe their emotional experience and identify the intensity of anger at various points in time (i.e., "What's your temperature rating?"). The anger hierarchy is revisited throughout the treatment course to also identify whether there has been a change in the assessment/rating of initial triggers and/or to determine if new triggers have emerged. It is also used to process and identify ways in which the child has learned to cope and respond to certain triggers.

Before exposures are introduced, the therapist collaborates with the child to identify potential tailored adaptive responses to anger to practice during the exposure. The "OK ways and NOT OK ways to show anger" worksheet (Fig. 3) is another activity critical in developing a foundation prior to the start of the exposure-focused phase of the treatment to help identify behaviors the child should attempt to inhibit during exposure to triggers and

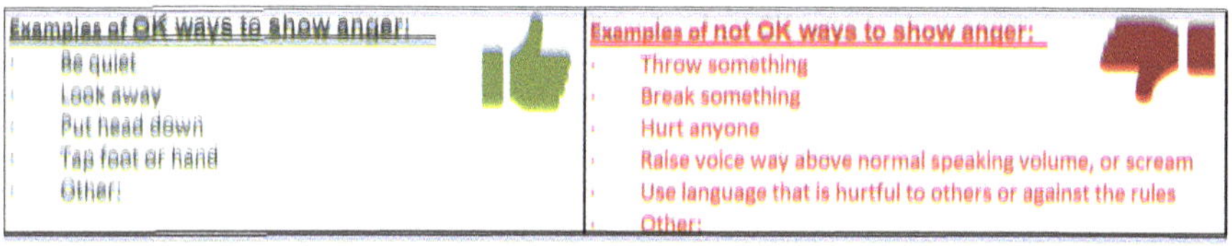

Fig. 3 OK ways and NOT OK ways to show anger worksheet

Table 2
Steps for exposure therapy for irritability in session

1	Prepare, select exposure collaboratively with child and identify target behavior
2	Child predicts highest level of anger (1–10) during exposure
3	Collect baseline anger rating before exposure starts
4	Engage in the exposure and rate anger every minute during exposure
5	Debrief exposure

to provide alternative, more adaptive, behaviors to engage in while angry. This exercise of identifying adaptive ways to express anger helps the child better understand the behavioral targets of the exposures.

One of the main goals of this exercise is to normalize anger while highlighting the importance of expressing it in a more adaptive way. Considering each family's way of interacting, it is often helpful to invite caregivers to be a part of developing the list of "OK ways to show anger" to ensure the responses are most relevant to their family's mode of engagement. For example, for Junior, some *ok* ways to express anger were staying silent, walking away, going to his room, breathing deeply, and frowning; and some *not ok* ways to express anger were yelling, name-calling, cursing, slamming doors, shoving, kicking, or hitting mom/dad/peers.

Most of this second phase is then spent conducting exposures with the child. Each exposure has a defined structure (Table 2). Prior to each exposure, the therapist and child determine the target behavior during the exposure (often based on the anger hierarchy). For example, for Junior, a target behavior during an exposure on having to stop playing a game to complete a complicated math homework was staying quiet during the entirety of the exposure. Once the target behavior(s) is identified, the child predicts the highest level of anger that they believe they may experience (for Junior, this was 8). At the start of the exposure, the child shares baseline anger rating at time 0 (for Junior, this was 1). Then the exposure begins. During the exposure, the therapist collects anger ratings in 1-minute intervals to understand the trajectory of the child's anger during the entirety of the exposure/situation. The

Exposure # within session (1) 2 3

Exposure Description: Having to do homework

Predicted Temperature Rating: 8

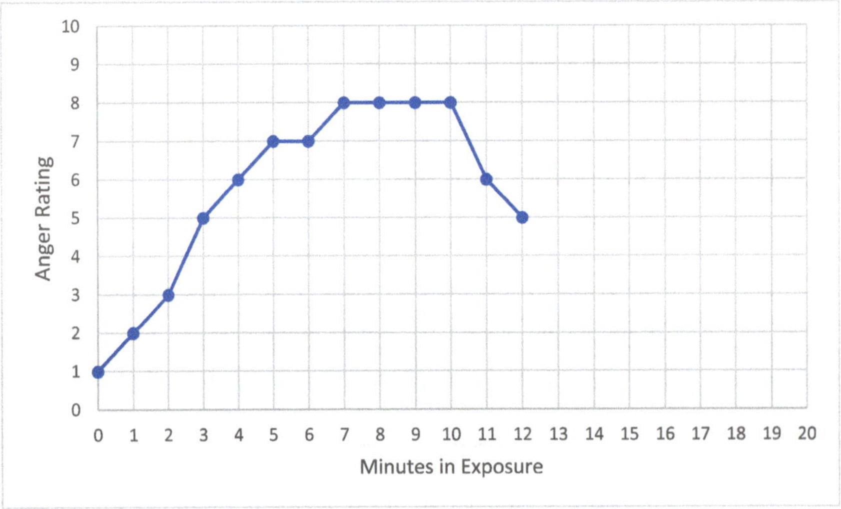

Fig. 4 Sample exposure graph of an exposure of child having to do homework

therapist avoids any distractions that would possibly disrupt the exposure or impact and/or dampen the child's anger response, such as talking about an unrelated topic, or fiddling with an object, toy, phone, or instrument. Once the exposure is completed, the therapist debriefs with the child to encapsulate learning and provides the graph collected during the exposure to show the trajectory of their anger (see Fig. 4 for an example of the exposure sheet for an exposure with Junior on completing his homework). A significant point of discussion during the debrief is the child's predicted highest anger rating compared to their actual highest anger rating, especially as treatment progresses. For some children, across multiple exposures, the trajectory begins to flatten out and becomes lower, suggesting an overall decrease in anger. At the end of each session, the therapist provides a review of what was learned and develops relevant practice homework in collaboration with child based on the exposures done in session. For example, a home practice exposure for Junior was to start his homework as soon as he arrived home before engaging with electronics; parents were encouraged to facilitate these exposures at home.

3.2.3 Phase 3

In the final phase of treatment for the child, there is emphasis on the encapsulation of learning, maintenance of skill development, and preparation for and processing of termination. For example, for Junior, one of the main takeaways that he shared was "you can be as angry as you would like, it's what you do with it that matters."

Therefore, the final phase would capitalize on this learning and assure that this learning generalizes across multiple future anger-inducing situations. In conjunction with the parent component, the final phase of the packaged treatment involves developing an optimal personalized aftercare plan for the child which includes making appropriate referrals, as needed.

3.3 Parent Component: Parent Management Training (PMT)

Parent sessions begin with assessment of how the family approaches the child's irritability, particularly, understanding the accommodations that the parents make to avoid the child's outbursts. Assessing the parents' accommodations is helpful in determining target behaviors to address with the parents. For example, as described in the Junior example, parents were making accommodations of not asking Junior to complete his homework or chores and allowing him unlimited access to electronics to "maintain the peace." Therefore, the PMT involved addressing with parents how to properly command child to complete chores and homework, how to set appropriate limits on electronics use, and how to address noncompliance and handling temper outbursts. Like the child's treatment, motivational interviewing is utilized in this initial phase (and throughout treatment) to increase buy-in from the parents to enhance their intrinsic motivation to change some of their parenting strategies and to maintain the parenting approaches that have proven to be helpful during the treatment.

The parent component of this novel treatment is modeled like established parent behavior management training programs [15], in which the parents are presented with learning modules (see Table 3) with the goal of teaching concrete skills to help manage their child's behaviors, increasing consistency, decreasing parental stress, increasing parents-child positive interactions, and, ultimately, decreasing the child's irritability.

The structure of each parent session includes learning a new skill, brainstorming on how to apply or tailor it to the individual child, role-playing the skill in session, addressing potential barriers, troubleshooting in advance, and assigning at-home practice worksheets and exercises related to the skills being targeted that week. The topics covered (see Table 3) in this treatment are based on principles of instrumental learning, specifically, contingencies for behaviors. In this approach, parents are first taught how antecedents for and consequences of child behaviors either increase or decrease reinforcement of those behaviors and, subsequently are taught how to implement more productive consequences for and/or responses to their child's behaviors. For example, for Junior's parents, exploring antecedents and consequences revealed a pattern of positively reinforcing outbursts by parents "giving in" to Junior. Parents also tended to be inconsistent with providing non-reinforcing consequences to noncompliance. For example, when parents asked Junior to stop using electronics to start the

Table 3
Topics for parent component sessions

Module	Topic
1	Assessment and motivational interviewing
2	Introduction to parent intervention
3	Instrumental learning
4	Praise/acknowledgment
5	Active ignore
6	Dealing with outbursts
7	Commands
8	Limit setting
9	Pleasant interactions and unexpected rewards

bedtime routine, Junior would become dysregulated and have a temper outburst, and parents in response would let him continue to use his electronics. Parents indicated that they had a hard time tolerating the child being dysregulated and preferred to "maintain the peace."

A key factor of the parent component is the parallel process of the child learning to tolerate their anger, while the parents similarly learn how to engage in their own emotion regulation and distress tolerance through the process of ***actively*** ignoring their child's temper outbursts that do not pose a safety concern. As treatment progresses (for both the child and parents), the child's exposures may begin to involve scenarios that are enhanced with parental participation for a dual purpose of having the child tolerate frustrating situations while also allowing the parents to practice how to better navigate their child's maladaptive behaviors utilizing the PMT skills they have learned. For example, when conducting the exposure of Junior stopping a game to start homework, the parents can participate in this exposure by providing the command that leads to the transition. This joint exposure can serve as an opportunity for parents to learn how to provide effective commands. If the child becomes dysregulated during the exposure, then the parents learn to tolerate the child's outburst through active ignoring. If the child complies, parents can be taught how to positively reinforce compliance through praise/acknowledgement of positive behaviors. This example shows how to both conduct an exposure that addresses the child's trigger while also coaching the parent through their learning of the PMT skills.

Like the child's final phase of treatment, the termination process for the parents includes discussion around maintenance of gains and how to troubleshoot issues in the future. In addition, referrals are given to parents, as appropriate, for the continuation of their own support (emotional and/or parent coaching related) posttreatment.

4 Discussion

The treatment presented in this chapter continues to be refined and tested. Critically, it is not recommended for parents to try this treatment on their own. Before any treatment can begin, therapists assess safety and any risks to safety and develop a safety plan for child with the family. Safety is assessed throughout the treatment. It is important that the therapist examines fit of the child with the treatment. Some questions that would need to be considered before starting this treatment are: "Would this child benefit from conducting anger-inducing exposures?" For example, if the child tends to engage in proactive aggression [30] or shows callous or unemotional traits [31], conducting exposures would be contraindicated. In these cases, the child may be deriving satisfaction from engaging in anger-inducing exposures which has the potential of increasing the irritability. Another question is "are there underlying issues that might make the treatment ineffective for this child?" For example, children with significant inflexibility, intolerance to change, significant deficits in verbal and nonverbal social communication skills, or significant intellectual disability may be limited in their ability to understand the rationale for intentionally angering the child for the purposes of muting their responses to irritability. "What would exposures look like, and could they benefit this child?" To answer this question, therapists need to consider the triggers and the therapist's ability to expose the child to these triggers in session and through home practice. For example, for a child whose main irritability trigger is their inability to understand the intentions of others, conducting exposures to increase perspective taking may be ineffective. "What is the source of irritability?" Irritability can develop from multiple sources. For some youth, the source of the irritability may be a specific stressor, for others, it may be directly related to family circumstances or family dynamics, whereas for others, the irritability may stem from an underlying depressive disorder or anxiety disorder. Therefore, the best approach to treat the child's irritability would be to address and treat the direct source or cause of the irritability through evidence-based treatments and practices that would better address the root source. "Are there other issues to be addressed (outside of treatment) before starting the exposure treatment?" For example, if the child shows significant challenges academically and most of the irritability is present in the school setting or when engaging in schoolwork at home, it might be best to consider academic testing and any recommendations that result from the testing to explore if the symptoms abate before starting a round of CBT plus parent management training. "Is there another issue that needs to be triaged first or that is currently more pressing or primary?" Examples for answers to this question are school refusal, or a primary diagnosis that is more impairing currently than the child's irritability.

Importantly, when considering the first line of defense to treat a child's most pressing clinical problem (e.g., anxiety, depression, PTSD, ADHD), it is critical to know the specific diagnosis to understand what is being targeted, and since there are already empirically supported treatments for many psychological disorders (e.g., anxiety, depression, PTSD), it's clinically indicated to explore those first and talk with a mental health professional.

4.1 On the Efficacy of the Treatment

All families that started the treatment completed the full course of the treatment without dropping out, providing support for the acceptability and feasibility of this approach to treating youth with severe irritability. The initial pilot study [32] and the more recent study with 40 participants [33] have shown significant results in reducing irritability symptoms, such as irritable mood and temper outbursts (all $\beta > -0.04$, $ps < 0.011$, Cohen's d range: -0.33 to -0.98), and improving overall level of functioning ($p = 0.035$) [33]. The significant reduction in irritability symptoms was maintained at the follow-up assessment (6 months after the treatment ended; all β (39) < -0.001, $ps > 0.400$) [33]. Overall, youth are learning new, more adaptive ways of responding to anger-inducing situations that do not result in yelling, screaming, or problematic physical behavior. Clinical trials of this treatment are ongoing with active recruitment of participants. While this treatment method has demonstrated promise as an effective treatment for severe irritability in youth, we have yet to define specific mechanisms leading to improvement and to test their efficacy in head-to-head trials.

5 Conclusions and Future Directions

As stated, the current trial is ongoing. In the current trial, mechanisms are being explored; psychophysiological data is being collected, specifically heart rate and heart rate variability, to determine changes in these metrics as exposures progress and how psychophysiological changes relate to changes in clinical outcomes [34]. Moreover, real-time, in vivo emotional and mood assessments provide naturalistic data to gather the daily dynamics of mood change over the course of the day and throughout the week during treatment [35]. Interviews are being conducted with the child and parents undergoing the treatment about their perspective on the causes of change in this treatment. Interviews with the families on their attributions for symptom change may help in understanding the parents and child perceived mechanisms for change, which can include extra therapeutic factors, to optimize the treatment. Finally, pre- and posttreatment neural data (e.g., fMRI, diffusion tensor imaging) is being collected to assess changes in the putative brain mechanisms implicated in irritability and targeted by the treatment (c.f., [36]). Further work is needed to understand the neural mechanisms mediating improvement to develop more targeted treatments. Future work should continue to examine the use of exposure-based treatments for irritability, and specifically the unique effects of exposure versus parent management training on the clinical outcomes of youth with irritability and determine if pretreatment clinical, behavior, or neural factors can predict what treatment will be most effective for a particular child struggling with irritability.

Conflicts of Interest

The development and scientific investigation of the novel treatment approach presented in this chapter is supported by the Intramural Research Program (IRP) of the National Institute of Mental Health, National Institutes of Health (ZIAMH002786-15, ZIAMH002778-17), and conducted under NIH Clinical Study Protocols 02-M-0021 and 15-M-0182 [ClinicalTrials.gov identifier: NCT02531893 (Brotman)].

References

1. Copeland WE et al (2013) Prevalence, comorbidity, and correlates of DSM-5 proposed disruptive mood dysregulation disorder. Am J Psychiatry 170(2):173–179
2. Brotman MA et al (2006) Prevalence, clinical correlates, and longitudinal course of severe mood dysregulation in children. Biol Psychiatry 60(9):991–997
3. Copeland WE et al (2014) Adult diagnostic and functional outcomes of DSM-5 disruptive mood dysregulation disorder. Am J Psychiatry 171(6):668–674
4. Stringaris A et al (2009) Adult outcomes of youth irritability: a 20-year prospective community-based study. Am J Psychiatry 166(9):1048–1054
5. Stringaris A et al (2010) Pediatric bipolar disorder versus severe mood dysregulation: risk for manic episodes on follow-up. J Am Acad Child Adolesc Psychiatry 49(4):397–405
6. American Psychiatric Association, D. and D. American Psychiatric Association (2013) Diagnostic and statistical manual of mental disorders: DSM-5, vol 5. American Psychiatric Association, Washington, DC
7. Leibenluft E, Stoddard J (2013) The developmental psychopathology of irritability. Dev Psychopathol 25(4pt2):1473–1487
8. Perepletchikova F et al (2017) Randomized clinical trial of dialectical behavior therapy for preadolescent children with disruptive mood dysregulation disorder: Feasibility and outcomes. J Am Acad Child Adolesc Psychiatry 56(10):832–840
9. Brotman MA et al (2017) Irritability in youths: a translational model. Am J Psychiatry 174(6): 520–532
10. Leibenluft E et al (2024) Irritability in youths: a critical integrative review. Am J Psychiatry 181(4):275–290
11. Sukhodolsky DG et al (2016) Behavioral interventions for anger, irritability, and aggression in children and adolescents. J Child Adolesc Psychopharmacol 26(1):58–64
12. Augimeri LK et al (2007) The SNAP™ under 12 outreach project: effects of a community

based program for children with conduct problems. J Child Fam Stud 16:799–807
13. Barkley RA (2013) Defiant children: a clinician's manual for assessment and parent training. Guilford Press
14. Eyberg SM, Robinson EA (1982) Parent-child interaction training: effects on family functioning. J Clin Child Adolesc Psychol 11(2): 130–137
15. Kazdin AE (2017) Parent management training and problem-solving skills training for child and adolescent conduct problems. In: Evidence-based psychotherapies for children and adolescents, vol 3. Guilford Press, pp 142–158
16. McCart MR et al (2006) Differential effectiveness of behavioral parent-training and cognitive-behavioral therapy for antisocial youth: a meta-analysis. J Abnorm Child Psychol 34:525–541
17. Menting AT, de Castro BO, Matthys W (2013) Effectiveness of the incredible years parent training to modify disruptive and prosocial child behavior: a meta-analytic review. Clin Psychol Rev 33(8):901–913
18. Thomas R, Zimmer-Gembeck MJ (2007) Behavioral outcomes of parent-child interaction therapy and triple P – positive parenting program: a review and meta-analysis. J Abnorm Child Psychol 35:475–495
19. Williams W (2018) Instrumental learning. In: Encyclopedia of animal cognition and behavior. Springer, Cham
20. Patterson GR, Oregon E (1982) A social learning approach, Volume 3: coercive family process.
21. Patterson GR (2016) Coercion theory: The study of change. The Oxford handbook of coercive relationship dynamics 1:7–22
22. Smith JD et al (2014) Coercive family process and early-onset conduct problems from age 2 to school entry. Dev Psychopathol 26(4pt1):917–932
23. Moed A (2024) An emotion-focused extension of coercion theory: emerging evidence and conceptualizations for parental experienced emotion as a mechanism of reinforcement in coercive parent–child interactions. Child Dev Perspect 18(2):82–87
24. Braungart-Rieker J, Garwood MM, Stifter CA (1997) Compliance and noncompliance: the roles of maternal control and child temperament. J Appl Dev Psychol 18(3):411–428
25. Snyder J, Schrepferman L, St. Peter C (1997) Origins of antisocial behavior: negative reinforcement and affect dysregulation of behavior as socialization mechanisms in family interaction. Behav Modif 21(2):187–215
26. Crick NR, Dodge KA (1994) A review and reformulation of social information-processing mechanisms in children's social adjustment. Psychol Bull 115(1):74
27. Tseng W-L et al (2019) Brain mechanisms of attention orienting following frustration: associations with irritability and age in youths. Am J Psychiatry 176(1):67–76
28. Craske MG et al (2014) Maximizing exposure therapy: an inhibitory learning approach. Behav Res Ther 58:10–23
29. Foa EB, McLean CP (2016) The efficacy of exposure therapy for anxiety-related disorders and its underlying mechanisms: the case of OCD and PTSD. Annu Rev Clin Psychol 12(1):1–28
30. Miller JD, Lynam DR (2006) Reactive and proactive aggression: similarities and differences. Personal Individ Differ 41(8): 1469–1480
31. Frick PJ, White SF (2008) Research review: the importance of callous-unemotional traits for developmental models of aggressive and antisocial behavior. J Child Psychol Psychiatry 49(4):359–375
32. Kircanski K et al (2018) Psychosocial treatment of irritability in youth. Curr Treat Options Psychiatry 5:129–140
33. Naim R et al (2024) An exposure-based cognitive–behavioral therapy for youth with severe irritability: feasibility and preliminary efficacy. J Clin Child Adolesc Psychol 53(2): 260–276
34. Grasser LR et al (2023) Can peripheral psychophysiological markers predict response to exposure-based cognitive behavioral therapy in youth with severely impairing irritability? A study protocol. BMC Psychiatry 23(1):926
35. Naim R et al (2021) Using ecological momentary assessment to enhance irritability phenotyping in a transdiagnostic sample of youth. Dev Psychopathol 33(5):1734–1746
36. Haller SP et al (2024) Cognitive behavioral therapy normalizes fronto-parietal activation in unmedicated patients with pediatric anxiety disorders. Am J Psychiatry 181(3):201

Chapter 12

Cognitive Behavioral Therapy with Eating Disordered Youth

Jamal H. Essayli

Abstract

Current guidelines recognize family-based treatment (FBT) as the leading therapy for children and adolescents with eating disorders (EDs), with cognitive-behavioral therapy (CBT) recommended as a second-line option. Both approaches offer distinct advantages, and integrating components of CBT and FBT may optimize outcomes. This chapter describes how to combine CBT interventions with calibrated family support to promote and maintain adaptive eating behaviors in youth with EDs. CBT for children and adolescents with EDs employs interventions similar to adult treatment, aiming to modify eating-, weight-, and body shape-related beliefs and behaviors through developing a clinical formulation, psychoeducation, motivation enhancement, open weighing, meal planning, self-monitoring, cognitive restructuring, exposure therapy, and relapse prevention. New to this edition, this chapter also presents innovative modifications to some of these CBT interventions to target beliefs and emotions about the consequences of weight gain. These interventions are augmented with flexible family involvement, adjusting the frequency and format of family sessions based on the patient's age, symptom severity, and clinical needs. Particularly during the early phase of treatment for underweight patients, family sessions focus on coaching the patient's caregivers to support their child with normalized eating and weight restoration. As patients become more active participants in treatment and behavior change, family involvement is reduced gradually.

Key words Cognitive-behavioral therapy (CBT), Eating disorders, Children, Adolescents, Anorexia Nervosa, Bulimia nervosa, Family therapy, Family-based treatment (FBT)

1 Introduction

Eating disorders (EDs), including anorexia nervosa (AN), bulimia nervosa (BN), and binge-eating disorder (BED), are complex psychological conditions associated with serious medical complications [1], frequent relapse [2–4], and high risk for suicide [5, 6]. This is particularly true of AN, which ranks among the most dangerous psychiatric disorders [7]. Patients with EDs also demonstrate distinctive features that further complicate the clinical picture and can thwart successful treatment, including positive valuation of core symptoms and ambivalence about the prospect of change, as well as the psychological consequences of semi-starvation [8, 9].

Robert D. Friedberg and Micaela A. Thordarson (eds.), *Cognitive Behavioral Therapy with Youth: Tradition and Innovation*, Neuromethods, vol. 231, https://doi.org/10.1007/978-1-0716-5134-6_12,

Individuals with AN and BN, and a substantial proportion of those with BED, hold core beliefs about the importance of weight and/or shape as measures of self-worth, which contribute to the development of stereotyped behaviors designed to control body weight [10, 11]. These behaviors typically include dietary restraint, binge eating, and/or compensatory behaviors such as self-induced vomiting, laxative abuse, and excessive exercise. These symptoms can emerge across ED diagnoses: While AN is the only ED characterized by subnormal body weight, purging may occur in individuals with AN or BN and binge eating in those with AN, BN, or BED [12]. The diagnostic categories of "other specified feeding or eating disorder" (OSFED) and "unspecified feeding or eating disorder" describe individuals who experience disordered eating but do not meet full criteria for AN, BN, or BED. Two examples of diagnoses subsumed under the OSFED category including purging disorder, which is characterized by recurrent purging behavior in the absence of binge eating, and atypical AN, which is characterized by the same symptoms of AN (e.g., restrictive eating, fear of weight gain) with the exception of low body weight [12].

In contrast to BED, AN, BN, and OSFED typically emerge during adolescence or young adulthood [13–15]. Although EDs can occur in prepubescent children, most pediatric patients with EDs are adolescents: The great majority of patients with AN are at least 13 years old, while those with BN tend to be 15 years or older [16–18]. The fifth edition of the Diagnostic and Statistical Manual of Mental Disorders (DSM-5) also includes several feeding and eating disorders that typically present during childhood, including avoidant/restrictive food intake disorder (ARFID), pica, and rumination disorder [12]. Due to space limitations, the later characteristic age of onset for BED, and the dissimilar psychological mechanisms underlying ARFID, pica, and rumination disorder, the current chapter will focus specifically on adapting CBT for children and adolescents with AN, BN, and their subthreshold presentations classified under OSFED.

2 Theoretical and Empirical Foundations

Cognitive-behavioral theorists have proposed that the overvaluation of weight and shape and their control represent the "core psychopathology" of EDs [10, 19]. According to this theory, individuals with EDs learn to evaluate themselves largely or exclusively on the basis of dietary restraint, weight, and/or body shape. These core beliefs result in disordered eating- and weight-related "safety" behaviors, including severe calorie restriction, food avoidance, rigid dietary rules and rituals, excessive body checking, and body avoidance [20]. Safety behaviors reduce anxiety in the short term but maintain anxiety over the long term, as the nonoccurrence of the feared outcome (e.g., weight gain) is attributed to safety behaviors such as avoidance of specific foods or conformity to dietary rituals [21]. Dietary restraint and rigid rules about eating in turn predispose individuals with EDs to subjective overeating and objective binge-eating episodes, which are commonly associated with intense distress and the subsequent employment of harmful compensatory measures such as further dietary restraint, fasting, self-induced vomiting, laxative abuse, diuretic misuse, and excessive exercise [10]. Compensatory behaviors are maintained through negative reinforcement, as they serve to reduce eating- and weight-related anxiety [20, 22]. Additionally, successful dietary restraint and weight loss can be positively reinforced by feelings of euphoria, accomplishment, pride, power, identity, and self-control [8, 9, 23–27].

There is a strong basis for postulating that CBT is well-matched to the prominent cognitive and behavioral features of EDs. Distorted beliefs about eating, weight, and shape play a central role, avoidance of anxiety-provoking situations and behaviors maintains ED symptoms, and convictions about the value of dietary restraint and thinness contribute to ambivalence about change. CBT employs cognitive, exposure-based, and motivational interventions to modify these maladaptive behaviors, beliefs, and assumptions about eating, weight, and body shape [9, 19, 22, 28, 29]. Randomized controlled trials (RCTs) support CBT's efficacy for adults with BN, demonstrating more reliable and durable symptom remission compared to alternative psychological and pharmacological interventions. However, outcomes are less encouraging for adults with AN. As with all examined interventions for adults with AN, CBT yields lower recovery rates compared to treatment outcomes for adults with BN [30, 31].

The leading treatment for adolescents with EDs is not CBT, but family-based treatment (FBT), originally known as the Maudsley Model [32–34]. FBT shares a few similarities with CBT: Both treatments emphasize that the family is not to blame for the child's illness, agree that parents are responsible for ensuring their child's safety, and incorporate family involvement. However, their approaches have fundamental differences. FBT requires full family participation in all therapy sessions, whereas CBT involves parents more flexibly in a subset of sessions [22, 35, 36]. Unlike CBT's cognitive-behavioral framework, FBT takes a theoretically agnostic stance, viewing the ED as an invading force that has overtaken the child's life, and initially directs parents to take full control over their child's eating behaviors [34]. While FBT relies on an external control model, CBT respects patient autonomy, values the adolescent's perspective, and encourages active participation in treatment decisions from the start [35].

RCTs support the use of FBT above and beyond other treatments for adolescents with EDs [37], including adolescent-focused individual therapy [37, 38] and supportive psychotherapy [39]. However, research directly comparing FBT to CBT is limited. One nonrandomized trial for adolescents with restrictive EDs (e.g., AN, atypical AN) found that FBT led to greater weight gain than CBT, though both treatments resulted in comparable improvements on measures of ED psychopathology [40]. The only RCT comparing these treatments for adolescents with BN demonstrated that FBT yielded higher rates of abstinence from binge eating and purging at both end of treatment and 6-month follow-up [41]. Based on the existing evidence, current guidelines recommend FBT as the first-line treatment for adolescent with EDs, with CBT indicated for cases when FBT is either unacceptable or ineffective [35, 42].

Initial evidence suggests that FBT holds some advantages over CBT for adolescents with AN and BN, particularly in restoring weight and reducing behavioral symptoms. These advantages appear to be linked to FBT's external control framework. Although treatment ambivalence is common across age groups [9], adolescents with EDs who are identified early in their illness frequently exhibit heightened resistance to treatment [43]. For many adolescents, their limited motivation, coupled with developing cognitive maturity and decreased exposure to the adverse long-term effects of EDs, may necessitate an external control approach to initiate behavior change. This approach parallels recommendations for treatment-resistant adults with EDs, who often require inpatient care where meal completion and weight restoration occur under direct staff supervision [44]. The initial phase of FBT effectively translates aspects of the hospital-based model for EDs to the outpatient setting, with parents assuming a role analogous to hospital staff in supervising meals and monitoring eating behaviors. The parents' effectiveness in this supervisory role appears to be critical for favorable outcomes, with research identifying parental self-efficacy as a predictor of positive response to FBT [45, 46].

Despite FBT's demonstrated advantages, more than half of adolescents in clinical trials fail to achieve full recovery [37, 38, 41]. Even among patients who show behavioral improvements (e.g., weight gain, decreased purging), many continue to experience persistent and impairing psychological ED symptoms (e.g., eating and weight-related anxiety) [47]. While CBT appears theoretically well-suited to address both the cognitive and behavioral features of EDs, initial evidence suggests that FBT may yield superior outcomes for adolescents. Moreover, even in populations where CBT demonstrates its greatest efficacy—such as adults with BN—approximately half of participants in RCTs remain symptomatic at follow-up [30, 31, 48].

These findings suggest the need for meaningful modifications to existing CBT approaches to improve recovery rates for both adolescents and adults with EDs. One potential modification involves shifting CBT's focus from beliefs about the *likelihood* of weight gain to beliefs about the *consequences* of weight gain (e.g., "I will permanently hate myself if I gain weight"). For instance, standard CBT protocols employ "open weighing" (described later in this chapter) to challenge anxious predictions about the likelihood of weight gain (e.g., the belief that regular eating will lead to significant weight gain) [29, 49, 50]. This narrow focus on the likelihood of weight gain presents at least three limitations. First, it proves particularly challenging to dispute weight gain predictions in underweight and weight-suppressed patients who are, in fact, gaining significant weight in treatment. Second, teaching patients that weight gain is less likely to occur than predicted (e.g., "it looks like eating desserts does not lead to weight gain like I feared it would") may inadvertently reinforce the underlying belief that weight gain is problematic. Third, this approach fails to help patients learn that *weight gain itself* is both tolerable and acceptable [51, 52].

Shifting CBT's focus to beliefs about the consequences of weight gain necessitates adaptations to at least three CBT interventions: open weighing, cognitive restructuring, and exposure therapy. These adaptations are detailed in this chapter's second edition. Given the apparent advantages of parental participation in treatment for adolescents with EDs, this chapter recommends that therapists combine these adapted CBT interventions with flexible levels of family involvement, calibrated to each adolescent patient's need for external support in initiating and maintaining behavior change. This integrated approach may capitalize on the strengths of both CBT and FBT, potentially achieving outcomes that match or exceed those associated with either treatment alone.

3 Intervention

The first part of this section describes individual CBT interventions for EDs, most of which are discussed in more detail elsewhere [19, 22, 28, 29, 35, 50]. Due to space limitations, this chapter places greater emphasis on material that is specific to the distinctive features of EDs and on treatment elements that have received less attention in traditional CBT protocols for EDs. The second part of this section provides practical recommendations and considerations for augmenting individual CBT for EDs with a flexible degree of family involvement.

3.1 Individual CBT Interventions for Children and Adolescents with EDs

3.1.1 Assessment, Medical Evaluation, and Determining Appropriate Level of Care

Prior to initiating treatment, individuals with EDs should receive a comprehensive psychological evaluation that includes assessment of ED features [22]. Before or during the psychological assessment process, the therapist should obtain parental consent to contact the child's pediatrician or family physician. A medical evaluation should be conducted by a physician familiar with the medical findings and complications of EDs, particularly if the child is underweight, rapidly losing weight, vomiting, and/or engaging in laxative abuse. Some of the physical and physiological variables that might be assessed include weight and height; heart rate and blood pressure; levels of phosphorus, magnesium, potassium, and glucose; thyroid function; estradiol and/or testosterone levels; bone mineral density; hydration; and gastrointestinal function [53, 54]. Alarming results may indicate hospitalization to ensure medical stability.

3.1.2 Discussion of Confidentiality

Although all therapists who treat minors should be familiar with ethical dilemmas related to confidentiality, the medical complications associated with EDs warrant special consideration. Specifically, dangerous compensatory behaviors such as laxative abuse and self-induced vomiting, as well as severe dietary restraint and weight loss, constitute risk of harm to self and, depending on the severity, may need to be disclosed to the patient's family.

3.1.3 Setting an Agenda

Consistent with CBT for other conditions, therapists treating EDs start sessions by setting an agenda [19, 29, 55]. Agendas help keep sessions focused and goal-oriented, prioritize topics by importance, and emphasize the collaborative nature of CBT as patients are encouraged to add and modify agenda items.

3.1.4 Clinical Formulation

Several prominent cognitive-behavioral therapists recommend prioritizing the development of a visual clinical formulation with each patient at the start of treatment [19, 29]. Developing a clinical formulation helps patients understand the interconnectedness of their eating-related beliefs and behaviors, underscores the importance of modifying eating behaviors, and provides a rationale for CBT. The therapist typically starts by describing the CBT model that conceptualizes the overevaluation of weight and shape as the "core psychopathology" of EDs. Next, using a blend of Socratic questioning and psychoeducation, the therapist and patient collaboratively identify previous experiences that may have contributed to the patient's current beliefs about eating, weight, and shape. The therapist writes down each of these hypothesized factors inside a box or oval and draws arrows to denote the ways in which these variables contributed to the overevaluation of weight and shape. The therapist and patient then explore behavioral, cognitive, physiological, and/or emotional consequences (including perceived benefits) related to the overevaluation of weight and shape, which are subsequently added to the visual clinical formulation. All components of the formulation are then connected with arrows that identify hypothesized antecedents and consequences [19, 28, 29, 35]. New information should be added to the visual formulation as the patient develops additional insight into their ED over the course of treatment.

While there are benefits to developing a joint formulation early in treatment, therapists should use their clinical judgment to consider whether the formulation should be proposed before or after other treatment components are initiated. For some patients, key interventions such as providing psychoeducation, meal planning, and introducing open weighing may take up all of the first few therapy sessions. In these instances, the development of a clinical formulation can be postponed to a later session.

3.1.5 Psychoeducation

ED-related psychoeducation provides the rationale for prioritizing interventions that facilitate regular eating, corrects misinformation about eating and weight, and sets the foundation for the motivational and cognitive work to follow. Psychoeducational topics are heavily emphasized during the first few CBT sessions and revisited frequently throughout treatment.

3.1.5.1 The Effects of Strict Dieting and Semi-Starvation

Patients with EDs who are underweight or weight suppressed should be provided with detailed information about the physiological, behavioral, and psychological effects of semi-starvation [19, 22, 29], most extensively documented in the Minnesota Starvation Study [56]. Patients (and uninformed providers) often formulate pathological explanations for what are in fact typical human responses to starvation. It is important that a patient with AN recognize that starvation explains why she is cold all the time, why she tears her bread into small pieces, why she devotes an hour to eating an apple, why she spends much of her day reading recipes online, why she is constantly irritable, and why she has lost interest in spending time with her friends and family [9]. The hardwired effects of starvation virtually guarantee that patients will continue to experience unpleasant symptoms (e.g., depression, low energy, food preoccupation, irritability, binge eating) as long as they maintain restrictive eating patterns. Although this information is especially relevant to highly restrictive, underweight patients, all individuals with EDs are likely to manifest attenuated starvation effects. Those who experience subjective loss of control over eating and episodes of binge eating may benefit from learning that these are expected responses to dietary restraint [57, 58] that are unlikely to resolve until they moderate dietary rules and consume regular meals and snacks throughout the day [19, 22, 29].

3.1.5.2 The Ineffectiveness of Self-Induced Vomiting, Laxative Abuse, and Diuretic Abuse

Patients who engage in self-induced vomiting, laxative abuse, and/or diuretic abuse should be provided with accurate information about the limited effectiveness of these behaviors for weight control. Many individuals with EDs hold the erroneous belief that these purging behaviors "cancel out" the calories eaten. In reality, research suggests that individuals with BN who induce vomiting likely eliminate less than 50% of the calories consumed [59]. Laxatives act on the large intestine after most calories have already been absorbed and are thus an ineffective weight loss strategy [60]. The sensation of weight loss and any decreases in body weight after laxative use are merely the temporary effects of dehydration. Similarly, diuretics facilitate the loss of fluids from the body, with no effect on calorie absorption.

3.1.5.3 The Medical Complications Associated with EDs

Although nonmedical therapists cannot provide medical management to patients, it is crucial that they be familiar with the medical complications associated with EDs, discuss these risks with patients and their families, and recognize when to consult with a physician. Accurate, detailed information about the health consequences of EDs can be accessed readily online [61], in treatment manuals for EDs [29], and in scholarly articles [62, 63].

3.1.5.4 Myths About Food, Dieting, and Weight

Individuals with EDs are avid consumers of messages related to nutrition and weight loss but are often uncritical and biased in interpreting this information. Moreover, many in the general population hold inaccurate beliefs about food and experience eating-related guilt [64, 65]. A quick online search for healthy eating advice yields warnings about the negative effects of gluten on the brain, the addictive properties of carbohydrates, the link between processed foods and poor sleep, and the impact of refined sugars on early aging. Unsurprisingly, patients are often puzzled when ED therapists recommend that they reduce their vegetable intake, avoid low-fat foods, and eat desserts more often. Clinicians must be prepared to correct dietary misinformation and provide both psychological and nutritional rationales for eating a varied diet of carbohydrates, fats, and protein [66].

Working with patients to discern nutritional truth from fiction is complicated by the fact that *certain* weight-related beliefs are valid [52]. For instance, an adolescent with AN may divulge that she dropped 20 lbs. by skipping breakfast, cutting out carbohydrates, and avoiding restaurants and report fear that changing these dietary rules and rituals will lead to uncontrollable weight gain. It is important that therapists recognize and effectively communicate that it is not these specific dietary rules and rituals per se that caused the patient's weight loss but rather her consistent decrease in energy intake relative to energy expenditure over time [67]. Similarly, the patient should understand that weight gain after deliberate weight suppression is the result of eating an appropriate amount of food rather than a direct effect of consuming her first meal earlier in the day, reintroducing pizza and pasta into her diet, or eating in restaurants more frequently.

3.1.5.5 Enhancing Motivation

Many children and adolescents with EDs, particularly AN, enter treatment against their wishes and report low intrinsic motivation to change [9]. Although eating behaviors may improve solely in response to external pressure from parents or hospital staff, individuals who do not increase their internal motivation to recover are prone to poor long-term outcomes and relapse [68–70]. While research has not consistently favored *separate* motivational interviewing sessions for EDs [71], CBT *integrates* motivational enhancement strategies with other cognitive-behavioral interventions [9, 29]. Therapists should assess motivation to change on an ongoing basis and implement motivational enhancement strategies as warranted throughout the course of treatment.

3.1.5.6 Pros and Cons of Maintaining Versus Changing Eating Behaviors

Several CBT manuals for EDs recommend exploring the pros and cons of recovery as a strategy to increase motivation for change [19, 29, 72]. Pros and cons include the perceived benefits and consequences of both maintaining and changing eating behaviors. After exploring pros and cons and providing the patient with relevant psychoeducation about EDs, therapists can discuss the "package deals" associated with various degrees of recovery [9]. An adolescent with AN who maintains extremely restrictive eating patterns and low body weight may continue to derive feelings of achievement and increased body satisfaction from her ED but must also endure depression, poor concentration, low energy, and food preoccupation. Partial recovery, which might be achieved by loosening dietary restraint and gaining some weight, characteristically yields improvement in some physical and psychological effects; however, weight preoccupation, avoidance of preferred foods, and anxiety in social eating situations are likely to persist, as well as heightened risk of relapse. While full recovery from an ED involves foregoing the positive feelings derived from controlled eating and weight loss, it also frees the patient from the multitude of negative consequences inextricably tied to an ED.

3.1.5.7 Projecting into the Future

Encouraging patients to envision what their future might look like if they were to maintain versus recover from an ED is another recommended strategy that may enhance intrinsic motivation to change [9, 29]. Therapists should use a collaborative and curious Socratic style to help patients reflect upon the degree to which an ED is compatible with their future goals. The sample dialogue below includes some of the questions one might use when asking patients to project into the near future.

Therapist: I want us to spend more time discussing what *you* think about changing your eating and getting back to a healthy weight. I'm curious—what do you think would happen if your parents said that you no longer had to come to therapy and that you could choose to eat in any way you wanted?

Patient: I'm pretty sure I would drop out of therapy and go back to the way I was eating before. I hate having to eat so much food and I feel better about myself when I'm losing weight.

Therapist: OK, so you anticipate that you would immediately cut down on your eating. In the 4 months before you and your parents came to treatment, you had lost a lot of weight. What do you think would happen to your weight if you returned to your old eating habits?

Patient: I think I would quickly lose all the weight that I've gained in treatment so far.

Therapist: That does seem likely. After getting back to what you weighed when treatment began, do you think you would keep trying to lose more weight?

Patient: I'm pretty sure I would. My parents stopped me before I could get to my goal weight.

Therapist: If you were able to reach your goal weight, do you think you would stop trying to lose weight at that point, or keep trying to lose more?

Patient: I'm not sure. I'd be really happy that I got to my goal weight. But I guess I might be tempted to lose a little more weight even after that. It's just that losing weight makes me feel better about myself than almost anything else.

Therapist: When do you think your temptation to lose weight might stop?

Patient: I don't know. It's hard for me to see that temptation ever going away.

Therapist: Even before reaching your goal weight, you started experiencing some medical issues, like feeling dizzy and weak, and blacking out that one time. You also noted that you often felt depressed and had trouble concentrating on things. Based on what we know about how the body is affected by starvation, it's likely that you would have even worse problems of that kind at lower weights. What do you think might happen if no one stopped you from continuing to lose weight?

Patient: I guess I could see it leading to more medical problems. It could get pretty bad. I might even have to be taken to a hospital.

Therapist: I think you're right. Hospitalization is usually what happens when people lose that much weight. To make sure you were safe, the hospital would refeed you until you got up to a somewhat healthier weight before letting you go. If that were to happen to you, what might you do after getting out of the hospital?

Patient: I honestly don't think that even going to the hospital could get me to stop losing weight. I think I would go right back to eating as little as I could and start losing weight again.

Therapist: And what might happen if you were to lose more weight again?

Patient: I might have to be put back in the hospital.

Therapist: So it sounds like not eating enough and losing weight would lead to a cycle of being in and out of the hospital.

Patient: Yeah, I guess so. Well, maybe I could get myself to lose *some* weight, but not so much that I would have to go to the hospital.

Therapist: How do you think you would feel about still not being at your preferred weight? You said that it's hard to imagine the temptation to lose going away.

Patient: I don't really know. I'd probably still be unhappy that I'm not at my goal weight.

Therapist: That sounds like a real dilemma. Right now, the pull toward eating less and losing weight is so strong that you can't imagine making the decision *yourself* to eat enough food to stay within a healthy weight range and avoid all those medical consequences, as well as the problems with depression and concentration—but you also hate it when other people take over your life in order to prevent those things from happening. Maybe if you could kind of split the difference, you think, you could keep your eating and weight *partly* restricted and have *fewer* of those negative effects—but you're not sure how you could give up your longing to weigh less.

Patient: Honestly, all those options sound bad. It seems like I lose out no matter what I do.

Another exercise encourages patients to imagine that they have been transported to some time (e.g., 5 years) in the future and are meeting their therapist again to discuss life. Patients are asked to role-play different future versions of themselves, including someone who has fully recovered from an ED and someone who continues to have an ED [29]. These role-plays begin with a lighter tone and questions unrelated to eating and weight to increase interest in the exercise, reduce defensiveness, and assist patients with vividly imaging their future. For example, patients can be asked to start by describing their age, place of residence, college major, job, roommates, friends, relationship with family, dating life, and hobbies from this future perspective. Questions then gradually shift to eating, weight, body image, and mental health, such as: "What do you eat on an average day?", "How often do you think about your weight and body shape?", "How do you feel when you go out to eat with friends or family?", "What were some of the most important changes that helped you achieve recovery?", "Is there any part of you that regrets recovering from an ED?", "How do you handle stressful aspects of life without ED symptoms?", "How has your mood been?", "Is anxiety any better or worse than it was before?", and "If you could go back in time and give your old self some advice, what might you say?". Therapists can use similar questions to assist patients with envisioning themselves as young adults who continue to have an ED. After being "transported" back to the present, patients reflect on the exercise, compare and contrast the different future versions of themselves, and discuss how this insight affects their decisions about maintaining or changing their eating behaviors.

3.1.5.8 Exploring the Patient's Values

Control of eating and weight is often *consonant* with the values reported by patients with AN, such as being disciplined, self-restrained, rule-oriented, hard-working, and favoring higher-order goals over primal human urges [9]. Therapists should explore the full range of patients' personal beliefs and values, gently highlighting both consistencies and inconsistencies between various values and behaviors, and encouraging the patient to consider ways to resolve incompatibilities. For example, a therapist might discuss whether values related to achievement are more compatible with maintaining ED symptoms or embarking on the difficult path to recovery. Collaborative and strategic Socratic questioning usually leads patients to reveal some ambivalence about selecting thinness as the ultimate measure of achievement, individuality, and self-worth. When asked, almost all patients state that they would not recommend that their own (hypothetical) children, close friends, or family members develop an ED, suggesting some incongruence between their values and the decision to maintain an ED. Encouraging patients to reflect on societal beauty standards and pressures to be thin may also increase cognitive dissonance between beliefs and behaviors [73, 74] and increase motivation to change.

3.1.6 Open Weighing

Weighing patients with EDs on a regular basis is necessary to monitor an important indicator of their health and treatment progress. While many therapists who treat EDs choose to keep patients blind to their weight [75, 76], CBT involves *open* weighing: Patients see their weight on the scale during therapy sessions once per week and are involved in direct discussion about their weight, including weight goals for patients who are underweight or weight suppressed [19, 22, 29]. For post-menarchal patients who are assigned female at birth and not using hormonal contraception, resumption of menses indicates reaching a *minimum* weight threshold. However, adolescents' premorbid body mass index-for-age percentile provides a better target weight estimate, assuming appropriate eating behaviors and growth prior to the onset of their ED [81] (growth charts from the Centers for Disease Control and

Prevention can be accessed online at www.cdc.gov/growthcharts). During the weight restoration phase of treatment for EDs, patients should gain an average of at least 1 lb. per week. Because of the serious medical complications associated with malnutrition in children and adolescents, underweight patients and their families should be informed from the start that any further weight loss or failure to gain weight adequately in outpatient treatment will signal a need for higher levels of care, such as partial hospitalization, residential, or inpatient treatment.

Open weighing aims to modify maladaptive beliefs, reduce anxiety, and increase self-efficacy related to eating and weight [49]. CBT manuals recommend that therapists provide the rationale for open weighing and psychoeducation about weight before the first in-session weigh-in, including information about variables that impact short-term weight fluctuations (e.g., water retention, hormonal changes, the weight of urine, stools, liquid, and food in the body). Additionally, therapists convey that weight must be measured over several weeks before drawing conclusions about weight trajectory [29]. Given the high likelihood that patients will misinterpret weight information, open weighing should occur during therapy sessions, and patients are discouraged from weighing themselves at home [19]. After discussing patients' concerns about seeing their weight, the therapist constructs a weight graph with the number of weeks (or the date of the therapy session) on the x-axis and weight (e.g., in pounds) on the y-axis. This graph is used to track patients' weight from week to week over the course of CBT.

Several CBT protocols incorporate open weighing as a form of exposure therapy that aims to modify exaggerated beliefs about one's *likelihood* of weight gain [29, 49, 50, 77]. The therapist begins by discussing patients' eating behaviors over the past week and then asks patients to predict their weight to the nearest pound. After marking this prediction on a graph, the therapist weighs the patient (shoes off, clothing on) using a standing scale, announces the weight rounded to the nearest pound, records it on the graph, and explores the patient's reaction, including potential reasons for any discrepancy between predicted and actual weight.

During each weekly weigh-in, the therapist follows a similar procedure, graphing both the patient's actual weight and *cumulative predicted* weight over time [29, 49]. For instance, if a patient predicts a 3-pound gain since her last weigh-in, the therapist adds a data point 3 pounds above the previous week's predicted weight. After several weeks, comparing these two trend lines—predicted versus actual weight—helps patients learn that regular eating does not lead to their feared degree of weight gain. This comparison can help modify inaccurate beliefs about food and weight, particularly by examining whether patients misinterpret certain experiences (e.g., feeling full, eating feared foods, poor body image, anxiety, depression) as indicators of weight gain. For example, a 15-year-old girl with BN might learn that maintaining regular eating patterns, incorporating previously "forbidden" foods into her diet, and ending her cycle of binge eating and vomiting does not result in notable changes to her weight.

However, this approach to open weighing does not explicitly help patients learn that *weight gain itself* is acceptable and tolerable [51, 52]. Additionally, while many assume that patients with EDs overestimate the impact of eating on their weight [49], data from one of our current projects suggest that a substantial portion of adolescents with EDs either accurately predict or underestimate weekly changes in weight. Furthermore, patients often experience intense emotional reactions during open weighing, yet existing CBT protocols offer minimal guidance on specific strategies clinicians should use to respond to these reactions.

In two ongoing clinical trials, we are evaluating a novel approach to open weighing that targets patients' beliefs about the *consequences* of weight gain rather than narrowly focusing on beliefs about the *likelihood* of weight gain. This approach uses patients' emotional responses during weighing sessions to identify, and then modify, underlying beliefs about the consequences of weight gain. Prior to initiating open weighing, clinicians provide psychoeducation about the CBT model and the relationship between thoughts, emotions, and behaviors [55, 78]. When patients experience weight increases during open weighing sessions, therapists help them identify their emotional responses and associated thoughts, validate these emotional reactions (e.g., anxiety, sadness) as logical within the context of their beliefs (e.g., "my life will become meaningless if I gain weight"), and then apply cognitive restructuring techniques (see "cognitive restructuring").

Most of our study participants have required assistance identifying their core negative beliefs about weight gain. The belief that "weight gain is bad" pervades industrialized societies [79], leading both patients and clinicians to accept rather than examine this assumption. The ED Fear Questionnaire [80] is one measure that clinicians can use to help patients identify their specific weight gain fears. Alternatively, therapists can provide patients with a list of common feared consequences of weight gain (see Fig. 1) and ask them to identify which beliefs align with their emotional responses. These beliefs are intentionally framed as catastrophic (e.g., "I will permanently hate my body if I gain weight") to distinguish them from more realistic concerns ("I will feel worse about my body if I gain weight") and increase their potential for disconfirmation. After identifying catastrophic beliefs triggered by weight increases, therapists can guide patients in documenting the evidence for and against these beliefs throughout the week (see Fig. 2). This

Feared Consequences of Weight Gain

- I will be judged, rejected, and/or abandoned by others
- I will look ugly, hideous, and/or disgusting
- I will permanently hate my body
- I will become so distracted by my body that I won't be able to focus on anything else
- I will lose control of my life
- My life will become meaningless and/or amount to nothing
- My life will be worse off
- I will become permanently unhappy, depressed, and/or lonely
- I will be unlovable
- I will permanently hate myself
- I will no longer have value, worth, confidence, and/or a sense of self

Fig. 1 A list of feared consequences of weight gain

Feared Outcome ("Worst Case Scenario")	Date	Belief in Feared Outcome 0 – 100%
I will permanently hate my body if I gain weight	*12/10*	*90%*

This week, how would you know if this belief is coming **true**?	This week, how would you know if this belief is **not** coming true?
• *Amount of time I spend having hateful or angry thoughts about my body throughout the week* • *Amount of time I spend feeling strong negative emotions (sad, angry, disgusted) because of my body* • *Number of times that hating my body stops me from doing something I want to do*	• *Amount of time I spend having neutral or positive thoughts about my body throughout the week* • *Amount of time I spend thinking about things other than my body* • *Amount of time I spend feeling neutral or positive emotions* • *Number of times I am able to do something I want to do*
What evidence did you find that **supports** this belief?	What evidence did you find that goes **against** this belief?
• *I had negative thoughts about my body at least twice every day this week* • *I have more negative thoughts about my body now than I did before gaining weight* • *On Saturday, I spent over 4 hours feeling sad about gaining weight, and cried for a few minutes* • *On Saturday, I felt too fat and ugly to go shopping with my friends, and stayed home instead*	• *There are many hours throughout the week that I am not thinking about my body* • *I was able to go to school and have fun with friends* • *Just because I have negative thoughts about my body now, doesn't mean that I will always have negative thoughts about my body* • *People who have recovered from an eating disorder say that you slowly start to feel better about your body over time*

Balanced Thought	Date	Belief in Balanced Thought 0 – 100%	Belief in Feared Outcome 0 – 100%
I have more negative thoughts about my body since I started gaining weight, but I am still able to live my life, and these negative thoughts might decrease over time	*12/17*	*80%*	*65%*

Fig. 2 A sample cognitive restructuring form completed by an adolescent patient with anorexia nervosa

transforms the process of weight gain into a behavioral experiment, with the aim of helping patients learn that weight gain does not lead to their feared catastrophic outcomes.

3.1.7 Meal Planning

An important CBT intervention for EDs is meal planning, during which the therapist and patient collaboratively decide what, when, and where the patient will eat over the next several days [19, 22, 29]. To help the patient maintain consistent blood sugar, reduce the likelihood of significant hunger and binge eating, increase control over eating, and/or restore weight to a healthy range, the meal plan should include three meals (i.e., breakfast, lunch, and dinner) and at least one snack per day. Meal plans should specify times and settings for all meals and snacks and note the foods and

Time	Place	Planned Food/Bev	Actual Food/Bev	B/V/L	Thoughts/Emotions/Other
7:00am	Home	1 bowl of cereal 1 banana	1 bowl of cereal 1 banana		I feel bad for choosing the sugary cereal. Why did I do that!?
10:00am	School	1 yogurt	1 yogurt		I'm still full from breakfast. I didn't really want to eat the yogurt but I did anyway.
12:00pm	School	1 peanut butter and jelly sandwich 1 small bag of chips 1 juice box	Nothing (threw lunch in trash)		I'm STILL full from breakfast! It's my fault for eating that sugary cereal.
3:00pm	Home	8 baby carrots with ranch dressing	Nothing		I feel hungry now but can wait a few more hours until dinner.
6:30pm	Home	1 bowl of alfredo pasta 2 pieces of garlic bread	1 bowl of alfredo pasta 4 pieces of garlic bread 5 bags of chips 10 cookies	B V	I ate more bread than I was supposed to! I left the table and went to my room to eat the chips and cookies I keep there. Felt a little better after throwing the food up.
8:00pm	Home	2 cookies	Nothing		

Fig. 3 A sample daily food record completed by an adolescent patient with bulimia nervosa

drinks to be consumed. This information is written down during therapy sessions on a self-monitoring form, also referred to as a food record [82]. Clinicians can use the self-monitoring form in Fig. 3, an alternative electronic variation [83], or a revised food record that better fits the individual patient. Figure 3 illustrates a daily food record completed by a 16-year-old female patient with BN. In this example, the adolescent wrote down the planned foods and beverages for each meal and snack in the "Planned Food/Bev" column, the time of day when she intended to eat in the "Time" column, and the anticipated location for all meals and snacks in the "Place" column.

Meal planning assists patients with developing regular eating behaviors, increasing food variety, modifying rigid dietary rules and rituals, and decreasing binge eating and purging. Moreover, CBT for patients with AN (and those with other EDs who are significantly weight suppressed) involves the additional treatment goal of weight restoration. Accordingly, providers treating children and adolescents with AN should be familiar with the general calorie content of foods that are selected for the patient's meal plan, as well as the recommended daily calorie intake for the patient based on age, gender, height, and activity level. If a 15-year-old male with AN is adherent to his meal plan but failing to gain weight, the therapist needs to assist him with modifying his meal plan to augment calorie intake. For instance, the patient might decide to increase the size of his breakfast, add a large afternoon snack, and eat more calorie-dense foods for dinner.

The first several weeks of therapy tend to focus intensively on meal planning, which is faded as patients' eating behaviors improve. Many young patients with EDs can be persuaded to begin improving their eating behaviors with little to no external control by family members. Other patients are initially unwilling to change their eating behaviors and/or gain weight on their own and may require frequent family participation in therapy sessions focused on meal planning during the early stages of treatment. Factors guiding the degree to which parents should assist with meal planning are discussed in more detail in a subsequent section.

3.1.8 Self-Monitoring

After meal planning with the patient, the therapist describes how to complete the rest of the self-monitoring form, which keeps track of what occurs between therapy sessions. Patients are asked to record the foods and beverages they consume throughout the week in the "Actual Food/Bev" column, and the times and locations where they eat in the "Time" and "Place" columns, respectively (see Fig. 3). Binge eating, self-induced vomiting, and laxative use are recorded in the "B/V/L" column: "B" represents a perceived binge-eating episode; "V" represents self-induced vomiting; and "L" represents laxative use. Patients are encouraged to record any notable thoughts or emotions they experienced throughout the week in the "Thoughts/Emotions/Other" column. Additionally, patients should document exercise and other behaviors related to eating and weight (e.g., diet pill and diuretic use) on the self-monitoring form.

To target binge eating and inappropriate compensatory behaviors noted on the self-monitoring form, the therapist works with the patient to uncover behavioral, situational, cognitive, emotional, and physiological antecedents and consequences of these symptoms. This discussion often identifies useful treatment components, such as psychoeducation about vulnerability to binge eating in response to dietary restraint, cognitive restructuring aimed at modifying "all or none" thinking around food, or behavioral strategies that attempt to interrupt the binge-purge cycle [19, 22, 29]. For example, after talking to her therapist about the binge/purge episode recorded in Fig. 3, the patient identified several contributing factors, including skipping lunch and afternoon snack, keeping food in her room, and evening time. Through this discussion, the patient recognized that skipping meals and snacks increases her vulnerability to binge eat, while she is unlikely to binge or purge during daytime, when she is around others, and when she is outside of her bedroom. To decrease binge eating and vomiting over the next week, the patient agreed to have lunch and an afternoon snack every day, move food out of her bedroom, and engage in a pleasant activity after dinner such as playing video games with her brother.

In general, delaying binge eating or purging for a predetermined amount of time (e.g., 30 min) can reduce anxiety and urges to binge or purge. During this "wait period," patients often find it helpful to participate in an enjoyable activity, read through or listen to reasoned conclusions that have been recorded in therapy [55], and/or engage in explicit decision-making about the choice to binge eat or purge, such as by writing down a pros and cons list. Keeping track of binge eating and/or purging frequency, transforming the data into a line graph, and reviewing the graph each week can also heighten motivation to reduce these behaviors. Any interventions planned for the week are written down on the patient's self-monitoring form or other document, and outcomes of these efforts are reviewed with the therapist during the next therapy session. While the amount of time in therapy spent reviewing the self-monitoring form tends to decrease as the patient's eating behaviors improve, patients are usually encouraged to continue with self-monitoring until the last phase of treatment.

3.1.9 Cognitive Restructuring

Cognitive restructuring involves identifying, evaluating, and modifying patients' thoughts, assumptions, and beliefs. Therapists maintain a curious, collaborative, and empirical stance while using Socratic questioning to help patients examine interpretations from different angles [9, 78]. Following established cognitive restructuring techniques for other disorders [55, 78, 84], CBT for EDs assists patients with understanding the relationship between thoughts, behaviors, and emotions; identifying "automatic" thoughts; examining the accuracy of thoughts; generating more reasoned conclusions; uncovering core beliefs; making behavioral changes that are consistent with reasoned conclusions; and evaluating the outcomes of behaving in different ways [22, 29].

Therapists can draw thoughts about eating, weight, and shape from food records (see Fig. 3) and therapy discussions. Once a relevant thought is identified, therapists can ask a series of cognitive restructuring questions, such as the following:

- What evidence supports this thought? What evidence challenges this thought?
- What might other people say about this thought?
- What would you tell a friend in this situation?
- Let's imagine that you are fully recovered from an ED in the future. What might you think if you were to look back at this thought?
- Let's think of this thought as one "extreme." What would the "opposite" extreme thought be? Now, what would be a "middle ground" thought between these two extremes?
- In what ways is this thought helpful? In what ways is this thought unhelpful?

Cognitive restructuring is integrated throughout the course of CBT for EDs. Whether reviewing psychoeducational material, planning meals for the upcoming week, exploring why the patient chose to skip breakfast over the past few days, or discussing the patient's reaction to seeing their weight, therapists should attend to patients' views and assist them with considering more accurate and adaptive interpretations and conclusions.

In one of our intervention studies, we are evaluating whether integrating cognitive restructuring with open weighing modifies beliefs about the consequences of weight gain among adolescents' with EDs. In this approach, therapists help patients identify their negative beliefs about weight gain (see "open weighing") and establish methods for evaluating whether or not these beliefs are being confirmed (e.g., "This week, how would you know if this belief is coming true?") (see Fig. 2). Between sessions, patients observe and record evidence from their daily experiences that supports and challenges their beliefs about the consequences of weight gain. During the next session, therapists and patients collaboratively review this evidence. If needed, therapists help patients strengthen their "evidence against" column by asking additional cognitive restructuring questions. Finally, therapists guide patients in developing a balanced thought that incorporates all documented evidence. For instance, in the sample form illustrated in Fig. 2, a 16-year-old female with AN shifted from the feared belief "I will permanently hate my body if I gain weight" to a more balanced perspective: "I have more negative thoughts about my body since I started gaining weight, but I am still able to live my life, and these negative thoughts might decrease over time."

3.1.10 Exposure Therapy

Inherent in many CBT interventions for EDs are elements of exposure: Therapists openly weigh patients despite their anxiety about this process; discourage safety behaviors such as purging, excessive exercise, frequent weighing, body checking, and dietary rituals; and encourage patients to eat regularly and incorporate feared foods into their diets [19, 29]. To target anxiety associated with eating certain foods, for example, therapists might assist patients with developing a hierarchy of feared foods with subjective units of distress ratings and encourage patients to choose progressively more anxiety-provoking foods to add to their meal plans each week [85, 86]. The way in which exposure is usually conducted for EDs, however, has been narrow in scope and conservative in approach [87]. Unlike CBT protocols for EDs, exposure therapy for anxiety disorders involves *overcorrection*: Exposure moves beyond simply "being around" the feared stimulus to more challenging sessions in which the patient's specific fears are amplified or exaggerated [88–90]. By adopting the principle of overcorrection, therapists have the potential to use exposure in imaginative ways to address a relatively wide range of issues, including ritualized eating, weight-related self-evaluation, body dissatisfaction, and eating-related social situations [87].

In recent years, scientist-practitioners have begun to explore more creative adaptations of exposure for EDs. Relatively novel exposure interventions include therapist-guided mirror exposure, which targets body image by instructing individuals to look at themselves in a full body mirror and describe what they see using objective terminology [91]; interoceptive exposure targeting distressing eating-related somatic sensations such as fullness by asking patients to drink a high volume of water [92]; imaginal exposure targeting fear of weight gain by encouraging patients to develop and repeatedly read scripts describing catastrophic weight gain [51, 93]; and in vivo exposure targeting anxiety-provoking eating situations such as portioning food out of a large container [94] and eating a high energy food in front of a romantic interest [95].

Extending the principle of overcorrection used in exposure treatment for anxiety disorders suggests even more challenging exposure scenarios designed to amplify patients' fears about a wide range of scenarios [87]. For example, to address a patient's fear of negative judgment from others about her food choices, therapists could develop an exposure session in which the patient eats a calorie-rich pasta with a therapist who orders a salad and makes critical comments about the patient's meal (e.g., "*That's* what you decided to eat?!", "Wow, that looks like it has a lot of fat in it," "I guess you aren't into 'clean' eating like I am," "Aren't you worried about gaining weight?") [96].

Advances in the field of exposure therapy for anxiety disorders—specifically the shift from habituation-based to "inhibitory learning" approaches [97]—also offer new possibilities for addressing weight gain fears in adolescents with EDs. The inhibitory learning model differs from the traditional habituation-based approach to exposure, which focuses on decreasing anxiety ratings during exposure sessions (i.e., within-session habituation) [78, 97, 98]. Instead, inhibitory learning recommends conducting exposures in flexible, unpredictable, and intensive ways to maximize "expectancy violation" (i.e., challenging anxious predictions and beliefs) and enhance tolerance of anxiety [88, 89]. We are currently evaluating this more contemporary approach in a clinical trial for individuals with EDs. Following inhibitory learning principles, participants in our trial are encouraged to do more anxiety-provoking exposures at a faster rate; make in-the-moment adjustments to exposure scenarios; frame the goal of exposure as "bravery building" rather than "learning that my anxiety will decrease"; and integrate various in vivo elements (e.g., weighing, mirror work, eating feared foods) with imaginal components (e.g., scripts, audio recordings, role-playing scenarios about feared weight gain consequences) during exposure sessions. We hypothesize that this approach will enhance exposure therapy's effectiveness in targeting weight gain fears among individuals with EDs.

Exposure therapy is the leading treatment for anxiety disorders [98] and is a logical choice to target fear responses in patients with EDs. In adapting the intervention to EDs, however, clinicians should consider several distinctive features that affect the timing, targets, and implementation of exposure. For example, some ritualistic patterns such as prolonged eating and excessive use of salt and spices are more accurately construed as starvation behaviors rather than ED symptoms and should be treated through weight restoration rather than response prevention. In addition, many ED patients have intentionally *cultivated* their fear of "fattening" or "unhealthy" foods and are invested in *retaining* their anxiety as a weight control tactic that facilitates dietary restraint [99]. Psychoeducation about the effectiveness of exposure is unlikely to be persuasive for individuals who are terrified of becoming less afraid. These considerations may need to be addressed through other therapeutic strategies before patients can be engaged as active collaborators in designing and carrying out exposure exercises.

3.1.11 Relapse Prevention

The final therapy sessions tend to focus on relapse prevention. Common relapse prevention interventions include summarizing the progress that the patient has made over the course of treatment, reviewing skills that were instrumental to the patient's positive outcome, identifying areas of continued vulnerability and warning signs of relapse, anticipating potential future difficulties and plans to address these challenges, assisting the patient with differentiating small setbacks from a full relapse, discussing any concerns about ending treatment, and developing a plan for appropriate weight monitoring [19, 22, 29].

4 Augmenting CBT for EDs with Family Involvement

4.1 Determining the Degree and Format of Family Involvement

CBT for both pediatric anxiety disorders and EDs typically incorporates family into treatment, as therapists encourage parents to help their child participate in between-session exposure to feared stimuli and reduce avoidance behaviors [100]. By definition, exposure-based treatment is anxiety-provoking, and many children require family support to initiate and maintain distressing behavior change. In addition, EDs are characterized by particularly dangerous avoidance and safety behaviors such as extreme dietary restraint and self-induced vomiting for which rapid intervention is indicated.

At the same time, long-term recovery requires patients to make the decision *themselves* to sustain repeated exposure to feared stimuli. Getting over AN depends on a daily commitment to eat normal meals and cope with body image concerns, just as surmounting social anxiety involves ongoing participation in social events. While CBT for children and adolescents often enlists the family to help initiate behavior change, an important treatment goal is to increase patients' own investment in working toward recovery without external pressure. Accordingly, we recommend that therapists depend on parental control to the minimum degree needed to facilitate behavior change and ensure the patient's safety and gradually decrease external direction as the patient makes progress.

Consistent with these principles, CBT for children and adolescents with EDs typically includes a blend of individual and family therapy sessions [22, 53]. The degree of family involvement varies according to the amount and forms of external support patients need to participate in treatment and change their eating behaviors. Each patient's age, symptom presentation and severity, motivation to recover, and progress over time should be used to determine the frequency and content of family sessions. In general, patients who are younger, underweight, and/or reluctant to change require substantially higher levels of family involvement. Weekly family sessions are indicated for most children and adolescents with AN in the initial stages of treatment; in contrast, CBT for adolescents with BN may involve only a few family sessions over the course of treatment.

Therapists should use clinical judgment to guide the optimal format of family involvement. Research on FBT suggests that there are advantages and disadvantages to the alternative formats examined, with individual patients responding better to conjoint, separate, and parent-only approaches [101, 102]. For example, families with higher levels of expressed emotion and criticism fare worse in conjoint family therapy [103, 104], perhaps because airing parental frustrations in the child's presence increases conflict during and between therapy sessions.

For the subset of patients who require substantial family involvement to participate, treatment typically begins with one individual and one family therapy session per week. In most instances, family sessions are split into two segments, meeting initially with the parents and then with parents and patient conjointly. Factors that influence the proportion of time dedicated to parent-only versus conjoint sessions include the preferences of the parents and patient as well as the clinician's observations of family interactions during sessions. Conjoint sessions may be initiated quickly when working with a 15-year-old with AN who has a strong relationship with her parents and prefers to participate in family discussions of the recovery plan; however, when treating a 13-year-old who becomes agitated during family therapy, it might be more appropriate to separate the patient and parents for most of the session.

4.2 The Family's Role in Supporting the Normalization of Eating Behaviors and Weight Restoration

4.2.1 Psychoeducation and Communicating Empathy

Parents should be provided with detailed material about the physical and psychological consequences of EDs, the hardwired effects of semi-starvation, and the reinforcement principles that maintain ED behaviors. Psychoeducation helps parents develop a more informed and compassionate understanding of their child's predicament and perspective and conveys the importance of immediate weight restoration and normalized eating. Weight restoration for children with AN, like chemotherapy for children with leukemia, is frightening, aversive, uncomfortable, and difficult—*and* an essential and nonnegotiable part of recovery. From the start of treatment, therapists should assist families with communicating and implementing this sympathetic yet firm message at home. A mother might repeatedly tell her 14-year-old daughter with AN that she loves her and is committed to her health, safety, and future—which is precisely why she cannot allow her to eat less than her body requires.

Therapists should also be prepared to provide families with accurate nutritional information. Myths about healthy eating, dieting, and weight are ubiquitous, and many parents inadvertently convey false or oversimplified messages about food to their children. Conscientious parents who know that relatively low-fat, low--sugar diets are recommended for the general public and who work to provide healthy, home-cooked meals to their families are often puzzled by the apparently contrary advice given in treatment. Family members deserve clear explanations about why their child with an ED should be encouraged to include fried foods and desserts in their diet and to eat occasional meals in fast-food restaurants.

4.2.2 Setting Clear Expectations and Increasing Supervision

Parents should communicate to their child that eating three meals per day is nonnegotiable and demonstrate an unwavering commitment to this expectation. Therapists should also help parents understand and convey that their child must eat enough food to support participation in daily life, including school, work, and extracurricular activities. A father may need to sit at the kitchen table with his 17-year-old daughter with AN during mealtimes and calmly remind her that she cannot leave to attend a social event until she has finished eating. Gradually regaining independence and reducing parental monitoring act as logical rewards for treatment progress, while further decreases in autonomy and reduced access to unsupervised or physically demanding activities are natural consequences of persistent ED behavior.

4.2.3 Collaborative Meal Planning

In contrast to FBT, which empowers families to make all eating-related decisions for their child [34], CBT actively involves the patient in meal planning from the outset. Most young patients with BN and some with AN can participate effectively in meal planning during individual sessions; for others, decisions are made by the parents during family sessions. In all cases, however, the therapist solicits input from the child as well as the parents. Some components of the meal plan are nonnegotiable, such as eating three meals and at least one snack per day and consuming a sufficient number of calories to support weight gain or maintenance. Within these boundaries, however, the patient is encouraged to make a number of decisions related to what, when, and where to eat. For example, a 14-year-old girl with AN may decide to eat one large snack rather than two smaller snacks each day. Although it is preferable for patients to eat the same foods as other family members, some flexibility may be appropriate in the early stages of recovery. For example, a therapist might encourage parents to allow their 13-year-old son with atypical AN to eat pasta on a night when the rest of the family has pizza, if pizza is ranked near the top of his hierarchy of feared foods.

Collaborative meal planning should accompany ongoing discussion about the kinds of comments the patient experiences as helpful and unhelpful, which can help parents use verbal reinforcement more effectively. While some children appreciate praise for improved eating behavior at meal time, others prefer to have parents recognize their efforts to change at other times or in different ways.

5 Conclusion

While FBT remains the leading treatment for adolescents with EDs, combining CBT with individualized levels of family involvement may enhance treatment outcomes by capitalizing on the strengths of both approaches. Because ambivalence about change is a prominent feature of EDs, especially AN, parental involvement often proves necessary to ensure treatment participation and prevent ED-related harm, with the degree of parental control varying from case to case. At the same time, CBT emphasizes modifying maladaptive beliefs about eating, weight, and shape as key to long-term recovery. This requires therapists to balance external support with patient autonomy, attending to patients' perspectives while working to develop internal motivation to change. As patients become more engaged in treatment, external control is faded and treatment shifts to self-regulation of eating behaviors, self-selected exposure, cognitive restructuring, and related CBT interventions. Although clinical experience suggests that augmenting CBT for EDs with family involvement in this way increases the treatment's effectiveness, this integrated approach awaits empirical validation. This chapter aims to serve as a valuable supplement to more comprehensive CBT protocols and to support further study of the efficacy of CBT for children and adolescents with EDs.

Conflicts of Interest

The author has no conflicts of interest to declare that are relevant to the content of this chapter.

References

1. Hambleton A, Pepin G, Le A, Maloney D, Touyz S, Maguire S (2022) Psychiatric and medical comorbidities of eating disorders: findings from a rapid review of the literature. J Eat Disord 10(1):132
2. Berends T, Boonstra N, Van Elburg A (2018) Relapse in anorexia nervosa: a systematic review and meta-analysis. Curr Opin Psychiatry 31(6):445–455
3. Khalsa SS, Portnoff LC, McCurdy-McKinnon D, Feusner JD (2017) What happens after treatment? A systematic review of relapse, remission, and recovery in anorexia nervosa. J Eat Disord 5:1–12
4. Sala M, Keshishian A, Song S, Moskowitz R, Bulik CM, Roos CR et al (2023) Predictors of relapse in eating disorders: a meta-analysis. J Psychiatr Res 158:281–299
5. Pompili M, Mancinelli I, Girardi P, Ruberto A, Tatarelli R (2004) Suicide in anorexia nervosa: a meta-analysis. Int J Eat Disord 36(1):99–103
6. Amiri S, Khan MA (2023) Prevalence of non-suicidal self-injury, suicidal ideation, suicide attempts, suicide mortality in eating disorders: a systematic review and meta-analysis. Eat Disord 31(5):487–525
7. Chesney E, Goodwin GM, Fazel S (2014) Risks of all-cause and suicide mortality in mental disorders: a meta-review. World Psychiatry 13(2):153–160
8. Vitousek KB, Hollon SD (1990) The investigation of schematic content and processing in eating disorders. Cogn Ther Res 14:191–214
9. Vitousek K, Watson S, Wilson GT (1998) Enhancing motivation for change in treatment-resistant eating disorders. Clin Psychol Rev 18(4):391–420
10. Fairburn CG, Cooper Z, Shafran R (2003) Cognitive behaviour therapy for eating disorders: a "transdiagnostic" theory and treatment. Behav Res Ther 41(5):509–528
11. Grilo CM, Masheb RM, White MA (2010) Significance of overvaluation of shape/weight in binge-eating disorder: comparative study with overweight and bulimia nervosa. Obesity 18(3):499–504
12. American Psychiatric Association D, American Psychiatric Association D (2013) Diagnostic and statistical manual of mental disorders: DSM-5, vol 5. American Psychiatric Association, Washington, DC
13. Hoek HW, Van Hoeken D (2003) Review of the prevalence and incidence of eating disorders. Int J Eat Disord 34(4):383–396
14. Hudson JI, Hiripi E, Pope HG Jr, Kessler RC (2007) The prevalence and correlates of eating disorders in the National Comorbidity Survey Replication. Biol Psychiatry 61(3): 348–358
15. Solmi M, Radua J, Olivola M, Croce E, Soardo L, Salazar de Pablo G et al (2022) Age at onset of mental disorders worldwide: large-scale meta-analysis of 192 epidemiological studies. Mol Psychiatry 27(1):281–295
16. Jaite C, Hoffmann F, Glaeske G, Bachmann CJ (2013) Prevalence, comorbidities and outpatient treatment of anorexia and bulimia nervosa in German children and adolescents. Eat Weight Disord 18:157–165
17. Smink FR, Van Hoeken D, Hoek HW (2012) Epidemiology of eating disorders: incidence, prevalence and mortality rates. Curr Psychiatry Rep 14(4):406–414
18. Stice E, Marti CN, Shaw H, Jaconis M (2009) An 8-year longitudinal study of the natural history of threshold, subthreshold, and partial eating disorders from a community sample of adolescents. J Abnorm Psychol 118(3):587
19. Fairburn C (2008) Cognitive behavior therapy and eating disorders. Guilford
20. Pallister E, Waller G (2008) Anxiety in the eating disorders: understanding the overlap. Clin Psychol Rev 28(3):366–386
21. Salkovskis PM (1991) The importance of behaviour in the maintenance of anxiety and panic: a cognitive account. Behav Cogn Psychother 19(1):6–19
22. Garner DM, Vitousek KM, Pike KM (1997) Cognitive-behavioral therapy for anorexia nervosa
23. Haynos AF, Koithan E, Hagan KE (2023) Learned industriousness as a translational

mechanism in anorexia nervosa. Nat Rev Psychol 2(2):112–126

24. Haynos AF, Lavender JM, Nelson J, Crow SJ, Peterson CB (2020) Moving towards specificity: a systematic review of cue features associated with reward and punishment in anorexia nervosa. Clin Psychol Rev 79: 101872
25. Gale C, Holliday J, Troop NA, Serpell L, Treasure J (2006) The pros and cons of change in individuals with eating disorders: a broader perspective. Int J Eat Disord 39(5): 394–403
26. Serpell L, Treasure J, Teasdale J, Sullivan V (1999) Anorexia nervosa: friend or foe? Int J Eat Disord 25(2):177–186
27. Skårderud F (2007) Shame and pride in anorexia nervosa: a qualitative descriptive study. Eur Eat Disord Rev 15(2):81–97
28. Gowers SG, Green L (2009) Eating disorders: cognitive behaviour therapy with children and young people. Routledge
29. Waller G, Cordery H, Corstorphine E, Hinrichsen H, Lawson R, Mountford V et al (2007) Cognitive behavioral therapy for eating disorders: a comprehensive treatment guide. Cambridge University Press
30. Waller G (2016) Treatment protocols for eating disorders: clinicians' attitudes, concerns, adherence and difficulties delivering evidence-based psychological interventions. Curr Psychiatry Rep 18:1–8
31. Grilo CM (2024) Treatment of eating disorders: current status, challenges, and future directions. Annu Rev Clin Psychol 20
32. Dare C, Eisler I (1992) Family therapy for anorexia nervosa. Harwood Academic Publishers, New York
33. Eisler I (2005) The empirical and theoretical base of family therapy and multiple family day therapy for adolescent anorexia nervosa. J Fam Ther 27(2):104–131
34. Lock J, Le Grange D (2015) Treatment manual for anorexia nervosa: a family-based approach, Guilford Press
35. Dalle Grave R, Calugi S (2024) Enhanced cognitive behaviour therapy for adolescents with eating disorders: development, effectiveness, and future challenges. BioPsychoSocial Med 18(1):18
36. Dalle Grave R, Calugi S, Conti M, Doll H, Fairburn CG (2013) Inpatient cognitive behaviour therapy for anorexia nervosa: a randomized controlled trial. Psychother Psychosom 82(6):390–398
37. Gorrell S, Loeb KL, Le Grange D (2019) Family-based treatment of eating disorders: a narrative review. Psychiatr Clin 42(2): 193–204
38. Lock J, Le Grange D, Agras WS, Moye A, Bryson SW, Jo B (2010) Randomized clinical trial comparing family-based treatment with adolescent-focused individual therapy for adolescents with anorexia nervosa. Arch Gen Psychiatry 67(10):1025–1032
39. Le Grange D, Crosby RD, Rathouz PJ, Leventhal BL (2007) A randomized controlled comparison of family-based treatment and supportive psychotherapy for adolescent bulimia nervosa. Arch Gen Psychiatry 64(9): 1049–1056
40. Le Grange D, Eckhardt S, Dalle Grave R, Crosby RD, Peterson CB, Keery H et al (2022) Enhanced cognitive-behavior therapy and family-based treatment for adolescents with an eating disorder: a non-randomized effectiveness trial. Psychol Med 52(13): 2520–2530
41. Le Grange D, Lock J, Agras WS, Bryson SW, Jo B (2015) Randomized clinical trial of family-based treatment and cognitive-behavioral therapy for adolescent bulimia nervosa. J Am Acad Child Adolesc Psychiatry 54(11):886–894. e2
42. UK NGA (2017) Eating disorders: recognition and treatment
43. Manwaring J, Blalock DV, Le Grange D, Duffy A, McClanahan SF, Johnson C et al (2021) Motivation to recover for adolescent and adult eating disorder patients in residential treatment. Eur Eat Disord Rev 29(4): 622–633
44. Martini M, Lepora M, Longo P, Amodeo L, Marzola E, Abbate-Daga G (2023) Anorexia nervosa in the acute hospitalization setting. In: Eating disorders. Springer, Cham, pp 623–640
45. Byrne CE, Accurso EC, Arnow KD, Lock J, Le Grange D (2015) An exploratory examination of patient and parental self-efficacy as predictors of weight gain in adolescents with anorexia nervosa. Int J Eat Disord 48(7): 883–888
46. Sadeh-Sharvit S, Arnow KD, Osipov L, Lock JD, Jo B, Pajarito S et al (2018) Are parental self-efficacy and family flexibility mediators of treatment for anorexia nervosa? Int J Eat Disord 51(3):275–280
47. Murray SB, Quintana DS, Loeb KL, Griffiths S, Le Grange D (2019) Treatment outcomes for anorexia nervosa: a systematic

review and meta-analysis of randomized controlled trials. Psychol Med 49(4):535–544
48. Linardon J, Wade TD (2018) How many individuals achieve symptom abstinence following psychological treatments for bulimia nervosa? A meta-analytic review. Int J Eat Disord 51(4):287–294
49. Waller G, Mountford VA (2015) Weighing patients within cognitive-behavioural therapy for eating disorders: how, when and why. Behav Res Ther 70:1–10
50. Waller G, Turner H, Tatham M, Mountford V, Wade T (2019) Brief cognitive behavioural therapy for non-underweight patients: CBT-T for eating disorders. Routledge
51. Levinson CA, Rapp J, Riley EN (2014) Addressing the fear of fat: extending imaginal exposure therapy for anxiety disorders to anorexia nervosa. Eating Weight Disord 19: 521–524
52. Murray SB, Loeb KL, Le Grange D (2016) Dissecting the core fear in anorexia nervosa: can we optimize treatment mechanisms? JAMA Psychiatry 73(9):891–892
53. Anderson LK, Reilly EE, Berner L, Wierenga CE, Jones MD, Brown TA et al (2017) Treating eating disorders at higher levels of care: overview and challenges. Curr Psychiatry Rep 19:1–9
54. Rock CL (1999) Nutritional and medical assessment and management of eating disorders. Nutr Clin Care 2(6):332–343
55. Beck Judith S (1995) Cognitive therapy: basics and beyond. Guilford, New York
56. Keys A, Brožek J, Henschel A, Mickelsen O, Taylor HL (1950) The biology of human starvation (2 vols)
57. Polivy J, Zeitlin SB, Herman CP, Beal AL (1994) Food restriction and binge eating: a study of former prisoners of war. J Abnorm Psychol 103(2):409
58. Tuschl RJ (1990) From dietary restraint to binge eating: some theoretical considerations. Appetite 14(2):105–109
59. Kaye WH, Weltzin TE, Hsu LG, McConaha CW, Bolton B (1993) Amount of calories retained after binge eating and vomiting. Am J Psychiatry 150:969
60. BO-LINN GW, SANTA ANA CA, MORAWSKI SG, FORDTRAN JS (1983) Purging and calorie absorption in bulimic patients and normal women. Ann Intern Med 99(1): 14–17
61. Association NED: Health Consequences (2024). https://www.nationaleatingdisorders.org/health-consequences/. Accessed 12/16/2024
62. Mehler PS, Brown C (2015) Anorexia nervosa–medical complications. J Eat Disord 3:1–8
63. Westmoreland P, Krantz MJ, Mehler PS (2016) Medical complications of anorexia nervosa and bulimia. Am J Med 129(1): 30–37
64. Gonzalez VM, Vitousek KM (2004) Feared food in dieting and non-dieting young women: a preliminary validation of the Food Phobia Survey. Appetite 43(2):155–173
65. King GA, Herman CP, Polivy J (1987) Food perception in dieters and non-dieters. Appetite 8(2):147–158
66. Committee DGA, HHS, Prevention OoD, Promotion H, USDA, Promotion CfNP (2015) Dietary guidelines for Americans 2015–2020. Government Printing Office
67. Casazza K, Fontaine KR, Astrup A, Birch LL, Brown AW, Bohan Brown MM et al (2013) Myths, presumptions, and facts about obesity. N Engl J Med 368(5):446–454
68. Carter JC, Kelly AC (2015) Autonomous and controlled motivation for eating disorders treatment: baseline predictors and relationship to treatment outcome. Br J Clin Psychol 54(1):76–90
69. Carter JC, Mercer-Lynn KB, Norwood SJ, Bewell-Weiss CV, Crosby RD, Woodside DB et al (2012) A prospective study of predictors of relapse in anorexia nervosa: implications for relapse prevention. Psychiatry Res 200(2–3): 518–523
70. Gregertsen EC, Mandy W, Kanakam N, Armstrong S, Serpell L (2019) Pre-treatment patient characteristics as predictors of drop-out and treatment outcome in individual and family therapy for adolescents and adults with anorexia nervosa: a systematic review and meta-analysis. Psychiatry Res 271:484–501
71. Dray J, Wade TD (2012) Is the transtheoretical model and motivational interviewing approach applicable to the treatment of eating disorders? A review. Clin Psychol Rev 32(6): 558–565
72. Vitousek KB, Orimoto L (1993) Cognitive-behavioral models of anorexia nervosa, bulimia nervosa, and obesity
73. Stice E, Marti CN, Spoor S, Presnell K, Shaw H (2008) Dissonance and healthy weight eating disorder prevention programs: long-term effects from a randomized efficacy trial. J Consult Clin Psychol 76(2):329
74. Stice E, Rohde P, Shaw H (2013) The body project: a dissonance-based eating disorder

prevention intervention. Oxford University Press
75. Forbush KT, Richardson JH, Bohrer BK (2015) Clinicians' practices regarding blind versus open weighing among patients with eating disorders. Int J Eat Disord 48(7): 905–911
76. Kelly-Weeder S, Kells M, Jennings K, Dunne J, Wolfe B (2018) Procedures and protocols for weight assessment during acute illness in individuals with anorexia nervosa: a National Survey. J Am Psychiatr Nurses Assoc 24(3):241–246
77. Golden NH, Jacobson MS, Sterling WM, Hertz S (2008) Treatment goal weight in adolescents with anorexia nervosa: use of BMI percentiles. Int J Eat Disord 41(4): 301–306
78. Becker CB, Farrell NR, Waller G (2019) Exposure therapy for eating disorders. Oxford University Press
79. Tolin DF (2024) Doing CBT. Guilford
80. Brown A, Flint SW, Batterham RL (2022) Pervasiveness, impact and implications of weight stigma. EClinicalMedicine 47
81. Levinson CA, Vanzhula IA, Christian C (2019) Development and validation of the eating disorder fear questionnaire and interview: preliminary investigation of eating disorder fears. Eat Behav 35:101320
82. Wilson GT, Vitousek KM (1999) Self-monitoring in the assessment of eating disorders. Psychol Assess 11(4):480
83. Juarascio AS, Manasse SM, Goldstein SP, Forman EM, Butryn ML (2015) Review of smartphone applications for the treatment of eating disorders. Eur Eat Disord Rev 23(1): 1–11
84. Greenberger D, Padesky CA (1995) Mind over mood: a cognitive therapy treatment manual for clients. Guilford
85. Hildebrandt T, Bacow T, Greif R, Flores A (2014) Exposure-based family therapy (FBT-E): an open case series of a new treatment for anorexia nervosa. Cogn Behav Pract 21(4):470–484
86. Steinglass JE, Albano AM, Simpson HB, Wang Y, Zou J, Attia E et al (2014) Confronting fear using exposure and response prevention for anorexia nervosa: a randomized controlled pilot study. Int J Eat Disord 47(2):174–180
87. Vitousek KM, Gray JA (2005) Eating disorders. In: Gabbard GO, Beck JS, Holmes J (eds) Oxford textbook of psychotherapy. Oxford University Press, New York, pp 177–202
88. Craske MG, Treanor M, Conway CC, Zbozinek T, Vervliet B (2014) Maximizing exposure therapy: an inhibitory learning approach. Behav Res Ther 58:10–23
89. Huppert JD, Siev J, Kushner ES (2007) When religion and obsessive–compulsive disorder collide: treating scrupulosity in ultra-orthodox Jews. J Clin Psychol 63(10): 925–941
90. Springer KS, Tolin DF (2020) The big book of exposures: innovative, creative, and effective CBT-based exposures for treating anxiety-related disorders. New Harbinger
91. Hildebrandt T, Loeb K, Troupe S, Delinsky S (2012) Adjunctive mirror exposure for eating disorders: a randomized controlled pilot study. Behav Res Ther 50(12):797–804
92. Reilly EE, Anderson LM, Gorrell S, Schaumberg K, Anderson DA (2017) Expanding exposure-based interventions for eating disorders. Int J Eat Disord 50(10): 1137–1141
93. Levinson CA, Ralph-Nearman C, Brown ML, Gardner T, Cardi V, Treasure J et al (2021) A pilot randomized control trial of online exposure for eating disorders and mechanisms of change delivered after discharge from intensive eating disorder care: a registered report. Int J Eat Disord 54(11):2066–2074
94. Glasofer DR, Albano AM, Simpson HB, Steinglass JE (2016) Overcoming fear of eating: a case study of a novel use of exposure and response prevention. Psychotherapy 53(2):223
95. Trottier K, Carter JC, MacDonald DE, McFarlane T, Olmsted MP (2015) Adjunctive graded body image exposure for eating disorders: a randomized controlled initial trial in clinical practice. Int J Eat Disord 48(5): 494–504
96. Schaumberg K, Reilly EE, Gorrell S, Levinson CA, Farrell NR, Brown TA et al (2021) Conceptualizing eating disorder psychopathology using an anxiety disorders framework: evidence and implications for exposure-based clinical research. Clin Psychol Rev 83:101952
97. Craske MG, Kircanski K, Zelikowsky M, Mystkowski J, Chowdhury N, Baker A (2008) Optimizing inhibitory learning during exposure therapy. Behav Res Ther 46(1):5–27
98. Abramowitz J (2019) Exposure therapy for anxiety: principles and practice. Guilford

99. Garner DM, Bemis KM (1982) A cognitive-behavioral approach to anorexia nervosa. Cogn Ther Res 6:123–150
100. Taboas WR, McKay D, Whiteside SP, Storch EA (2015) Parental involvement in youth anxiety treatment: conceptual bases, controversies, and recommendations for intervention. J Anxiety Disord 30:16–18
101. Hughes EK, Sawyer SM, Loeb KL, Le Grange D (2015) Who's in the room? A parent-focused family therapy for adolescent anorexia nervosa. Eat Disord 23(4):291–301
102. Le Grange D, Hughes EK, Court A, Yeo M, Crosby RD, Sawyer SM (2016) Randomized clinical trial of parent-focused treatment and family-based treatment for adolescent anorexia nervosa. J Am Acad Child Adolesc Psychiatry 55(8):683–692
103. Eisler I, Dare C, Hodes M, Russell G, Dodge E, Le Grange D (2000) Family therapy for adolescent anorexia nervosa: the results of a controlled comparison of two family interventions. J Child Psychol Psychiatry 41(6):727–736
104. Le Grange D, Eisler I, Dare C, Russell GF (1992) Evaluation of family treatments in adolescent anorexia nervosa: a pilot study. Int J Eat Disord 12(4):347–357

Chapter 13

Mindfulness-Based Cognitive Therapy for Children (MBCT-C): A Clinical Approach for Teaching Mindfulness to Children

Randye J. Semple and Laila A. Madni

Abstract

Mindfulness-based cognitive therapy for children (MBCT-C) is a downward and sideways adaptation of the adult mindfulness-based cognitive therapy (MBCT) program. Downward from adults to children ages 8–13 years and sideways from focus as a treatment for depressive relapse to focus on managing stress and anxiety disorders. Both are structured group psychotherapies that have similar aims and goals. MBCT-C engages children in short, experiential group activities and offers a variety of interactive, age-appropriate activities that explore mindfulness using child-friendly language. The model is based on the concept that hands-on activities and creative repetition promote engagement and learning. Activities include breath meditations, still and movement-based body awareness practices, and sensory-based activities (i.e., mindful seeing, touching, tasting, smelling, and hearing). One or two therapists typically lead groups of six to eight children in 12 weekly, 90-min sessions. Like the adult MBCT program, children are expected to practice at home between sessions. Parents or caregivers attend two (adult-only) sessions that encourage participation in their child's home practices. Goals of MBCT-C are to:

- Enhance social-emotional resiliency in a safe group environment.
- Promote metacognitive insight, which often results in positive changes in how children relate to their own thoughts, emotions, and body sensations.
- Distinguish thoughts that are judgmental from those that describe experiences.
- Recognize that judgments can escalate anxiety or mood disturbances, which may then trigger maladaptive behaviors.
- Expand awareness of personal emotional and behavioral choices.
- Cultivate self-acceptance and acceptance of those things that cannot be changed.

Key words Mindfulness, Children, MBCT-C, MBCT, Anxiety, Meditation, Depression, Acceptance, Awareness, Group therapy

Robert D. Friedberg and Micaela A. Thordarson (eds.), *Cognitive Behavioral Therapy with Youth: Tradition and Innovation*, Neuromethods, vol. 231, https://doi.org/10.1007/978-1-0716-5134-6_13,

1 Introduction

The latest nationally representative data from the National Survey of Children's Health (NSCH) indicate that depression, anxiety, and behavioral problems are prevalent among US children and adolescents [1]. Anxiety disorders are the most common psychiatric diagnoses for children and adolescents, with lifetime prevalence rates as high as 32% for youth aged 13–17 years [2]. Up to 50% of adolescents' anxiety symptoms begin by age 6 [3]. Further exacerbating the problem, co-occurring anxiety or depressive disorders are present in many of these youth. Anxious children are also more likely to develop additional problems in the future. Data from the National Comorbidity Survey suggest that anxiety disorders are the most common precursors to major depression [3]. Anxiety can be considered a gateway disorder that is associated with both short- and long-term sequelae, including an increased risk of substance use [4]; academic [5], occupational, family, social, and conduct problems; and overall lower quality of life [6]. Effective early interventions have been shown to mitigate some of these sequelae and improve overall functioning [7].

Cognitive behavioral therapy (CBT) and psychotropic medications (mainly antidepressants and anxiolytics) are the current "best practice" treatments for childhood anxiety [8]. While psychotherapy and medications independently can be effective, the combination of CBT and medication for severe anxiety has been shown to have an additive effect [9]. However, many parents prefer not to medicate their children, and, although helpful for some children, CBT is not accessible or effective for many. A randomized clinical trial of CBT for children (aged 8–14 years) with anxiety disorders found that up to one half of children in a CBT group still met criteria for an anxiety disorder posttreatment [10]. A recent meta-analysis found that CBT for children resulted in remission only 49.4% of the time [11]. CBT is more effective than no treatment; however, it's clear that additional effective interventions are still needed.

Williams, Teasdale, Segal, and Kabat-Zinn [12] have defined mindfulness in the mindfulness-based cognitive therapy (MBCT) program as "the awareness that emerges though paying attention, on purpose, in the present moment, and nonjudgmentally to things as they are" (p. 47). Mindfulness-based cognitive therapy for children [13] has adopted this definition. MBCT and MBCT-C include some components of CBT; however, the theoretical bases and interventions are significantly different.

2 Theoretical and Empirical Foundations

CBT is a structured, present-focused, time-limited treatment aimed at reducing symptoms primarily through cognitive restructuring and development of emotion and behavior regulation strategies. The cognitive model suggests that holding inaccurate or distorted thoughts about a situation triggers emotional distress. Also structured, present-focused, and time-limited, MBCT and MBCT-C are based on a different premise—that thoughts, although related to emotions, are not the primary trigger for emotional distress. Rather, this model assumes that the source of emotional distress is the way we relate to our own thoughts. Instead of changing thoughts, when practicing mindfulness, clients learn to recognize that thoughts are simply events in the mind and cultivate acceptance of all their experiences—including thoughts and feelings—just as they are. Decentering is the metacognitive awareness of thoughts and emotions as transient "mind stuff" and is believed to be what allows us to relate to our thoughts in more wholesome ways [14]. Although acceptance and change are seemingly paradoxical goals, acceptance may serve as a catalyst for change. Additionally, Kuyken et al. [15] suggested that MBCT could be helpful in reducing symptoms of depression and depressive relapse by increasing self-compassion and decreasing reactivity. Schoenberg and Speckens [16] identified neurological changes potentially related to participation in MBCT, including recalibration of cortical circuitry, which allows individuals to disengage from rumination and associated negative cognitive and emotional processes.

3 Mindfulness-Based Cognitive Therapy

Mindfulness-based cognitive therapy (MBCT), the adult version of MBCT-C, was developed by Segal, Williams, and Teasdale [17] in the late 1990s to prevent depressive relapse in remitted patients with previous episodes of major depression. MBCT now has significant research support for overall health and well-being [18] as well as a variety of other mental health conditions such as severe depression and suicidality [19–21]; anxiety spectrum disorders [22] and obsessive-compulsive disorders [23]; posttraumatic stress disorder [24]; bipolar disorder [25]; and attentional disorders [26]. MBCT is also being used as an adjunctive treatment for cancer and other serious medical conditions [27–30].

MBCT is an 8-week group program that consists of 150-min weekly classes plus one all-day retreat. The classes teach mindful breath and body practices. Insights are facilitated by guided inquiries that follow each exercise. Mindfulness training is integrated with cognitive behavioral methods, which includes psychoeducation about depressive symptoms, awareness of changing mood states, and planning pleasant activities. Daily home practice is an important part of the program. Participants cultivate mindful awareness of thoughts, emotions, and body sensations with an open, curious, and a nonjudgmental stance. A primary aim of MBCT is to change how the clients relate to their own internal experiences. The 8-week program focuses on learning to shift from "automatic pilot" mode, which is an absence of present moment awareness to mindfulness. We do this by practicing giving attention to thoughts, emotions, and body sensations, which cultivates present-focused awareness, acceptance, and compassion toward internal and external experiences. Repeated mindfulness practice is aimed at developing the ability to decenter from thoughts and feelings.

4 Adapting Mindfulness Training for Children

Children differ from adults in three ways that are relevant to teaching mindfulness. First, children's cognitive and affective development is at an earlier stage than adults, which limits their abstract reasoning skills. To accommodate these differences, MBCT-C has adopted child-friendly language that is more concrete and somewhat more directive than MBCT. For example, we explain that mindfulness is "being curious and paying attention to whatever is happening right now, even if we don't like it." Initial sessions focus on teaching children to observe and differentiate their thoughts, feelings (emotions), and body sensations while engaged in short mindful breath and body practices. Activities are simple, not competitive, fun, and interesting.

Children also have less capacity for sustained attention than most adults and may get restless trying to stay engaged in one activity for a long time. You need only imagine trying to get a 9-year-old child to practice a 45-min seated breath meditation. For this reason, the program was extended from 8 to 12 weeks, and each session was shortened to 90 min. MBCT-C includes breath meditations, body scan, and mindful movement activities that are similar to those used in the adult program, albeit considerably shortened. Each weekly session includes opportunities for children to practice mindful movements or stretching, alternating with breath and body awareness practices, sensory-based activities, visualizations, reading, writing, or drawing activities, and interactive group discussions.

Lastly, children are dependent on their families in ways that most adults are not. Not only do they depend on caregivers to provide the basic needs required for survival, but they also depend on adults for emotional, social, and spiritual support.

Consequently, parents and caregivers are integrated into the program in four ways. We encourage parents to (a) attend an initial orientation session, (b) maintain contact with the therapist throughout the program, (c) attend a review session at the end of the program, and (d) participate with their child in the home practice activities.

Given that MBCT-C groups are intended for children with social, emotional, or behavioral challenges, the groups are generally smaller than adult groups and smaller than most classrooms. Depending on the age and dynamics of the group, and the specific needs of the children, we suggest that a single therapist limit the group size to 8 or 10; two co-therapists may feel comfortable working with up to 14 or 15 children. Having a second therapist present allows one to attend to a child who might need individual attention, while the other continues to facilitate the group.

In addition to child-friendly activities, the language used in MBCT-C should be accessible to children and appropriate to their developmental level. Language, concepts, and activities can be further simplified to respond to the emotional and cognitive needs of the group as a whole. Practice activity examples should be relevant to the social or emotional concerns of each group. A group of young teens may want to focus on exploring peer relationships and interpersonal dynamics, for example, whereas a younger group may be more interested in examining hobbies, sports, or family issues. MBCT-C has also been adapted to be a school-friendly intervention for teachers and school-based mental health providers. The *Mindfulness Matters Program* is modeled on MBCT-C but provides a more flexible framework along with a host of engaging activities for teaching mindfulness to children and adolescents with differing needs [31].

5 Research Support for MBCT-C

MBCT-C is an evidence-based intervention with growing research support [32]. An early study provided evidence for the overall acceptability and feasibility of mindfulness training adapted for children [33]. Children understood mindfulness concepts, practiced mindful awareness activities, and used them appropriately in their daily lives. The first randomized controlled trial of MBCT-C reported reductions in internalizing and externalizing symptoms, and fewer attentional problems as compared to a waitlisted control group [34, 35]. Other studies of MBCT-C have reported reductions in anxiety [36–43]; depressive symptoms [44, 45]; PTSD symptoms [46, 47]; externalizing behaviors, oppositional defiant, and conduct problems [48]; bullying behaviors [49]; anger, physical aggression, and hostility [50]; and inattention, conduct problems, and peer relations problems [51]. Improvements in mindfulness and resiliency were reported in a randomized, wait-listed controlled trial of 65 children ages 12–14 [52] and in a separate study of resiliency in children of divorced parents [53]. In a randomized trial comparing MBCT-C to the gold standard CBT, equivalent improvements were found on symptoms of anxiety and depression, quality of life, and attention control [54]. A randomized clinical trial of MBCT-C conducted with children hospitalized with cancer reported reductions in internalizing and attention problems [55].

Strawn et al. [56] conducted an fMRI study of MBCT-C for anxious children at risk for bipolar disorder and found increased activation of brain structures (insula and anterior cingulate) that are associated with fear processing and emotional regulation. Another fMRI study found topological changes which suggest that MBCT-C can help to balance the brain's ability to specialize in various types of information processing [57]. Yang and his colleagues proposed that such changes could enhance emotion regulation and increase the capacity for complex cognitive functions, two changes that may underlie the therapeutic benefits of MBCT-C treatment.

Although the research support for MBCT-C is growing, many of these studies suffer from methodological limitations. These include insufficiently powered designs, failure to use active control conditions, and not including a long-term follow-up assessment [58]. Studies have been conducted in a variety of settings ranging from school classrooms, outpatient clinics, inpatient psychiatric and medical units, and community centers with widely heterogenous groups of children. Heterogeneity, however, may be helpful at this stage of the research, since the positive findings across many studies support the effectiveness of MBCT-C to improve the mental health and well-being of children.

6 Intervention

6.1 Program Overview

MBCT-C [13] consists of 12, 90-min group sessions that use breath, body, and sensory-based activities to teach mindful awareness skills. Rather than sitting still for prolonged breath and body practices, children are encouraged to develop mindfulness using their five senses—engaging with the world around them with "eyes wide open."

Parents or caregivers are encouraged to cultivate personal mindfulness practices and participate in home practice assignments with their child. Written session summaries familiarize parents with the concepts, language, and practices of MBCT-C. They are invited to attend an Introduction to Mindfulness session prior to the beginning of the groups. During this 2-hour session, they are given a definition of mindfulness and its benefits and discuss expectations for participating in the program. They engage in guided meditation practices to introduce them to some of the activities and skills their child will be learning over the next few months. At the conclusion of the program, parents meet again with the therapist in a group to discuss their observations of the children's experiences, progress over the previous 12 weeks, or any concerns. Parents and therapists discuss strategies that may help each child continue his or her mindfulness practice after the program ends and explore child-specific modifications that might be needed.

MBCT-C is conducted in three phases. Sessions 1–3 are devoted to orienting the children to the concept of mindfulness, identifying program expectations and parameters, and participating in short, experiential mindfulness activities. Early activities focus on establishing awareness of the breath and body. Sessions 4–10 expand the practice of mindfulness using the five senses. Children learn to distinguish thoughts, feelings, and body sensations; understand how these internal experiences relate to each other; and recognize how thoughts and anxiety are connected. These middle sessions also teach ways to differentiate judging from observing (or noting), identify "choice points," and discover new ways to respond to situations—with mindful awareness rather than impulsive, emotional reactivity. Sessions 11 and 12 focus on generalization and continuation of mindfulness practices after the program ends. Children often develop strong connections with other group members, so we offer opportunities in these closing sessions for each child to explore and share his or her thoughts and feelings about being a part of the group and the ending of the group. To conclude the program, we celebrate our shared journey with a small "graduation" ceremony and party.

6.1.1 Materials and Supplies

When working with children, some supplies and materials are generally needed. These are inexpensive items, many of which may be found in your own home. Yoga or exercise mats are useful for mindful movement and mindfulness of body activities but are not required. A clean, carpeted floor or bath towels work fine. Child-sized meditation cushions can be used for seated meditations, but you can choose to sit in chairs or use small firm pillows. We use meditation bells (bells of mindfulness) to signal the beginning and end of each mindfulness activity. A variety of chimes, bells, gongs, or musical instruments such as triangles may be used for this purpose. Meditation bell apps are available for downloading to smartphones. Since children often benefit from organizational assistance, we give them binders or folders to store their session summaries, handouts, and home practice assignments. You will need an assortment of art supplies (markers, colored pencils, construction paper, tape, etc.) and, optionally, cheerful stickers to encourage attendance and home practice assignments.

Table 1
Partial list of supplies needed to conduct MBCT-C

Session supplies needed
1. A nonbreakable cup and water
2. One small box of raisins
3. One clementine orange per child and hand wipes or paper towels
4. Audio player with five or six short segments of music
5. Small musical instruments (e.g., finger symbols, castanets, tambourine, maracas, or shaker eggs). Many household items can be used as musical instruments (e.g., wooden chopsticks, water bottle with sand or pebbles, or pot lids) and a "conductor's baton" (e.g., a pencil or short stick)
6. Several optical illusions
7. An assortment of blocks or other items to build a three-dimensional abstract structure
8. Small items with interesting textures (e.g., lava rock, sea shell, piece of velvet cloth, a pine cone)
9. Scents in small containers (e.g., vinegar, vanilla extract, camphor, cinnamon, garlic, aromatic lotions)

Most of the sensory-based mindfulness activities require some materials. Many of these items can be found at home or purchased inexpensively. Each session includes written session summaries, poems, stories, or information handouts, and home practice assignments. These printable session materials and a session-by-session list of suggested materials can be found in the MBCT-C treatment manual [13]. A few examples of the supplies that can be used are listed in Table 1.

6.1.2 Guiding the Learning Process

An essential component of teaching mindfulness is the post-activity inquiry. Therapist-led discussions encourage participants to explore and deepen their understanding of mindfulness. The inquiries begin with an exploration of the thoughts, emotions, and body sensations experienced during the mindfulness activity. The aim of this is to shift the child away from thinking *about*, analyzing, or critiquing the experience to a nonjudgmental description of the actual experience. The second part of the inquiry explores what might be different about that experience when practiced with mindful awareness than the same activity done in everyday life and what might be learned from that awareness. The third part of the inquiry explores ways in which these understandings might be applied in everyday life. The inquiries aim to help children (a) learn to describe present moment experiences—in thoughts, feelings, and body sensations; (b) understand how living with mindful awareness differs from living on "autopilot;" and (c) explore how this knowledge might be integrated into everyday life.

Helping children notice when they are judging or describing must be done with care. Children often have no qualms offering blunt opinions, however, can easily take feedback about those opinions as criticism. For example, you might respond to a child who says, "That stuff (camphor) smells disgusting." with "I noticed you used a judgment word there. Would a Martian know what disgusting means? Can we find a different word to describe how this smells to a Martian who doesn't know what this stuff is?"

The aim is not to stop the children from making judgments but rather learn to see them for what they are—just thoughts, beliefs, desires, or expectations—that are not always consistent with reality and recognize how they influence our felt experiences. All of us have tendencies to cling to things we like and push away things we do not like. MBCT-C is based on the idea that unhappiness arises from holding onto beliefs and expectations about an experience that are incongruent with the reality of experience itself. For example, an adult who believes that all drivers should be polite and courteous will be unhappy when (inevitably) a rude driver cuts in front of them. Belief that a new love relationship will last forever will lead to unhappiness when that relationship inevitably ends—be it 3 months or 75 years later. What *is* will always prevail over what *should be*.

It is important to validate and normalize the challenges that often arise during mindfulness activities. Examples include finding it hard to focus on the object of attention, feeling unable to remain still during the quiet activities, and forgetting to do the home practice assignments. Validating the reality of these common issues encourages open, authentic dialogue and helps each child feel comfortable participating (at whatever level they can) in the group and home-based activities. A child who complains, "I'm terrible at this. My mind is all over the place." can be praised for noticing how busy her mind is and pointing out that observing all this busyness is how we practice mindfulness. Next, we briefly describe the 12 sessions of the program. Table 2 describes the theme, key points, and mindfulness practices for each session.

7 The 12-Session Program

Session 1

Children participate in an "ice-breaker" activity to get to know each other and are introduced to the structure of each session. One optional component is the Feely Faces Scale, which is a 7-point self-rating scale with visual representations of feelings that range from very unhappy to very happy. Using the Feely Faces Scale reminds children to "check in" with themselves at the beginning and the end of each session, becoming mindfully aware of their

Table 2
Overview of the theme, key points, and mindfulness activities for the 12-session program[a]

	Theme	Key points	In-session practices
1	Being on Automatic Pilot	We live much of our lives on automatic pilot Mindfulness exists, and it is a different, more helpful way of being in the world	Getting to Know You Discovering Awareness in a Cup What Mindfulness Means to Me Taking Three Mindful Breaths
2	Being Mindful Is Simple, But It Is Not Easy!	Living with awareness isn't easy, so why are we doing this anyway? We give attention to the barriers to practice Understanding the importance of practice Bringing awareness to the breath and body	Taking Three Mindful Breaths Raisin Mindfulness Mindfully Moving Slowly Taking Three Mindful Breaths
3	Who Am I?	Thoughts arise in the present but are often about the past or future Thoughts may not be accurate to the present reality Thoughts are not facts	Taking Three Mindful Breaths Mindfulness of the Body Hey, I Have Thoughts, Feelings, and Body Sensations! Listening to the Sounds of Silence Taking Three Mindful Breaths
4	A Taste of Mindfulness	We have thoughts, feelings, and body sensations, but these are not who we are Thoughts, feelings, and body sensations are not exactly the same as the events they describe	Introduction to Three-Minute Breathing Space Opening to One Orange Mindful Yoga Movements Three-Minute Breathing Space
5	Music to Our Ears	Thoughts, feelings, and body sensations often color how we experience the world With our thoughts, we create individual and unique relationships and experiences Awareness holds it all	Three-Minute Breathing Space Do You Hear What I Hear? Mindfulness of the Body Three-Minute Breathing Space
6	Sound Expressions	Practicing mindful awareness helps us recognize that thoughts, feelings, and body sensations influence how we express ourselves We can choose to express ourselves with mindful awareness	3-Minute Breathing Space Sounding Out Emotions—Mindfully Mindful Yoga Movements 3-Minute Breathing Space
7	Practice Looking	Judging is not the same as noting If we simply observe experiences rather than judge them, the experience may change We can choose to observe or note our experiences instead of judging them	3-Minute Breathing Space Visualizing with Clarity Mindful Yoga Movements Seeing What Is in the Mind's Eye 3-Minute Breathing Space

(continued)

Table 2
(continued)

	Theme	Key points	In-session practices
8	Strengthening the Muscle of Attention	Judging often changes how we experience the world Becoming more aware of judgments may change how we relate to thoughts and feelings Discovering "choice points"	3-Minute Breathing Space Seeing Through Illusions Moving Mindfully Seeing What Is Not There 3-Minute Breathing Space
9	Touching the World with Mindfulness	We have little control over most events that occur We do have choices in how we respond to events Choice points exist only in the present moment Bringing greater awareness to this moment, we may see more choice points	Three-Minute Breathing Space Being in Touch Mindfulness of the Body Three-Minute Breathing Space
10	What the Nose Knows	We often react to events by moving toward things we like or judge as "good" and moving away from things we don't like or judge as "bad" Judging an experience may interfere with seeing clearly what is present in each moment We have choices in how we respond to events	Three-Minute Breathing Space Judging Stinks! Mindful Yoga Movements 3-Minute Breathing Space
11	Life Is Not a Rehearsal	Mindfulness is available in everyday life We can practice mindful awareness using all our senses	3-Minute Breathing Space Thoughts Are Not Facts Feelings Are Not Facts Either Raisin Mindfulness Mindfulness Is… 3-Minute Breathing Space
12	Living With Presence, Compassion, and Awareness	Mindful awareness can be helpful in our daily lives Bringing greater awareness to our lives is a personal choice Living with awareness requires commitment, compassion, and continued daily practice	Three-Minute Breathing Space Exploring Everyday Mindfulness Program Evaluation (optional) Three-Minute Breathing Space Graduation Ceremony Graduation Party! Three-Minute Breathing Space
	3-Month Follow-Up	Support for maintaining a daily practice of mindful awareness	No session Therapist mails Letter to My Self and Daily Practice Calendar to each child

[a]Table excerpt from Semple and Lee (2011). Mindfulness-Based Cognitive Therapy for Anxious Children. New Harbinger Publications, Oakland, CA. Reprinted by permission

emotional state in that moment. Therapists then describe the concept of automatic pilot (autopilot) and engage children in a discussion to elicit examples from their own lives. To experience the differences between functioning on autopilot vs. functioning with mindful awareness, children pass a cup of water hand to hand around the group circle, first half-full, then completely full, and then again after darkening the room or asking them to close their eyes. Each time around the circle, the task demands increase and require giving greater attention to smaller details. During the post-activity inquiry, children explore how the experience differed with each new challenge. For example, one child might mention that passing a cup filled to the brim needed more focused attention than passing the half-full cup. Another might have noticed needing to move more slowly. When the room was darkened, some might notice that they were relying more on the other senses, perhaps being attuned to sounds, or even noticing the movement of air as the person seated next to them moved the cup closer. Many comment on the increasing difficulty. The therapist can use this activity as a metaphor for the development of greater mindfulness. Emphasizing the importance of practice and patience in cultivating mindfulness is useful as well. After this introduction to mindfulness, children are invited to draw a picture that represents their understanding of mindfulness. This drawing becomes the front cover of their binder, which will consist of Feely Faces Scales, session summaries, home practice assignments, poems, information handouts, and a personal journal. In concluding this first session, the therapist guides children in taking three mindful breaths and invites them to participate in a group poetry reading. At the end, each child can again check in with themselves and record their mood state on the Feely Faces Scale. Children are given a written summary of the session and home practices to complete before the next session. Every session ends with distributing a session summary and home practice handouts. These are the general components of each session. For the sake of brevity, the weekly mood check-in, home practice review, opening and closing breathing practices, poetry reading, session summary, and new home practice assignment will not be noted in the individual session descriptions below.

Session 2

Now that children have dipped their toes into the sea of mindfulness, they learn that although practicing mindfulness may be simple, it is often not easy. And not every child will take to mindfulness right away. When told that most of the mindfulness activities are done without any talking, one child replied, “That won’t work for me, I can’t shut up for that long.” This is where starting with short practices, offering gentle encouragement, and having parental support for home practices will be helpful.

Learning mindfulness takes practice, so each session consists mainly of experiential activities. Therapists invite children to participate in a few core activities to start this process. Children practice taking three mindful breaths while focusing on the inflow and the outflow of breath throughout their bodies. During the post-activity inquiry, the therapist elicits descriptions of thoughts, feelings, and bodily sensations observed during the breathing practice. Children then learn to incorporate mindfulness into an ordinary daily activity—mindful eating. Each child practices eating one raisin with full, mindful awareness of the experience—observing the qualities of the raisin—its texture, weight, appearance, taste, and smell. They may notice how the raisin transforms in the mouth and how the body responds to the raisin. Normally active children are not comfortable sitting still for 90 min, so each session includes one movement activity. Mindfully Moooving Slooowly is a slow movement activity aimed at bringing awareness to the many body sensations in each small movement. Children are invited to walk as slowly and mindfully as they can. The goal is not to get from one place to another but rather to experience each inbreath and each outbreath; to feel the movement of the muscles and the changing sensations in each foot as it contacts the ground; and bring awareness to the moments of stillness between each step. During the inquiry, therapists help children describe discrete sensations and begin to identify when the mind may be judging the activity, lost in another moment, or comparing it to other experiences. The session closes with another practice of taking three mindful breaths.

Session 3

Being mindful of restlessness or disengagement, therapists may wish to extend the three mindful breaths practice for a few additional breaths. Stop the activity if the group becomes restless. Mindful breathing is practiced in every session. Three mindful breaths may be added between any activity to provide a transitional space or a moment of self-calming. Session 3 introduces children to mindfulness of the body through a brief (approximately 10 min), guided body scan, which may be done sitting or lying down. During this activity, children are prompted to direct attention to one part of their motionless body at a time, starting with the toes of one foot, moving up the leg, shifting to the other foot and leg, and exploring the feeling of the breath moving through the body while bringing awareness to sensations in each part of the body. Children then have an opportunity to describe their experiences and explore what made a body scan different than just sitting or lying down with eyes closed.

This also lays the groundwork for the next practice, which is connecting thoughts, feelings, and body sensations. The CBT model suggests that changing thoughts can change feelings and behaviors. For example, a CBT therapist might encourage a child to identify anxiety-related distorted thoughts, challenge their accuracy and/or helpfulness, and then change the maladaptive thoughts to be more realistic or helpful in managing the anxiety. Alternatively, the MBCT-C model suggests that increasing awareness of thoughts, feelings, and body sensations changes how we respond to them and that decentering from the unhelpful thoughts reduces their emotional impact and thereby reduces anxiety. For example, a MBCT-C therapist might encourage the same child to sit and breathe with the anxious feelings, bring attention to the associated thoughts and body sensations, recognize them as internal experiences, cultivate nonjudgmental acceptance of these experiences as "just thoughts," "just feelings," or "just body sensations," and then help the child make a conscious choice about how to respond to them.

A brief mindful listening activity is an opportunity to adopt a different perspective by focusing attention on the silence between the sounds of a bell being rung several times. Then a visualization activity encourages children to closely observe their thoughts, feelings, and body sensations. A short inquiry after each of these activities deepens the children's understandings of their internal experiences.

Session 4

In this session, children are introduced to the 3-Min Breathing Space, which is a core activity borrowed unchanged from MBCT. The 3-Min Breathing Space will be practiced at the beginning and end of each remaining session. Building on the raisin activity that was done in Session 2, we now focus on mindfulness in the sense of taste. Mindful eating is practiced with an experience of touching, holding, smelling, and then peeling and eating an orange—while bringing mindful awareness to thoughts, feelings, and body sensations. The therapist invites the children to compare the experience of mindfully eating this orange to how they might typically eat (i.e., on autopilot). A post-activity inquiry follows. After being seated for 15–20 min, the next activity is a mindful movement practice with simple yoga postures. The aim is to cultivate mindful awareness of the body in movement rather than achieving perfect form. Competing with other group members is also discouraged. An inquiry and discussion of the mindful movement activity follows.

Session 5

In the previous session, children practiced mindfulness of taste. The next two sessions focus on mindfully engaging with sounds. Session 5 focuses on receptive listening while Session 6 focuses on expressively creating sounds. During Session 5, children practice listening with mindful awareness to five or six segments of music. Select a wide variety of instrumental music (without lyrics). For example, you might choose segments from a Jazz rift, African drumming, a Bach sonata, Bossa Nova, Calypso, and Mariachi.

After checking in with themselves to identify their thoughts, feelings, and body sensations, they create a title for each piece of music that reflects their current emotional state. Titles for a single piece can vary widely—from "wedding march" to "angry monster" to "funeral service." The post-activity inquiry explores how each child experienced something different while listening to the same music. The therapist helps them explore the concept of subjectivity and how their own thoughts and feelings might influence how they experience their worlds. They learn that they can transform an experience simply by attending to it with nonjudgmental awareness. They also begin to learn that awareness is the container within which all experiences are created. This concept is emphasized as one key point of this session: "awareness holds it all." The second mindfulness activity for this session is another brief body scan, which is followed by the second 3-Min Breathing Space.

Session 6

This session also attends to mindfulness of sounds but is now focused on the mindful expression of sounds. We offer a caveat here; this session can get somewhat loud. Based on children's previous experiences identifying thoughts, feelings, and body sensations arising in response to listening to sounds, they will now use their thoughts, feelings, and body sensations to guide the creation of sound by conducting their own symphony. Each child is asked to create a short piece of music that expresses his or her present-moment experience—in thoughts, feelings, and body sensations. Emphasize that each child will have a turn being the "conductor." As conductor, each child assigns each of the other children a musical instrument (or repurposed household item). Then, without words, the conductor guides his or her "orchestra" in creating a piece of music, instructing the musicians to play their instruments fast or slow, loud or soft—without words—simply by pointing or using a baton. The conductor invites the other children to create sounds that express the conductor's internal experiences. The inquiry also explores the subjectivity of how we interpret our experiences but mainly focuses on examining the ways we communicate our internal experiences to others. The second mindfulness activity for Session 6 is another mindful movement practice.

Session 7

Sessions 7 and 8 involve mindfulness of seeing. This session focuses on seeing clearly—practicing observing the world with mindful awareness of small details. Two visualization activities offer practice in using the mind's eye to see clearly. The first visualization helps them observe the different experiences of judging vs. observing or noting. The therapist guides the children in visualizing a pleasant scene. Afterward, the group inquiry explores the thoughts, feelings, and body sensations of the experience and identifies what specific elements defined the experience as being pleasant. The aim is to elicit understanding that thoughts about the experience (positive judgments) define the experience. Following a mindful yoga activity, the therapist guides the second visualization, which is combined with a drawing task. The task now is to first use the "mind's eye" to see clearly a familiar object in their home. This might be the kitchen stove, a piece of furniture, or a favorite toy. The therapist prompts the children to see clearly all the details of the object and then draw it. The children will then compare their drawing with the actual object. The aim is to notice ways the image in the mind's eye might distort the reality of a familiar object. They will explore elements they accurately recalled through the drawing, as well as components they may have changed or omitted entirely. During the home practice review at the beginning of Session 8, we review parts of the drawing that are not true to the reality of the object—drawn from vivid yet often inaccurate, memories. For some children, the drawing activities may bring out their lack of self-confidence or self-criticism more than many of the other activities. It is important to emphasize that the quality of the drawings is not important. The response we offer to a child's self-criticism is different than the often recommended cognitive response of challenging the negative self-statement and developing a more helpful statement. Instead, we gently identify the self-critical thought as simply being another judging thought and explore how that thought affects their emotions. We teach that all thoughts are just thoughts—even the thoughts that tell us that they're real. And that we don't have to believe everything we think.

Session 8

Mindfully looking at optical illusions helps children learn how their experiences might change when they intentionally shift their attention. Children discover that what they "see" is the actual object in combination with a cognitive interpretation of that object. Practice in redirecting attention is the aim of the Seeing Through Illusions activity. One thing they may discover with the rabbit–duck optical illusion, for example, is that both images cannot be seen simultaneously. Most children will initially see only one image, until the other is pointed out to them. Then, it becomes a mindfulness activity to shift attention back and forth to see first the rabbit, then the duck. This is followed by a mindful movement activity. The last mindfulness activity, Seeing What is Not There, is a negative-space drawing activity. The therapist creates a three-dimensional abstract structure from blocks or other objects in the room. The structure should have interior angles and open spaces.

The children are invited to sit in a circle around the structure and give mindful attention to drawing the spaces between (e.g., spaces where air, light, or shadows pass through). Each child sees the structure from a different angle, and consequently, each drawing will be different. This and the optical illusion activity demonstrate clearly that our felt experiences are influenced by what we choose to attend to and that attending in a different way—adopting a different perspective—changes how we experience the world. Variations of this activity include drawing only the straight lines, or drawing just the colors, or textures. The idea is to shift off autopilot by encouraging the children to look at the world in a different way. The post-activity inquiry can explore the idea that each of us sees the world differently and that no one perspective is the absolute truth. In this session, children are also introduced to "choice points," which arise in the moments between a trigger event and the response to that event. These are the points at which conscious choices can be made. The therapist also continues to cultivate the children's understanding the distinction between descriptive (objective) observations and thoughts and feelings (subjective) about the experience.

Session 9

In this session, children practice bringing mindful awareness to the sense of touch. The activity we call Being in Touch involves an assortment of small objects that have a variety of textures. Examples might include a hairbrush, bubble wrap, cotton balls, pine cones, or a fuzzy cat toy. The children take turns being blindfolded, and then one object is placed in his or her hand (or they may prefer to hold the object behind their back). Without seeing or naming it, the task is simply to describe the object. They can explore and describe the temperature, size, shape, texture, and weight of the object using only their sense of touch. The other children in the group can see the object, so they may add color, shading, and pattern to the description. All children help each other identify when a word becomes a label or a subjective judgment rather than an objective description. Each child gets the opportunity to practice mindfully touching and describing one object. This inquiry introduces the idea that, although we often have little control over things that happen, we always have choices in how we respond to events. Learning to stay present with what is happening, separating the objective from the judgmental, we may discover different ways to respond—different "choice points" that may be more skillful or wholesome than their habitual choices. Should moments of stress, frustration, or anger arise during a session, take the opportunity to remind the child that he or she doesn't need to allow themselves to be caught up in their emotions—they can choose how they want to respond—moment by moment. The body awareness activity for this session is a 10-min body scan.

Session 10

Our olfactory sense is one of the tools that humans developed to determine if something is safe to eat. Hence, smells tend to elicit rapid judgments, which are usually emotionally reactive "like" or "dislike" responses. Because this is an instinctive, nearly automatic reaction, mindfulness of smells can be considered an advanced practice. In this session, children practice smelling and describing different scents while observing their judgments about the scents. For this activity, the therapist offers five or six different scents, passing each scent, one at a time, around the circle to each child. The task is for the child to describe each scent using objective words while bringing awareness to the thoughts, feelings, and body sensations that arise in response to the scent. They may discover how difficult it is to find descriptive words (e.g., strong, sour, or sweet) rather than judgmental words (e.g., nice, gross, or yucky), particularly for strong scents. However, it can be easy to observe the emotional "like" or "dislike" response, particularly when the child physically moves closer to or away from the scent. This point can be emphasized during the inquiry—that we frequently react to events by moving toward the ones we judge positively and by moving away from those we judge negatively. Neither reaction, often driven by emotions or quick judgments, will reduce anxiety or increase happiness. The alternative to emotional reactivity is to use mindful awareness to first become aware of the available "choice points" and then consciously select how to respond. The second activity for this session is a mindful yoga movement activity.

Session 11

The last two sessions focus on maintaining and generalizing mindfulness as a way of life. Although practicing mindfulness in daily life is emphasized throughout the program, the aim of these last sessions is to reinforce the idea that practicing mindfulness can change how we relate to the world and to ourselves in significant ways. Practicing mindfulness can be done with formal breath or body practices, or informally, by remembering to bring mindful awareness to the things we see, hear, touch, smell, and taste. To promote decentering, the mindfulness activities for Session 11 again explore thoughts and feelings as internal events— mind stuff—not facts. Two visualization activities reinforce the idea that thoughts are not facts and that emotions are not facts either. Relationships between thoughts, emotions, and body sensations are also reexamined with a repeat of the same raisin activity that was done in Session 2. Each child then creates a drawing that becomes the back cover of his or her notebook. The drawing prompt is simply, "Mindfulness is" This activity gives children an opportunity to see how the meaning of mindfulness may have changed for them over the past months. We also take a few minutes to plan the Session 12 graduation party. The home practice activity for this week is to write a letter to themselves. Practicing mindfulness is not particularly difficult.

Remembering to practice is. So, these letters will be mailed back to the children 3 months after the end of the program. The written home practice handout includes a number of prompts to focus the writing. Examples of these include the following:

- What does mindfulness mean to me now?
- Has mindfulness changed the way I interact with other people?
- If so, how?
- What skills have I learned that might be helpful to me?
- How would I like to continue cultivating mindful awareness in my everyday life?
- What could I do that will help me remember to practice?

Session 12

The first goal of Session 12 is to reflect and explore ways to continue the practice of mindfulness. Following the opening 3-Min Breathing Space, the children are invited to share the Letter to Myself that they wrote after Session 11. By this time, the children tend to be very cohesive and supportive of each other, and although sharing is always optional, we have found that nearly all the children do choose to share what they wrote. This discussion is mostly guided by the children themselves. They discuss what influence the group may have had on their lives or their relationships; they explore everyday mindfulness—what has changed, and how mindfulness might bring other changes to their lives; and share strategies for continuing the practices beyond the weekly group sessions. They may also share what they have learned from other group members and often express appreciation for another child's presence in the group. A few of the pearls of wisdom we've heard from children in our groups are the following:

- "I've learned that I don't *have* to get mad at my little brother."
- "It's easier to have a good day when you start by smiling and taking three mindful breaths."
- "When I get worried about something, I tell myself that it's just thoughts."
- "Mindfulness makes me realize when I'm going to get in trouble and just breathe instead."
- "I taught my mom how to be mindful and now we don't fight so much."
- "Mindfulness is not letting thoughts get in your way when you can be doing something better."
- "Mindfulness helps me see my choice points."
- "I don't hate cleaning my room as much."
- "Mindfulness makes my life easier."

The discussion ends with another 3-Min Breathing Space. Then comes the graduation ceremony and farewell party. The graduation ceremony is a way of creating closure by marking the dissolution of the group, celebrating the shared journey, their achievements, and exchanging goodbyes. As therapists, we may choose to write a short farewell letter and prepare congratulatory "graduation from MBCT-C" certificates for each child. In addition, we share one more mindfulness activity in which we distribute a small remembrance of the program to each child (usually a brightly colored polished pebble). The children often bring cookies, drinks, balloons, or other party goods to the party. We generally bring pizza as our contribution.

8 When the Program Ends

Mindfulness as a way of life is emphasized throughout MBCT-C, and more so as the group nears its end. Children have considered ways to bring mindfulness into their everyday lives. Therapists have encouraged the children to explore the benefits of using mindfulness skills in daily activities and emphasized the importance of ongoing practice. Children, like adults, may find themselves skipping daily practices and moving back into old habits. The therapist has fostered motivation for ongoing practice and worked with the children to devise practical strategies for integrating mindfulness in everyday life. Three months after the program ends, letters the children wrote to themselves after Session 11 should be mailed back to them. In this way, we allow each child's own "words of wisdom" to motivate their ongoing practice.

Adapting MBCT-C for Individual Therapy

Individual therapy adaptations of MBCT-C interventions are not difficult, and the treatment manual offers suggestions for adapting each session to this modality. The key considerations are the reduced time for individual sessions and the ways that the therapist engages with the child. Some activities will naturally take less time. Others can be abbreviated. Most do not require major modifications. We recommended shortening an activity rather than eliminating it altogether. Session review handouts may be taken home and read with the parent rather than reading them during the session. A few activities, for example, the expressive sounds activity (Session 6) will need significant modifications for an individual therapy format. The treatment manual suggests starting by prompting the child to identify thoughts, feelings, and body sensations associated with different situations that have triggered strong emotional responses (e.g., an argument with a sibling or not being invited to a party). Then, rather than assigning musical instruments to others, the child creates sounds that represent each identified emotion. The therapist participates by responding with sounds that represent his or her emotional response to the child's music.

More important is how the therapist engages with the child when a therapist–child dyadic relationship replaces the group interactions. The therapist will need to participate more fully in each of the activities, with mindful attention that their own contributions do not come across as being judgmental or critical of the child. Children are commonly corrected by adults and are likely to be less comfortable relating to an adult on an equal basis. Authenticity is important—you are on a shared journey and should expect to learn as much as your young client. Closure is important. Children in individual therapy often develop strong attachments to their therapists. Each child should have a small graduation, perhaps choosing to read the letter he or she wrote, reflect and discuss his or her experience of MBCT-C, and receive a small remembrance and graduation certificate from the therapist. Together, the therapist and child can plan a fun activity or share food to celebrate completing the program.

Adapting MBCT-C for Special Populations

No one treatment, including MBCT-C, suits the needs of all children. The treatment manual discusses adaptations as needed for children with special needs. Examples include eliminating the mindful tasting activities for children with food allergies or eating disorders and modifying the mindful movement activities for children with physical challenges. Children with severe anxiety frequently prefer to keep their eyes open during the breathing activities or body scan and may feel more comfortable seeing the item they are touching rather than being blindfolded. Children with autism spectrum disorder or other developmental disabilities may have verbal, intellectual, or sensory issues that need to be addressed. Activities and language may need to be simplified, shortened, or the program extended to allow more frequent repetition to match the learning pace of the children. Children with reading difficulties may prefer not to read aloud to the group. Therapeutic sensitivity, mindful attention, and understanding the needs of each child in your group will allow you to modify the interventions appropriately.

9 Discussion

It is important to keep in mind that mindfulness is not simply a set of techniques one can learn and then sit back and enjoy the fruits [59]. There are a number of personal and clinical training considerations for therapists wishing to conduct mindfulness-based interventions. First, basic clinical competencies must be mastered. Given that MBCT-C is generally conducted in groups, therapists should be knowledgeable about the theory and practice of group facilitation. Group therapy adds the complexity of managing multiple interpersonal dynamics while cultivating the wholesome therapeutic alliance that is essential to any therapy. When implementing any child therapy, some understanding of child development and family dynamics is helpful. Recognizing how depression, anxiety, trauma, and other issues manifest in children is essential, as well as understanding the precepts and practices of CBT.

In CBT, as in most traditional psychotherapies, effecting change is the primary aim. Therapists and clients tend to focus on what changes in thinking or behaviors might be beneficial. Seasoned therapists know that promoting change can be complex and difficult. Particularly in longer-term therapies, motivation and engagement can wax and wane. Alternatively, mindfulness-based interventions foster acceptance rather than change. Mindful therapists may become aware of desires in themselves to promote change in their clients. MBCT-C encourages observing rather than judging, participating with authenticity, accepting the client as they are, and letting go of the desire to push for change. Moving from a change-oriented model to an acceptance and awareness-based approach may be the biggest challenge for conventionally trained therapists. The development of mindfulness is a process. Learning to trust the process will motivate the changes you seek.

As the field of child mindfulness continues to develop, mindful therapists and practitioners are evaluating what constitutes competency in teaching mindfulness. The use of mindfulness-based interventions in clinical settings has strengthened the call for unified standards and specialized clinical training, but no consensus has yet been reached. The British Association of Mindfulness-Based Approaches has established *Good Practice Guidelines for Teaching Mindfulness-Based Courses* [60]. These guidelines emphasize the need to be a trained mental health provider with experience providing evidence-based treatments or have a background in education, with experience in group-based teaching. Familiarity with the population being served is recommended. Personal participation in the mindfulness-based curriculum one plans to facilitate and having a regular practice of mindfulness are also strongly encouraged. Other recommendations include completion of a mindfulness-based teacher training program, regular supervision, consultation, and collaboration with colleagues, and maintaining awareness of current empirical evidence. Saltzman's [61] *Qualifications for Teachers of Mindfulness-Based Stress Reduction Youth Programs* suggests qualifications for mindfulness group facilitation with children. These include 5 years of mindfulness experience, attendance at mindfulness retreats, experience teaching yoga or another movement practice in a mindfulness context, professional experience working with children, facilitating process groups, and creating a network of mindfulness colleagues.

Many mindfulness programs, including MBCT-C, emphasize the importance of cultivating a personal mindfulness practice. As noted in both sets of guidelines discussed above, we strongly encourage therapists interested in facilitating MBCT-C to learn and practice mindfulness in their own lives. Participating in a Mindfulness-based Stress Reduction (MBSR; [62]) or MBCT program and developing a regular personal mindfulness practice can be a good beginning. Although MBCT-C does not yet have a formal therapist-training program, we agree with Saltzman's model, with one important addition. Since MBCT-C is intended as a clinical intervention, we believe that facilitators must have clinical training to conduct child therapy with understandings of both CBT and child psychopathology.

Mindful therapists teach through their own embodiment of mindfulness. Embodying mindfulness includes cultivating present-focused awareness, acting and speaking with intention, and maintaining a nonjudgmental stance while incorporating acceptance and compassion [59]. Additional qualities include being able to establish a sense of safety in the group, convey genuine interest, open curiosity and compassion for all aspects of the human experience, "unshockability," and profound empathy [61]. However, it may not be realistic to expect oneself to embody all these traits at once. Crane [59] noted that becoming a mindful therapist requires "the development of awareness (clear seeing) cultivated during formal and informal practice in daily life; the intention to underpin one's practice and life with a particular attitudinal framework characterized by warm acceptance, non-striving and curiosity; and a willingness to engage in an alive exploration of what it is to be human and of how our suffering is caused" (p. 160). Development of these qualities takes practice, patience, and a tremendous amount of self-compassion. Although this personal journey can be demanding, it also holds the possibility of a lifetime of enrichment.

Conflicts of Interest

The authors have no conflicts of interest to declare that are relevant to the content of this chapter.

References

1. Ghandour RM, Sherman LJ, Vladutiu CJ, Ali MM, Lynch SE, Bitsko RH et al (2019) Prevalence and treatment of depression, anxiety, and conduct problems in US children. J Pediatr 206:256–67.e3
2. Kessler RC, Avenevoli S, McLaughlin KA, Green JG, Lakoma MD, Petukhova M et al (2012) Lifetime co-morbidity of DSM-IV disorders in the US National Comorbidity Survey Replication Adolescent Supplement (NCS-A). Psychol Med 42(9):1997–2010
3. Merikangas KR, He JP, Burstein M, Swanson SA, Avenevoli S, Cui L et al (2010) Lifetime prevalence of mental disorders in U.-S. adolescents: results from the National Comorbidity Survey Replication-Adolescent

Supplement (NCS-A). J Am Acad Child Adolesc Psychiatry 49:980–989
4. Goodwin RD, Fergusson DM, Horwood LJ (2004) Association between anxiety disorders and substance use disorders among young persons: results of a 21-year longitudinal study. J Psychiatr Res 38(3):295–304
5. Mojtabai R, Stuart E, Hwang I, Eaton W, Sampson N, Kessler RC (2015) Long-term effects of mental disorders on educational attainment in the National Comorbidity Survey ten-year follow-up. Soc Psychiatry Psychiatr Epidemiol 50(10):1577–1591
6. Swan AJ, Kendall PC (2016) Fear and missing out: youth anxiety and functional outcomes. Clin Psychol Sci Pract 23(4):417–435
7. Swan AJ, Kendall PC, Olino T, Ginsburg G, Keeton C, Compton S et al (2018) Results from the Child/Adolescent Anxiety Multimodal Longitudinal Study (CAMELS): functional outcomes. J Consult Clin Psychol 86(9): 738–750
8. Sigurvinsdóttir AL, Jensínudóttir KB, Baldvinsdóttir KD, Smárason O, Skarphedinsson G (2020) Effectiveness of cognitive behavioral therapy (CBT) for child and adolescent anxiety disorders across different CBT modalities and comparisons: a systematic review and meta-analysis. Nord J Psychiatry 74(3):168–180
9. Asarnow JR, Rozenman MS, Carlson GA (2017) Medication and cognitive behavioral therapy for pediatric anxiety disorders. JAMA Pediatr 171(11):1038
10. Flannery-Schroeder EC, Kendall PC (2000) Group and individual cognitive-behavioral treatments for youth with anxiety disorders: a randomized clinical trial. Cogn Ther Res 24: 251–278
11. James AC, Reardon T, Soler A, James G, Creswell C (2020) Cognitive behavioural therapy for anxiety disorders in children and adolescents. Cochrane Database Syst Rev 11
12. Williams JMG, Teasdale JD, Segal ZV, Kabat-Zinn J (2007) The mindful way through depression: freeing yourself from chronic unhappiness. Guilford, New York
13. Semple RJ, Lee J (2011) Mindfulness-based cognitive therapy for anxious children: a manual for treating childhood anxiety. New Harbinger Publications, Oakland
14. Lenz AS, Hall J, Bailey Smith L (2016) Meta-analysis of group mindfulness-based cognitive therapy for decreasing symptoms of acute depression. J Spec Group Work 41(1):44–70
15. Kuyken W, Watkins E, Holden E, White K, Taylor RS, Byford S et al (2010) How does mindfulness-based cognitive therapy work? Behav Res Ther
16. Schoenberg PL, Speckens AE (2014) Modulation of induced frontocentral theta (Fm-theta) event-related (de-)synchronisation dynamics following mindfulness-based cognitive therapy in major depressive disorder. Cogn Neurodyn 8(5):373–388
17. Segal ZV, Williams JMG, Teasdale JD (2013) Mindfulness-based cognitive therapy for depression, 2nd edn. Guilford, New York
18. Querstret D, Morison L, Dickinson S, Cropley M, John M (2020) Mindfulness-based stress reduction and mindfulness-based cognitive therapy for psychological health and well-being in nonclinical samples: a systematic review and meta-analysis. Int J Stress Manage 27:394–411
19. Zhang B, Fu W, Guo Y, Chen Y, Jiang C, Li X et al (2022) Effectiveness of mindfulness-based cognitive therapy against suicidal ideation in patients with depression: a systematic review and meta-analysis. J Affect Disord 319:655–662
20. Williams K, Hartley S, Langer S, Manandhar-Richardson M, Sinha M, Taylor P (2022) A systematic review and meta-ethnographic synthesis of mindfulness-based cognitive therapy for people with major depression. Clin Psychol Psychother 29(5):1494–1514
21. Tseng HW, Chou FH, Chen CH, Chang YP (2023) Effects of mindfulness-based cognitive therapy on major depressive disorder with multiple episodes: a systematic review and meta-analysis. Int J Environ Res Public Health 20(2):1555
22. Li J, Cai Z, Li X, Du R, Shi Z, Hua Q et al (2021) Mindfulness-based therapy versus cognitive behavioral therapy for people with anxiety symptoms: a systematic review and meta-analysis of random controlled trials. Ann Palliat Med 10(7):7596612–7597612
23. Pseftogianni F, Panagioti M, Birtwell K, Angelakis I, Nezu AM (2023) Mindfulness interventions for obsessive-compulsive and related disorders: a systematic review and meta-analysis of randomized controlled trials. Clin Psychol 30(3):233–243
24. Caceres-Melillo R, Wagner C (2023) Mindfulness-Based Cognitive Therapy (MBCT) and Mindfulness-Based Stress Reduction (MBSR) in the treatment of Post-Traumatic Stress Disorder (PTSD): a literature review. Salud Ment (México) 46(1):35
25. Perich T, Manicavasagar V, Mitchell PB, Ball JR, Hadzi-Pavlovic D (2013) A randomized controlled trial of mindfulness-based cognitive

therapy for bipolar disorder. Acta Psychiatr Scand 127:333–343

26. Schoenberg PL, Hepark S, Kan CC, Barendregt HP, Buitelaar JK, Speckens AE (2014) Effects of mindfulness-based cognitive therapy on neurophysiological correlates of performance monitoring in adult attention-deficit/hyperactivity disorder. Clin Neurophysiol 125(7):1407–1416
27. Tamura N, Park S, Sato Y, Sato Y, Takita Y, Ninomiya A et al (2022) Predictors and moderators of outcomes in mindfulness-based cognitive therapy intervention for early breast cancer patients. Palliat Support Care 20(2): 159–166
28. Haller H, Winkler MM, Klose P, Dobos G, Kummel S, Cramer H (2017) Mindfulness-based interventions for women with breast cancer: an updated systematic review and meta-analysis. Acta Oncol (Stockh) 56(12): 1665–1676
29. Abbott RA, Whear R, Rodgers LR, Bethel A, Thompson Coon J, Kuyken W et al (2014) Effectiveness of mindfulness-based stress reduction and mindfulness based cognitive therapy in vascular disease: a systematic review and meta-analysis of randomised controlled trials. J Psychosom Res 76:341–351
30. Metcalf CA, Dimidjian S (2014) Extensions and mechanisms of mindfulness-based cognitive therapy: a review of the evidence. Aust Psychol 49(5):271–279
31. Semple RJ, Willard C (2019) The mindfulness matters program for children and adolescents: strategies, activities, and techniques for therapists and teachers. Guilford Press, New York
32. CEBC (2024) Mindfulness-based cognitive therapy for children: california evidence-based clearinghouse for child welfare. https://www.cebc4cw.org/program/mindfulness-based-cognitive-therapy-for-children-mbct-c/
33. Semple RJ, Reid EF, Miller L (2005) Treating anxiety with mindfulness: an open trial of mindfulness training for anxious children. J Cogn Psychother 19:379–392
34. Semple RJ, Lee J, Rosa D, Miller LF (2010) A randomized trial of mindfulness-based cognitive therapy for children: promoting mindful attention to enhance social-emotional resiliency in children. J Child Fam Stud 19(2): 218–229
35. Lee J, Semple RJ, Rosa D, Miller L (2008) Mindfulness-based cognitive therapy for children: results of a pilot study. J Cogn Psychother 22:15–28
36. Cotton S, Luberto CM, Sears RW, Strawn JR, Stahl L, Wasson RS et al (2016) Mindfulness-based cognitive therapy for youth with anxiety disorders at risk for bipolar disorder: a pilot trial. Early Interv Psychiatry 10:426–434
37. Cotton S, Kraemer KM, Sears RW, Strawn JR, Wasson RS, McCune N et al (2020) Mindfulness-based cognitive therapy for children and adolescents with anxiety disorders at-risk for bipolar disorder: a psychoeducation waitlist controlled pilot trial. Early Interv Psychiatry 14(2):211–219
38. Shetty R, Kongasseri S, Rai S (2020) Efficacy of mindfulness based cognitive therapy on children with anxiety. J Cogn Psychother 34(4): 306–318
39. Kalnciema B, Svence G (2020) Changes in school anxiety and emotional self-regulation of an adolescence experimental group following the MBCT-C Programme. Sabiedrība Integrācija Izglītība (Society Integration Education) 3:253–265
40. Syeda MM, Andrews JJW (2021) Mindfulness-based cognitive therapy as a targeted group intervention: examining children's changes in anxiety symptoms and mindfulness. J Child Fam Stud 30(4):1002–1015
41. Bruns K, La Guardia A, Brubaker M, Farrow J, Cotton S, DelBello M (2021) Effectiveness of mindfulness-based cognitive therapy-child with youth who have a parent diagnosed with bipolar I disorder. J Ment Health Couns 43(1): 59–74
42. LaGue A, Eakin G, Dykeman C (2019) The impact of mindfulness-based cognitive therapy on math anxiety in adolescents. Prev Sch Fail Altern Educ Child Youth 63(2):142–148
43. Jang J-Y, Shin H-K (2019) The effects of Mindfulness-based Cognitive Therapy (MBCT-C) on middle school students anxiety and mindfulness. Korean J Youth Stud 26(8): 53–78
44. Esmaeilian N, Tahmasian K, Dehghani M, Mootabi F (2013) Effectiveness of mindfulness-based cognitive therapy on depression symptoms in children with divorced parents. J Clin Psychol 3:47–57
45. Ames CS, Richardson J, Payne S, Smith P, Leigh E (2014) Mindfulness-based cognitive therapy for depression in adolescents. Child Adolesc Mental Health 19(1):74–78
46. Kostova Z, Sarmiento I, Fulwiler C, Semple RJ (2019) Implementing Mindfulness-based Cognitive Therapy for Children (MBCT-C) with history of trauma in inpatient settings: a pilot study. In: European Society for Trauma and Dissociation conference: the legacy of trauma and dissociation – body and mind in a new perspective, Rome, 25 Oct 2019

47. Kostova Z, Semple RJ (2020) Implementing Mindfulness-based Cognitive Therapy for Children (MBCT-C) with history of trauma and severe psychopathology in inpatient psychiatric settings: a pilot study. In: International Society for Traumatic Stress Studies (ISTSS) 36th annual meeting: bridging science and practice to reach underserved communities; virtual conference, 12 Nov 2020
48. Haydicky J, Wiener J, Badali P, Milligan K, Ducharme JM (2012) Evaluation of a Mindfulness-Based intervention for adolescents with learning disabilities and co-occurring ADHD and anxiety. Mindfulness 3:151–164
49. Faraji M, Talepasand S, Boogar I (2019) Effectiveness of mindfulness-based cognitive therapy for child on bullying behaviors among children. Int Arch Health Sci 6(1):52
50. Milani A, Nikmanesh Z, Farnam A (2013) Effectiveness of Mindfulness-Based Cognitive Therapy (MBCT) in reducing aggression of individuals at the Juvenile Correction and Rehabilitation Center. Int J High Risk Behav Addict 2(3):126
51. Haydicky J, Shecter C, Wiener J, Ducharme JM (2015) Evaluation of MBCT for adolescents with ADHD and their parents: impact on individual and family functioning. J Child Fam Stud 24(1):76–94
52. Peter A, Srivastava R, Agarwal A, Singh AP (2022) The effect of mindfulness-based cognitive therapy on anxiety and resilience of the school going early adolescents with anxiety. J Indian Assoc Child Adolesc Ment Health 18(2):176–185
53. Esmaeilian N, Dehghani M, Dehghani Z, Lee J (2018) Mindfulness-based cognitive therapy enhances emotional resiliency in children with divorced parents. Mindfulness 9(4): 1052–1062
54. Wright KM, Roberts R, Proeve MJ (2019) Mindfulness-Based Cognitive Therapy for Children (MBCT-C) for prevention of internalizing difficulties: a small randomized controlled trial with Australian primary school children. Mindfulness 10:2277–2293
55. Abedini S, Habibi M, Abedini N, Achenbach TM, Semple RJ (2020) A randomized clinical trial of a modified Mindfulness-Based Cognitive Therapy for children hospitalized with cancer. Mindfulness 12:141–151
56. Strawn JR, Cotton S, Luberto CM, Patino LR, Stahl LA, Weber WA et al (2016) Neural function before and after Mindfulness-Based Cognitive Therapy in anxious adolescents at risk for developing bipolar disorder. J Child Adolesc Psychopharmacol 26(4):372–379
57. Yang J, Lei D, Suo X, Tallman MJ, Qin K, Li W et al (2022) A preliminary study of the effects of mindfulness-based cognitive therapy on structural brain networks in mood-dysregulated youth with a familial risk for bipolar disorder. Early Interv Psychiatry 16(9): 1011–1019
58. Semple RJ, Burke C (2019) State of the research: physical and mental health benefits of mindfulness-based interventions for children and adolescents. OBM Integr Complement Med 4(1):1–31
59. Crane R (2017) Mindfulness-based cognitive therapy: distinctive features. Routledge
60. British Association of Mindfulness-Based Approaches Good practice guidelines for teaching mindfulness-based courses 2023 Available from: https://bamba.org.uk/
61. Saltzman A (2014) A still quiet place: a mindfulness program for teaching children and adolescents to ease stress and difficult emotions. New Harbinger Publications, Oakland
62. Kabat-Zinn J (1994) Wherever you go there you are: Mindfulness meditation for everyday life. Hyperion, New York

Chapter 14

Cognitive Behavior Therapy with Diverse Youth

Alayna L. Park, Clarissa V. Velez, and Rachel E. Kim

Abstract

Cognitive behavioral therapy (CBT) produces robust effects in treating a variety of youth mental health concerns, yet there have been concerns about its cultural compatibility for ethnic-racial minoritized (ERM) youths. Cultural adaptation of CBT can address these concerns by systematically modifying treatment to better align with the cultural beliefs, values, and practices of ERM youth clients. Clinicians working with ERM youth clients must make conscientious decisions about whether, when, and how to culturally adapt CBT. This chapter describes a dynamic, five-step process for culturally adapting CBT: (1) planning to use unadapted CBT, unless otherwise indicated; (2) continuously assessing whether/when to culturally adapt CBT; (3) determining how to culturally adapt CBT; (4) implementing the cultural adaptation; and (5) evaluating the effect of the cultural adaptation. A fictionalized case example is provided to illustrate how this process can be applied in clinical practice, highlighting the importance of treatment integrity and cultural responsiveness in achieving successful treatment outcomes.

Key words Cultural adaptation, Ethnic-racial minority, Youth of color, Culture and mental health, Cultural competence, Cognitive behavioral therapy, Clinical decision-making

1 Cognitive Behavior Therapy with Diverse Youth

Cognitive behavior therapy (CBT) is a time-limited psychosocial treatment that focuses on changing maladaptive thinking patterns and behaviors to alleviate mental health concerns [1]. To date, CBT has been tested in thousands of randomized clinical trials and has demonstrated efficacy in treating various youth mental health concerns including anxiety, depression, traumatic stress, inattention and hyperactivity, eating, and substance use disorders [2–7]. However, most of these randomized clinical trials do not include significant numbers of youths from minoritized groups [8]. Given that cultural values, beliefs, and practices influence mental health and help-seeking [9], the significant underrepresentation of minoritized youths in the empirical CBT literature raises questions about the cultural compatibility of CBT.

Robert D. Friedberg and Micaela A. Thordarson (eds.), *Cognitive Behavioral Therapy with Youth: Tradition and Innovation*, Neuromethods, vol. 231, https://doi.org/10.1007/978-1-0716-5134-6_14,

This chapter describes whether, when, and how CBT should be culturally adapted for ethnic-racial minoritized (ERM) youths. Although there are several approaches to enhance the cultural compatibility of CBT [10]—including increasing clinician awareness of cultural beliefs, values, and practices through training and consultation [11] and addressing cultural similarities and differences between youths and their clinicians [12]—this chapter focuses on culturally adapting the therapeutic content and processes of CBT. Cultural adaptations (i.e., systematic modifications intended to enhance treatment compatibility with a minoritized group's cultural beliefs, values, and practices) have promising empirical research support for producing positive treatment engagement and outcomes among ERM youth [10, 13]. Because empirical studies on cultural adaptation typically focus on mitigating ethnic-racial mental health disparities, this chapter provides recommendations for culturally adapting CBT for ERM youths. That said, clinicians should consider their client's intersecting identities—including but limited to age, disability status, religion and spirituality, ethnicity and race, socioeconomic status, sexual orientation, indigenous heritage, national origin, and gender [14, 15]—when conceptualizing the case, and planning and delivering CBT.

2 Should CBT Be Culturally Adapted?

Although culturally adapted CBT produces positive youth mental health outcomes, the relative advantage of culturally adapted CBT over unadapted CBT for ERM youths is less clear [10, 16]. On one hand, culturally adapted CBT is designed to be more congruent with an ethnic-racial group's cultural beliefs, values, and practices, and this cultural compatibility can help ERM clients feel understood and empowered [12]. On the other hand, unadapted CBT produces robust effects [17], and untested modifications may reduce the effectiveness of CBT [18, 19]. Modifying treatment solely based on a client's ethnic-racial group also risks making them feel misunderstood and stereotyped, particularly if they hold neutral attitudes about their ethnic-racial group or suffer from internalized racism [20]. Therefore, clinicians deciding *whether* to culturally adapt CBT should carefully weigh these potential benefits and risks.

The benefits of culturally adapting CBT usually outweigh the risks in three scenarios [21, 22]. First, cultural adaptations might be indicated when an ERM group has needs that are not covered by unadapted CBT. Although unadapted CBT is efficacious for treating various mental health concerns experienced by ethno-racially diverse youths [23], ERM youths face minority stressors that are not typically addressed in CBT protocols. As many as 90% of ERM youths report experiencing ethnic-racial discrimination [24–26]. These experiences of discrimination can manifest into symptoms of racism-based stress and trauma such as negative thoughts about oneself, others, or the world, hypervigilance, or strong affective or physiological reactions to internal or external cues that resemble an aspect of the discriminatory event(s) [27, 28]. While these symptoms largely overlap with symptoms of posttraumatic stress disorder, trauma-focused CBTs require cultural adaptation to successfully address racism-based stress and trauma [29, 30]. For example, one element of trauma-focused CBTs involves gradually exposing youths to internal or external reminders of the traumatic event. To address a youth's posttraumatic stress reaction of avoiding car rides after being in a bad car accident, a clinician delivering a trauma-focused CBT might construct a set of objectively harmless situations that gradually reintroduces the youth to riding in cars. Similarly, to address a youth's racism-based stress and trauma, a clinician might construct a set of interpersonal situations where the youth is unlikely to receive a racist response (e.g., interacting with friendly congregants at their church). However, given the pervasiveness of prejudice and discrimination, a clinician addressing racism-based stress and trauma might also augment their delivery of trauma-focused CBT with racial socialization practices (e.g., role-playing how their client could respond to microaggressions or overt discrimination, encouraging their client to generate positive self-statements related to their ethnicity/race). These cultural adaptations can meet the unique needs of ERM youths while maintaining the core elements of CBT.

Second, cultural adaptations might be indicated for an ERM group that is disproportionately at risk for low treatment engagement. Most youths in need do not receive mental health services [31], and ERM youths face unique barriers to engaging in mental health treatment. One such barrier is stigmatized beliefs about mental health concerns and help-seeking. Stigma is greater among Latine, Black, and Asian Americans than White Americans [32]. Additionally, Latine, Black, and Asian and Pacific Islander parents are less likely to believe that their child's mental health concerns are caused by biopsychosocial factors (e.g., physical health problems, personality, relational issues, trauma exposure) compared with White parents [33]—and these ethnic-racial differences in etiological beliefs can contribute to disparities in treatment engagement [34, 35]. Therefore, cultural adaptations, such as assessing ERM families' etiological beliefs or addressing the myth of equating mental health concerns with "being crazy," might be indicated for ERM groups expressing stigma. Another barrier to engaging in mental health treatment is cultural mistrust. Historical misdiagnosis and maltreatment have made some ERM groups understandably apprehensive to engage in mental health treatment [36, 37]. To address this apprehension, clinicians can augment

CBT with therapeutic practices to promote ERM family engagement such as dedicating extra time to rapport building, sharing more about themselves, or providing families with detailed verbal and written information about CBT [38, 39]. A more practical barrier to engaging in mental health treatment is the language gap that exists between some ERM families and clinicians. While nearly one-third of Latine and Asian Americans have limited English proficiency [40], few mental health facilities employ multilingual staff. Loho and Rosenheck [41] examined the reported language skills of staff employed across 11,682 mental health facilities in the United States. Their results showed that 33% of mental health facilities employed staff who spoke Spanish, but fewer than 3% of facilities employed staff who spoke any other non-English language (e.g., Chinese, French, Russian). Therefore, cultural adaptations for delivering CBT through an interpreter might be necessary for many ERM families with limited English proficiency.

Third, cultural adaptations might be indicated for an ERM group that is disproportionately at risk for poor treatment outcomes with unadapted CBT. Although most clinical trials indicate no ethnic-racial differences in treatment outcomes [42], approximately 15% of studies show stronger effects for White youths than ERM youths [16, 43]. For example, Rohde and colleagues [44] randomized 114 adolescents with major depressive disorder to receive unadapted CBT or a life skills-tutoring intervention. Results from this randomized clinical trial showed that depression recovery time was significantly faster in the unadapted CBT condition than the life skills-tutoring control condition but only for White adolescents; recovery time did not differ by condition for ERM adolescents. As another example, Lochman and Wells [45] randomly assigned 183 fourth- and fifth-grade boys at risk for aggressive behaviors to receive the Coping Power Program (i.e., an unadapted CBT and behavioral parent training program designed to increase social competence, self-regulation, and positive parental involvement) or usual school behavioral health services. Ethnic-racial identity moderated 1-year follow-up outcomes, such that the Coping Power Program had larger effects on parent-rated substance use and school behavior functioning for White boys than for ERM boys. In cases such as these, culturally adapting CBT might be a promising strategy for mitigating any ethnic-racial disparities in treatment outcomes.

3 *When* Should CBT Be Culturally Adapted?

Cultural adaptation should be considered not only for ERM *groups* at risk of minority stressors, low treatment engagement, and poor treatment outcomes but also for *individual* ERM clients [46–49]. For example, a Black youth who experiences performance anxiety due to fears of reinforcing negative stereotypes about their ethnic-racial group might benefit from culturally adapted CBT, whereas a Black youth who experiences performance anxiety due to fears of being ridiculed might benefit from unadapted CBT. An individual ERM client might also benefit from some cultural adaptations and some elements of unadapted CBT. Expanding upon the previous example, cultural adaptation might be indicated for assessing a client's fears of reinforcing negative ethnic-racial stereotypes, constructing exposures that are as harmless as possible, and problem-solving how the client can respond to discriminatory remarks. However, therapeutic practices such as providing psychoeducation about anxiety and CBT or identifying thoughts, feelings, and behaviors in distressing situations might not necessitate cultural adaptation. Therefore, in addition to deciding whether to culturally adapt CBT, clinicians should make informed decisions about *when* to culturally adapt CBT.

Clinicians should consider culturally adapting CBT when an ERM client reports a minority stressor. Minority stressors can be assessed throughout treatment using interviews (e.g., Cultural Formulation Interview [50]) and/or questionnaires (e.g., Perceptions of Racism in Children and Youth [24]; Perceived Discrimination Scale [51]; Index of Race Related Stress [52]; and Youth-Racism-Based Traumatic Stress Symptoms Scale [53]) [54]. For example, the Cultural Formulation Interview (CFI) is a free, 16-item semi-structured interview that assesses the following:

1. Cultural definition of the problem (e.g., *Sometimes people have different ways of describing their problem to their family, friends, or others in their community. How would you describe your problem to them?*)
2. Cultural perceptions of cause, context, and supports (e.g., *Sometimes aspects of people's background or identity can make their problem better or worse. Are there any aspects of your background or identity that make a difference to your problem?*)
3. Cultural factors affecting self-coping and past help-seeking (e.g., *Often people look for help from many different sources, including different kinds of doctors, helpers, or healers. In the past, what kinds of treatment, help, advice, or healing have you sought for your problem?*)
4. Cultural factors affecting current help-seeking (e.g., *Sometimes clinicians and clients misunderstand each other because they come from different backgrounds or have different expectations. Is there anything that we can do to provide you with the care you need?*).

The CFI can be administered in as little as 20 minutes, with practice, and can supplement measures assessing client symptoms and functioning [55, 56]. If an ERM client endorses a minority stressor, then cultural adaptation might be indicated; otherwise, unadapted CBT might be sufficient for meeting the client's needs. Notably, an ERM client might initially conceal minority stressors due to cultural mistrust or concerns that their experience will be invalidated. Therefore, clinicians should assume responsibility for broaching topics of ethnicity, race, and culture with their clients [57–60]. Once the minority stressor is addressed, then the decision of whether to proceed with culturally adapted CBT or unadapted CBT should be reevaluated.

Clinicians should also consider culturally adapting CBT when an ERM client is at risk of low treatment engagement. Treatment engagement is commonly assessed by observing session attendance [61]—such that late arrivals, cancellations, or no-shows might indicate an engagement problem. Treatment engagement can also be assessed using questionnaires. For example, My Thoughts about Therapy is a free, 35-item questionnaire that prompts youths and their parents to rate the extent to which they agree with statements pertaining to five dimensions of treatment engagement [62, 63]:

1. Relationship, or therapeutic alliance (e.g., *My counselor understands my culture and values*)
2. Expectancy, or beliefs that treatment will be helpful (e.g., *I believe my counselor knows how to help other people who are like me*)
3. Attendance, or presence at treatment sessions (e.g., *I am able to attend appointments even when there are other important things going on in my life*)
4. Clarity, or understanding about the treatment approach and the roles of each person involved in treatment (e.g., *The work I do with my clinician fits with my goals*)
5. Homework, or client participation in sessions and completion of homework between sessions (e.g., *When I learn something new in counseling, I try to use it right away at home or at school*)

Treatment engagement should be recurrently assessed throughout treatment, so that clinicians can intervene in time to prevent premature dropout. If a client is at risk of low treatment engagement, then their clinician should consider the compatibility between CBT and their client's cultural beliefs, values, and practices. Low treatment engagement alone does not justify the need for cultural adaptation, as treatment engagement can be hindered by factors unrelated to culture (e.g., transportation issues, childcare issues, unrealistic treatment expectations). Instead, the decision to culturally adapt CBT should be informed by both the ERM client's treatment engagement and the fit between CBT and their culture. Relatedly, improvements in an ERM client's treatment engagement should prompt clinicians to consider whether continued cultural adaptation is needed or whether to transition to unadapted CBT.

Additionally, clinicians should consider culturally adapting CBT when an ERM client is not progressing as expected. Routine outcome monitoring—or the process of continuously gathering and reflecting on client outcome data—is an evidence-based practice for promoting high-quality youth mental health care [64, 65]. By monitoring a client's progress throughout the course of treatment, clinicians can determine when a client might not be progressing as expected and adapt the treatment plan to better meet their client's needs. Clinicians can practice routine outcome monitoring by gathering data on their client's symptoms and functioning at each session using brief standardized questionnaires (e.g., Screen for Child Anxiety-Related Disorders, [66]; Patient Health Questionnaire-Adolescent version [67]), or ideographic measures (e.g., Top Problems Assessment [68]). For a list of psychometrically sound, free, brief measures of youth symptoms and functioning, refer to Becker-Haimes et al. [69]. To interpret their client's progress, these outcome data can then be compared with that measure's clinical cutoff scores, CBT benchmarks, or the outcomes of other clients in their mental health organization [70]. For example, if clients at a mental health organization typically display reductions in anxiety symptoms by the fourth session, but a particular client's symptoms remain unchanged, then their clinician should consider the cultural compatibility of their treatment plan. As with ERM clients at risk of low treatment engagement, the decision to culturally adapt CBT should be guided not solely by treatment progress but by a holistic evaluation of treatment progress using multi-informant standardized measures and behavioral observation, the congruence between CBT and a client's culture, and any other factors that might be impacting a client's outcomes (e.g., baseline symptom severity, comorbid conditions, emergent life events, treatment fidelity). This decision should also be reevaluated when improvements in client outcomes are observed.

4 *How* Should CBT Be Culturally Adapted?

If cultural adaptation for an ERM youth client is indicated, then clinicians must consider *how* to culturally adapt CBT. Cultural adaptations should directly address the specific treatment fit issue they are intended to resolve. For example, dispelling harmful myths about mental health concerns might be helpful for an ERM youth expressing stigma but might feel irrelevant to an ERM youth disclosing minority stressors. To improve the cultural compatibility of CBT for specific ERM youth clients, clinicians should consider surface versus deep structure adaptations as well as various ways to adapt the content and process of CBT.

4.1 Surface and Deep Structure Adaptations

Cultural adaptations to CBT can be categorized into surface structure adaptations and deep structure adaptations [71]. *Surface structure adaptations* are intended to enhance the fit between CBT and observable, "superficial" characteristics of an ERM group. Examples of surface structure cultural adaptations include delivering CBT in the client's primary (non-English) language, using examples that are familiar to the ERM group (e.g., role models who share the same ethnic-racial identity, cultural foods or music), or delivering CBT in local or cultural institutions such as public parks, churches, or barbershops. Surface structure adaptations are often used to enhance an ERM youth client's initial "buy-in" [72].

Deep structure adaptations focus on incorporating "cultural, social, historical, and psychological" considerations into assessment, case conceptualization, and treatment planning. Examples of deep structure adaptations include assessing for minority stressors, augmenting CBT with racial socialization practices, or addressing intergenerational differences in acculturation among family members. Deep structure adaptations are used to elicit a cultural understanding of an ERM client's mental health concern and fundamentally alter the client's engagement with or response to CBT.

4.2 CBT Content Adaptations

In addition to using surface and/or deep structure adaptations, clinicians can also consider adapting the content of CBT, or *what* they work on with clients [73]. For instance, clinicians can adapt their *conceptualization*, or how information is framed. Conceptualization adaptations can be made when assessing an ERM client's mental health concern. For example, a clinician working with an ERM client who is being bullied at school might hypothesize about the role of prejudice and discrimination. To test this hypothesis, the clinician could ask their client whether they feel different from other students—and more broadly, in what way, or more specifically, whether they perceive these differences to be related to ethnicity and race. Conceptualization adaptations can also be made when delivering CBT. To expand upon the previous example, if the ERM client confirms the clinician's hypothesis that the client believes they are being bullied because of their ethnicity or race, then the clinician might frame cognitive restructuring around challenging the client's perceived inability to cope rather than challenging the perceived discrimination.

Clinicians can also adapt their *message*, or what they say to clients. For example, when working with an ERM youth client who expresses stigma, a clinician might refer to themselves as a "teacher" or "coach." Relatedly, to address stigmatized beliefs about mental health concerns and treatment while harnessing the advantages of strengths-based approaches [74], particularly for ERM youths [75], a clinician might refer to psychotherapy as a "skill-building program." Alternatively, a clinician working with a family who uses corporal punishment and is reluctant to try time-out might refer to the time-out chair as a "punishment chair" [39]. Message adaptations also include the use of metaphors or *dichos* (e.g., "el que persevera alcanza/*he who perseveres achieves*" [76] to help ERM youths and their families understand CBT content or promote skill use.

Additionally, clinicians can adapt their *procedures*, or what they do with clients. Cultural adaptations to procedures might involve augmenting CBT with culturally congruent therapeutic practices, modifying CBT practices to address cultural needs, or reducing or reordering CBT practices that are less compatible with an ERM client's culture [77]. For example, a clinician working with an ERM youth client experiencing racism-based stress and trauma might augment CBT with racial socialization practices, which are not included in most CBT protocols. As another example, a clinician working with an ERM family with intergenerational acculturation differences might modify their use of cognitive restructuring—which is typically used to address a youth's maladaptive thoughts—to help parents challenge beliefs that their child's growing independence is a sign of disrespect. As an example of a reordering cultural adaptation, a clinician might introduce the skill of mindful awareness before providing psychoeducation to an ERM client with a daily meditation practice to address any of the client's concerns about the cultural compatibility of psychotherapy. While reducing adaptations might be conceptually appealing and indicated in some cases, they should be made sparingly due to the risk of omitting core CBT elements that could compromise treatment effectiveness.

4.3 CBT Process Adaptations

Clinicians can also adapt the therapeutic process, or modify *how* they work with their clients [73]. For instance, clinicians can adapt their *therapeutic style*, or how they behave with clients. A collaborative therapeutic relationship is often considered a hallmark of high-quality psychotherapy [78], and many families appreciate when their clinician involves them in the treatment planning process [79, 80]. However, some Asian American families, particularly those who are less acculturated, might view their clinician as the person who knows best and consequently prefer a more directive approach [81]. Relatedly, some Latine families might value *personalismo* (i.e., warm interpersonal interactions [82]) and expect to engage in casual conversation with their clinician at the beginning of each treatment session.

Clinicians can also adapt their *communication*, or how they deliver the message (i.e., what they say to clients). CBT content is traditionally communicated using a combination of didactics and active learning techniques such as modeling or role-playing. However, some ERM clients might prefer to practice applying skills to a situation described in a vignette as opposed to their own situation for various reasons, including stigmatized beliefs about mental health concerns, desires to maintain psychological distance from distressing thoughts or emotions, or cultural beliefs about not disclosing personal struggles to anyone outside of the family. CBT can also be adapted by communicating through storytelling, theater, visual media (e.g., drawings, cartoons), or play.

Further, clinicians can adapt the *change agent*, or who is involved in treatment. Except for behavioral parent training which focuses on changing parent's behaviors, youths are often the primary participant in CBT. While parents play a crucial role in supporting and sustaining treatment gains [83, 84], the extent to which they or other individuals in a youth's life should be involved in CBT might vary. For example, parent involvement in CBT might be particularly important for some Latine families who value respeto (i.e., showing appropriate courtesy and decorum to other people of a particular age, gender, and social status [85]) and expect for parents to not only be involved in but also approve all treatment planning decisions.

4.4 Steps for Culturally Adapting CBT

The following five steps can help clinicians decide whether, when, and how to culturally adapt CBT for ERM youths: (1) plan to use unadapted CBT, unless otherwise indicated; (2) continuously assess whether/when to culturally adapt CBT; (3) determine how to culturally adapt CBT; (4) culturally adapt CBT; and (5) reevaluate whether to culturally adapt CBT. This dynamic cultural adaptation process can be likened to the scientific method [86]. That is, clinicians are encouraged to first *observe* an ERM youth client's response to unadapted CBT and then *generate a hypothesis* about whether cultural adaptation might improve, worsen, or have no effect on the client's response. Next, they would *design* a treatment plan, *test the effect* of that treatment plan, and *analyze their results.* Positive client outcomes would reflect either a true positive (i.e., cultural adaptation was indicated, and treatment was culturally adapted) or a true negative (i.e., cultural adaptation was not indicated, and treatment was not culturally adapted), while stagnant or deteriorating outcomes might reflect a type I error (i.e., cultural adaptation was not indicated, and treatment was culturally adapted) or a type II error (i.e., cultural adaptation was indicated, and treatment was not culturally adapted) [87]. Clinicians would then draw a conclusion based on their results and reinitiate the cultural adaptation process by generating a new hypothesis about whether cultural adaptation is indicated at that point in treatment.

4.4.1 Step 1: Plan to Use Unadapted CBT, Unless Otherwise Indicated

Unadapted CBT produces robust effects and is likely to be effective for many ERM youths with various mental health concerns [10, 23]. Given potential barriers to implementing an empirically tested, culturally adapted CBT protocol that matches an ERM client's unique characteristics—if such a protocol even exists—unadapted CBT might be preferable to many other available treatment options. Therefore, unless an evidence-based culturally compatible CBT protocol can be implemented or unadapted CBT is viewed as culturally unacceptable, clinicians should default to delivering and monitoring their ERM client's response to unadapted CBT.

4.4.2 Step 2: Continuously Assess Whether/When to Culturally Adapt CBT

Throughout treatment, clinicians should assess and be prepared to address minority stressors, low treatment engagement, or poor treatment outcomes. If any of these signals to culturally adapt CBT arise, then clinicians should weigh the potential benefits and risks of culturally adapting their treatment plan. Notably, minority stressors, treatment engagement problems, and interruptions to treatment progress can occur at any point in treatment. Therefore, the absence of these signals does not imply that CBT should never be culturally adapted for a given ERM client, only that it is not warranted at this time. If a clinician plans to continue delivering unadapted CBT, then they should routinely assess for changes in minority stress, treatment engagement, and treatment progress. If a clinician determines that cultural adaptation is indicated for an ERM youth client, then they should proceed with this dynamic process.

4.4.3 Step 3: Determine How to Culturally Adapt CBT

After deciding whether/when to culturally adapt CBT, clinicians must also decide how to culturally adapt CBT. Any cultural adaptations should address the perceived cultural incompatibility. Surface structure adaptations might be used to elicit client "buy-in," and deep structure adaptations might be used to contextualize the cultural, social, historical, and psychological factors causing or maintaining an ERM client's mental health concern [71]. The content of CBT, or what a clinician works on with their ERM client, can be adapted by modifying the conceptualization, message, or procedures [73]. The process of CBT, or how a clinician works with their ERM client, can also be adapted by modifying therapeutic style, communication, or change agents. If a clinician is uncertain about how to culturally adapt CBT, then they might evaluate the applicability of various cultural adaptations or a combination of cultural adaptations to address any minority stressors, treatment engagement problems, or interruptions to treatment progress. Given the influence of a clinician's identities, lived experiences, and biases on clinical decision-making and practice [88], consulting with a supervisor or colleague is encouraged.

4.4.4 Step 4: Implement the Cultural Adaptation

After deciding whether/when and how to culturally adapt CBT, clinicians should prepare to implement the cultural adaptation. This preparation phase might involve reviewing relevant treatment protocols, consulting with cultural adaptation experts, or role-playing delivery of the cultural adaptation. Clinicians should then implement the cultural adaptation with their ERM youth client.

4.4.5 Step 5: Evaluate the Effect of the Cultural Adaptation

After implementing a cultural adaptation, clinicians should evaluate whether it achieved the desired and expected effect. If the ERM youth client responds, then their clinician must decide whether to revert to unadapted CBT (**Step 1**) or proceed with additional cultural adaptations (**Steps 2–5**). Conversely, if the ERM youth client is unaffected or negatively impacted, then their clinician should consider whether the cultural adaptation was necessary (**Step 1**) or whether alternative or additional cultural adaptations are required to achieve the desired and expected effect (**Steps 2–5**).

4.5 Case Example

This dynamic cultural adaptation process can be seen with the fictionalized case of Carlos Álvarez, a 9-year-old Mexican American boy who was referred to psychotherapy by his mother for throwing tantrums, not obeying his parents' commands, and being "disrespectful" to his parents. After gathering multi-informant reports on his symptoms and functioning and observing his behaviors at home and in school, Carlos was diagnosed with attention-deficit/hyperactivity disorder (ADHD), combined type.

4.5.1 Step 1: Plan to Use Unadapted CBT, Unless Otherwise Indicated

Carlos' clinician planned to use unadapted behavioral parent training to address functional impairments caused by his inattention and hyperactivity. Behavioral parent training has strong research support for reducing behavior problems among youths with ADHD [4]. Consistent with many behavioral parent training protocols, the clinician planned to provide psychoeducation about ADHD and behavioral parent training, teach Carlos' parents how to better reinforce desired behaviors by providing labeled praise and strengthening the parent-child relationship, and teach his parents how to decrease problematic behaviors by ignoring temper tantrums or punishing noncompliant and destructive behaviors.

4.5.2 Step 2: Continuously Assess Whether/When to Culturally Adapt CBT

Although unadapted behavioral parent training effectively reduces behavior problems among Mexican-American youths [89], Carlos' clinician hypothesized that several cultural factors might interfere with the Álvarez family's treatment engagement and outcomes. First, Carlos' mother had limited English fluency and predominantly spoke Spanish, which limited her ability to communicate with others including English-speaking mental healthcare providers. Second, Carlos' father—who was bilingual in English and Spanish—worked long hours as a construction worker, which prevented him from regularly attending treatment sessions. Third, Carlos' parents held values of *respeto*, *machismo* (which emphasizes a man's role as the head of the family and as exuding strength), and *marianismo* (which emphasizes a woman's primary role as a mother responsible for raising respectful children). These pan-Latine cultural values shaped perceptions about Carlos' behaviors at home and in school. Given the potential influence of these cultural factors on Carlos' behaviors and response to treatment, his clinician decided to make cultural adaptations.

4.5.3 Steps 3–4: Determine How to Culturally Adapt CBT and Implement the Cultural Adaptations

Before treatment began, a surface structure adaptation was implemented by assigning a bilingual clinician to work with the Álvarez family. This case assignment relieved Carlos from having to interpret for his mother and allowed his mother to speak openly and privately about her experiences—while maintaining the core elements of behavioral parent training. Given the benefit of father engagement on child development and behavioral parent training outcomes [90], Carlos' clinician explored the possibility of rescheduling treatment sessions to better fit the family's schedule. When this proved unfeasible, the clinician used a communication adaptation of providing Carlos' father with handouts summarizing behavioral parent training skills. Carlos' clinician also responded to the Álvarez family's pan-Latine cultural values using various cultural adaptations. Because Carlos' parents viewed his behaviors as disrespectful, they believed that the clinician should work directly with Carlos. The clinician consequently adapted procedures by dedicating additional treatment sessions to providing psychoeducation to ensure that Carlos' parents understood why their participation was essential. During the skill-building phase of treatment, the clinician adapted messaging around labeled praise to emphasize its effect on promoting "respectful" behaviors. The clinician also used another procedure adaptation by implementing cognitive restructuring with Carlos' mother to help her challenge self-critical thoughts—rooted in cultural beliefs about gender roles—that Carlos' need for psychotherapy signified that she had failed as a mother.

4.5.4 Step 5: Evaluate the Effects of the Cultural Adaptations

The Álvarez family responded positively to each cultural adaptation. After helping Carlos' mother restructure her maladaptive thoughts about her son's psychotherapy, the Álvarez family became more engaged and began showing great treatment gains. Based on these improvements, the clinician hypothesized that no additional cultural adaptations were needed and returned to delivering unadapted behavioral parent training. The initial cultural adaptations made by the clinician helped establish a strong therapeutic alliance, which allowed the Álvarez family to trust the clinician's unadapted guidance on reducing Carlos' problematic behaviors by ignoring temper tantrums or punishing noncompliant and destructive behaviors. Had the clinician used unadapted behavioral parent training from the outset, then the Álvarez family might have disengaged from treatment before their goals were met. The clinician's use of this dynamic cultural adaptation process enabled the Álvarez family to receive treatment that was not only supported by research but also responsive to their needs and lived experiences.

5 Conclusion

There is no one-size-fits-all approach to culturally adapting CBT for ERM youths. While modifying elements of CBT carries the risk of compromising its well-established effectiveness, failing to address the unique needs and experiences of ERM youths can perpetuate long-standing ethnic-racial mental health disparities. Although unadapted CBT might be effective for many ERM youths, cultural adaptations might be necessary when minority stressors are present or when cultural incompatibilities hinder treatment engagement or progress. In these situations, clinicians can adapt the structure, content, or process of CBT to better align with the needs of their ERM youth client. By following a dynamic process—starting with a plan to use unadapted CBT, continuously assessing the need for cultural adaptation, determining the appropriate cultural adaptation, implementing the cultural adaptation, and evaluating its effect—clinicians can ensure that they are delivering effective and culturally responsive mental health care to their ERM youth clients.

Conflicts of Interest

The authors have no conflicts of interest to declare that are relevant to the content of this chapter.

References

1. Beck JS (2021) Cognitive behavior therapy: basics and beyond. The Guilford Press, New York
2. Datta N, Matheson BE, Citron K et al (2023) Evidence based update on psychosocial treatments for eating disorders in children and adolescents. J Clin Child Adolesc Psychol 52:159–170
3. Dorsey S, McLaughlin KA, Kerns SEU et al (2017) Evidence base update for psychosocial treatments for children and adolescents exposed to traumatic events. J Clin Child Adolesc Psychol 53(46):303–330
4. Evans SW, Owens JS, Bunford N (2014) Evidence-based psychosocial treatments for children and adolescents with attention-deficit/hyperactivity disorder. J Clin Child Adolesc Psychol 43:527–551
5. Higa-McMillan CK, Francis SE, Rith-Najarian L et al (2016) Evidence base update: 50 years of research on treatment for child and adolescent anxiety. J Clin Child Adolesc Psychol 45: 91–113
6. Hogue A, Henderson CE, Ozechowski TJ et al (2014) Evidence base on outpatient behavioral treatments for adolescent substance use: updates and recommendations 2007–2013. J Clin Child Adolesc Psychol 43:695–720
7. Weersing VR, Jeffreys M, Do M-CT et al (2017) Evidence base update of psychosocial treatments for child and adolescent depression. J Clin Child Adolesc Psychol 46:11–43
8. Polo AJ, Makol BA, Castro AS et al (2019) Diversity in randomized clinical trials of depression: a 36-year review. Clin Psychol Rev 67:22–35
9. Sanchez AL, Jent J, Aggarwal NK et al (2022) Person-centered cultural assessment can improve child mental health service engagement and outcomes. J Clin Child Adolesc Psychol 51:1–22
10. Huey SJ, Park AL, Galán CA et al (2023) Culturally responsive cognitive behavioral therapy for ethnically diverse populations. Annu Rev Clin Psychol 19:51–78

11. Chu W, Wippold G, Becker KD (2022) A systematic review of cultural competence trainings for mental health providers. Prof Psychol Res Pract 53:362–371
12. Chu J, Leino A, Pflum S et al (2016) A model for the theoretical basis of cultural competency to guide psychotherapy. Prof Psychol Res Pract 47:18–29
13. Chu J, Leino A (2017) Advancement in the maturing science of cultural adaptations of evidence-based interventions. J Consult Clin Psychol 85:45–57
14. Hays PA (2022) Addressing cultural complexities in counseling and clinical practice: an intersectional approach. American Psychological Association, Washington, DC
15. Hays PA (2024) Four steps toward intersectionality in psychotherapy using the ADDRESSING framework. Prof Psychol Res Pract 55: 454–462
16. Huey SJ, Tilley JL, Jones EO et al (2014) The contribution of cultural competence to evidence-based care for ethnically diverse populations. Annu Rev Clin Psychol 10:305–338
17. Weisz JR, Kuppens S, Ng MY et al (2017) What five decades of research tells us about the effects of youth psychological therapy: a multilevel meta-analysis and implications for science and practice. Am Psychol 72:79–117
18. Wiltsey Stirman S (2022) Implementing evidence-based mental-health treatments: attending to training, fidelity, adaptation, and context. Curr Dir Psychol Sci 31:436–442
19. Wiltsey Stirman S, Baumann AA, Miller CJ (2019) The FRAME: an expanded framework for reporting adaptations and modifications to evidence-based interventions. Implement Sci 14:58
20. Sue DW, Sue D, Neville HA et al (2022) Counseling the culturally diverse: theory and practice. Wiley, Hoboken
21. Barrera M, Castro FG (2006) A heuristic framework for the cultural adaptation of interventions. Clin Psychol Sci Pract 13:311–316
22. Lau AS (2006) Cultural adaptations of evidence-based treatments: examples from parent training. Clin Psychol Sci Pract 13:295–310
23. Pina AA, Polo AJ, Huey SJ (2019) Evidence-based psychosocial interventions for ethnic minority youth: the 10-year update. J Clin Child Adolesc Psychol 48:179–202
24. Pachter LM, Coll CG (2009) Racism and child health: a review of the literature and future directions. J Dev Behav Pediatr 30:255–263
25. Priest N, Paradies Y, Trenerry B et al (2013) A systematic review of studies examining the relationship between reported racism and health and wellbeing for children and young people. Soc Sci Med 95:115–127
26. Tobler AL, Maldonado-Molina MM, Staras SAS et al (2013) Perceived racial/ethnic discrimination, problem behaviors, and mental health among minority urban youth. Ethn Health 18:337–349
27. Saleem FT, Anderson RE, Williams M (2020) Addressing the "myth" of racial trauma: developmental and ecological considerations for youth of color. Clin Child Fam Psychol Rev 23:1–14
28. Williams MT, Metzger IW, Leins C et al (2018) Assessing racial trauma within a DSM–5 framework: the UConn racial/ethnic stress & trauma survey. Pract Innov 3:242–260
29. Metzger IW, Anderson RE, Are F et al (2021) Healing interpersonal and racial trauma: integrating racial socialization into trauma-focused cognitive behavioral therapy for African American youth. Child Maltreat 26:17–27
30. Williams MT, Malcoun E, Sawyer B et al (2014) Cultural adaptations of prolonged exposure therapy for treatment and prevention of posttraumatic stress disorder in African Americans. Behav Sci 4:102–124
31. Merikangas KR, He J, Burstein M et al (2011) Service utilization for lifetime mental disorders in U.S. Adolescents: results of the national comorbidity survey–adolescent supplement (NCS-A). J Am Acad Child Adolesc Psychiatry 50:32–45
32. Misra S, Jackson VW, Chong J et al (2021) Systematic review of cultural aspects of stigma and mental illness among racial and ethnic minority groups in the United States: implications for interventions. Am J Community Psychol 68:486–512
33. Yeh M, Hough RL, McCabe K et al (2004) Parental beliefs about the causes of child problems: exploring racial/ethnic patterns. J Am Acad Child Adolesc Psychiatry 43:605–612
34. Yeh M, McCabe K, Hough RL et al (2005) Why bother with beliefs? Examining relationships between race/ethnicity, parental beliefs about causes of child problems, and mental health service use. J Consult Clin Psychol 73: 800–807
35. Yeh M, Aarons GA, Ho J et al (2014) Parental etiological explanations and longitudinal medication use for youths with attention deficit hyperactivity disorder. Adm Policy Ment Health Ment Health Serv Res 41:401–409

36. David EJR (2010) Cultural mistrust and mental health help-seeking attitudes among Filipino Americans. Asian Am J Psychol 1:57–66
37. Whaley AL (2001) Cultural mistrust and mental health services for African Americans: a review and meta-analysis. Couns Psychol 29: 513–531
38. Chavira DA, Bustos CE, Garcia MS et al (2017) Delivering CBT to Rural Latino children with anxiety disorders: a qualitative study. Community Ment Health J 53:53–61
39. McCabe KM, Yeh M, Garland AF et al (2005) The GANA program: a tailoring approach to adapting parent child interaction therapy for Mexican Americans. Educ Treat Child 28: 111–129
40. Haldar S, Pillai D, Published SA (2023) Overview of health coverage and care for individuals with Limited English Proficiency (LEP). https://www.kff.org/racial-equity-and-health-policy/issue-brief/overview-of-health-coverage-and-care-for-individuals-with-limited-english-proficiency/
41. Loho H, Rosenheck RA (2022) Provision of Non-English language services in U.S. Mental health facilities. Psychiatr Serv 73:339–342
42. Cougle JR, Grubaugh AL (2022) Do psychosocial treatment outcomes vary by race or ethnicity? A review of meta-analyses. Clin Psychol Rev 96:102192
43. Huey SJ, Polo AJ (2008) Evidence-based psychosocial treatments for ethnic minority youth. J Clin Child Adolesc Psychol 37:262–301
44. Rohde P, Seeley JR, Kaufman NK et al (2006) Predicting time to recovery among depressed adolescents treated in two psychosocial group interventions. J Consult Clin Psychol 74:80–88
45. Lochman JE, Wells KC (2004) The coping power program for preadolescent aggressive boys and their parents: outcome effects at the 1-year follow-up. J Consult Clin Psychol 72: 571–578
46. Hall GCN, Berkman ET, Zane NW et al (2021) Reducing mental health disparities by increasing the personal relevance of interventions. Am Psychol 76:91–103
47. McCabe KM, Yeh M, Zerr AA (2020) Personalizing behavioral parent training interventions to improve treatment engagement and outcomes for culturally diverse families. Psychol Res Behav Manag 13:41–53
48. Sanchez AL, Comer JS, LaRoche M (2022) Enhancing the responsiveness of family-based CBT through culturally informed case conceptualization and treatment planning. Cogn Behav Pract 29:750–770
49. Yeh M, Zerr A, McCabe K (2022) Personalizing PCIT for culturally diverse families: outcomes from a pilot trial utilizing the PersIn framework. Behav Res Ther 159:104204
50. Lewis-Fernández R, Aggarwal N, Hinton L et al (eds) (2016) The DSM-5 handbook on the cultural formulation interview. American Psychiatric Publishing, Washington, DC
51. Phinney JS (1992) The multigroup ethnic identity measure: a new scale for use with diverse groups. J Adolesc Res 7:156–176
52. Utsey SO, Ponterotto JG (1996) Development and validation of the Index of Race-Related Stress (IRRS). J Couns Psychol 43:490–501
53. Galán CA, Willis HA, Polanco-Roman L et al (2024) Initial development and psychometric properties of the Youth Racism-Based Traumatic Stress Symptom Scale
54. Kahhale I, Farrise K, Das A et al (2024) Recognizing the impact of racism-based traumatic stress on youth: implications for research and clinical practice. J Am Acad Child Adolesc Psychiatry. S0890856724002429
55. Jarvis GE, Kirmayer LJ, Gómez-Carrillo A et al (2020) Update on the cultural formulation interview. Focus 18:40–46
56. Lewis-Fernández R, Aggarwal NK, Lam PC et al (2017) Feasibility, acceptability and clinical utility of the cultural formulation interview: mixed-methods results from the DSM-5 international field trial. Br J Psychiatry 210:290–297
57. Day-Vines NL, Wood SM, Grothaus T et al (2007) Broaching the subjects of race, ethnicity, and culture during the counseling process. J Couns Dev 85:401–409
58. Day-Vines NL, Cluxton-Keller F, Agorsor C et al (2020) The multidimensional model of broaching behavior. J Couns Dev 98:107–118
59. Day-Vines NL, Cluxton-Keller F, Agorsor C et al (2021) Strategies for broaching the subjects of race, ethnicity, and culture. J Couns Dev 99:348–357
60. King KM (2021) "I want to, but how?" Defining counselor broaching in core tenets and debated components. J Multicult Couns Dev 49:87–100
61. Lakind D, Bradley WJ, Patel A et al (2022) A multidimensional examination of the measurement of treatment engagement: implications for children's mental health services and research. J Clin Child Adolesc Psychol 51: 453–468
62. Becker KD, Boustani M, Gellatly R et al (2018) Forty years of engagement research in children's mental health services:

multidimensional measurement and practice elements. J Clin Child Adolesc Psychol 47:1–23
63. Chorpita BF, Becker KD (2022) Dimensions of treatment engagement among youth and caregivers: Structural validity of the REACH framework. J Consult Clin Psychol 90:258–271
64. Bickman L (2008) A Measurement Feedback System (MFS) is necessary to improve mental health outcomes. J Am Acad Child Adolesc Psychiatry 47:1114–1119
65. Bickman L, Douglas SR, De Andrade ARV et al (2016) Implementing a measurement feedback system: a tale of two sites. Adm Policy Ment Health Ment Health Serv Res 43:410–425
66. Birmaher B, Khetarpal S, Brent D et al (1997) The Screen for Child Anxiety Related Emotional Disorders (SCARED): scale construction and psychometric characteristics. J Am Acad Child Adolesc Psychiatry 36:545–553
67. Johnson JG, Harris ES, Spitzer RL et al (2002) The patient health questionnaire for adolescents. J Adolesc Health 30:196–204
68. Weisz JR, Chorpita BF, Frye A et al (2011) Youth top problems: using idiographic, consumer-guided assessment to identify treatment needs and to track change during psychotherapy. J Consult Clin Psychol 79:369–380
69. Becker-Haimes EM, Tabachnick AR, Last BS et al (2020) Evidence base update for brief, free, and accessible youth mental health measures. J Clin Child Adolesc Psychol 49:1–17
70. Chorpita BF, Daleiden EL, Bernstein AD (2016) At the intersection of health information technology and decision support: measurement feedback systems...and beyond. Adm Policy Ment Health Ment Health Serv Res 43:471–477
71. Resnicow K, Baranowski T, Ahluwalia JS et al (1999) Cultural sensitivity in public health: defined and demystified. Ethn Dis 9:10–21
72. Simons-Morton BG, Donohew L, Davis Crump A (1997) Health communication in the prevention of alcohol, tobacco, and drug use. Health Educ Behav 24:544–554
73. Park AL, Rith-Najarian LR, Saifan D et al (2023) Strategies for incorporating culture into psychosocial interventions for youth of color. Evid-Based Pract Child Adolesc Ment Health 8:181–193
74. Zimmerman MA (2013) Resiliency theory: a strengths-based approach to research and practice for adolescent health. Health Educ Behav 40:381–383
75. Onyeka O, Richards M, Tyson McCrea K et al (2022) The role of positive youth development on mental health for youth of color living in high-stress communities: a strengths-based approach. Psychol Serv 19:72–83
76. Domenech Rodríguez MM, Baumann AA, Schwartz AL (2011) Cultural Adaptation of an evidence based intervention: from theory to practice in a latino/a community context. Am J Community Psychol 47:170–186
77. Lau A, Barnett M, Stadnick N et al (2017) Therapist report of adaptations to delivery of evidence-based practices within a system-driven reform of publicly funded children's mental health services. J Consult Clin Psychol 85:664–675
78. Spencer J, Goode J, Penix EA et al (2019) Developing a collaborative relationship with clients during the initial sessions of psychotherapy. Psychotherapy 56:7–10
79. Langer DA, Holly LE, Wills CE et al (2022) Shared decision-making for youth psychotherapy: a preliminary randomized clinical trial on facilitating personalized treatment. J Consult Clin Psychol 90:29–38
80. Langer DA, Jensen-Doss A (2018) Shared decision-making in youth mental health care: using the evidence to plan treatments collaboratively. J Clin Child Adolesc Psychol 47:821–831
81. Pan D, Huey SJ, Heflin LH (2019) Ethnic differences in response to directive vs. non-directive brief intervention for subsyndromal depression. Psychother Res 29:186–197
82. Añez LM, Paris M, Bedregal LE et al (2005) Application of cultural constructs in the care of first generation Latino clients in a community mental health setting. J Psychiatr Pract 11:221–230
83. Dippel N, Szota K, Cuijpers P et al (2022) Family involvement in psychotherapy for depression in children and adolescents: systematic review and meta-analysis. Psychol Psychother Theory Res Pract 95:656–679
84. Dowell KA, Ogles BM (2010) The effects of parent participation on child psychothera outcome: a meta-analytic review. J Clin Child Adolesc Psychol 39:151–162
85. Calzada EJ, Fernandez Y, Cortes DE (2010) Incorporating the cultural value of respeto into a framework of Latino parenting. Cultur Divers Ethnic Minor Psychol 16:77–86
86. López SR, Grover KP, Holland D et al (1989) Development of culturally sensitive psychotherapists. Prof Psychol Res Pract 20:369–376
87. Bernal G, Bonilla J, Bellido C (1995) Ecological validity and cultural sensitivity for outcome

research: Issues for the cultural adaptation and development of psychosocial treatments with Hispanics. J Abnorm Child Psychol 23:67–82

88. FitzGerald C, Hurst S (2017) Implicit bias in healthcare professionals: a systematic review. BMC Med Ethics 18:19

89. McCabe K, Yeh M (2009) Parent–child interaction therapy for Mexican americans: a randomized clinical trial. J Clin Child Adolesc Psychol 38:753–759

90. Gonzalez JC, Klein CC, Barnett ML et al (2023) Intervention and implementation characteristics to enhance father engagement: a systematic review of parenting interventions. Clin Child Fam Psychol Rev 26:445–458

Chapter 15

Unified Protocols for Transdiagnostic Treatment of Emotional Disorders in Children and Adolescents

Elizabeth R. Halliday and Jill Ehrenreich-May

Abstract

Treatment for emotional disorders in children and adolescents is increasingly being applied in a transdiagnostic manner to address the need for more broadly applicable yet mechanism-focused psychotherapy. The Unified Protocols for Transdiagnostic Treatment of Emotional Disorders in Children and Adolescents (UP-C/A) are flexible, evidence-based treatment manuals that utilize evidence-based, largely cognitive and behavioral strategies to treat anxiety, mood, and other emotional disorders. In this chapter, we describe the rationale and empirical support for the use of the UP-C/A, provide a pragmatic overview of its contents, and briefly discuss application of core intervention components.

Key words Transdiagnostic, Unified protocol, Children, Adolescents, Anxiety, Depression, Emotional disorders, Internalizing disorders, Cognitive-behavioral therapy

1 Introduction

The Unified Protocols for Transdiagnostic Treatment of Emotional Disorders in Children and Adolescents (UP-C/A; [1–3]) are evidence-based interventions designed purposefully to treat a wide variety and/or combination of emotional disorders. Originally developed in concert with the Unified Protocol for Transdiagnostic Treatment of Emotional Disorders for adults (UP; [4]), the UP-C/A includes an array of cognitive and behavioral treatment (CBT) strategies in a general emotion-focused language to promote change across transdiagnostic targets. The UP-C/A uses elements of emotion education (e.g., emotion labeling), traditional cognitive techniques (e.g., cognitive reappraisal), mindful awareness practices, and a full range of behaviorally driven activation and exposure strategies to address underlying mechanisms that are theorized to predispose and maintain emotional disorders [5–8]. The UP-C and UP-A can be delivered to children and adolescents, approximately 6–12 and 12–17 years old, respectively, with

Robert D. Friedberg and Micaela A. Thordarson (eds.), *Cognitive Behavioral Therapy with Youth: Tradition and Innovation*, Neuromethods, vol. 231, https://doi.org/10.1007/978-1-0716-5134-6_15,

any emotional disorder, including, but not limited to, any combination of anxiety disorders (i.e., generalized anxiety, social anxiety, specific phobias, separation anxiety, panic disorder, illness anxiety disorder, agoraphobia), depressive disorders (persistent depressive disorder and major depressive disorder), and obsessive-compulsive and related disorders [9, 10]. There is also emergent evidence for the benefits of the UP-C/A for related concerns (see [9]), including posttraumatic stress disorder(PTSD; [11]). The goals of this chapter are to review the theoretical foundations for the development and use of the UP-C/A, summarize the empirical support for the UP-C/A, and provide readers with an overview of the concepts, activities, and materials included in the intervention.

2 Theoretical Foundations

Anxiety disorders and depression share genetic, neurobiological, and environmental risk factors, which may lead these concerns to co-occur at high rates [12–14]. Comorbidity is the rule rather than the exception regarding such emotional disorders [12], and youth are likely to progress between emotional disorder conditions in a predictable course (e.g., separation anxiety leading to generalized anxiety, then depression, etc.) over time without effective intervention [13]; therefore, transdiagnostic approaches offer significant benefits compared to single-disorder-focused treatment. Instead of focusing on symptoms of a particular disorder, the UP-C/A use examples applicable to the experiences of anger, sadness, and anxiety/fear for most techniques, making it particularly useful for children presenting with multiple clinical and/or subclinical emotional disorders.

The UP, UP-C, and UP-A (UPs) specifically focus on the experience of neuroticism, postulated as a core dysfunction across emotional disorders [5, 6, 8, 15, 16]. Individuals high in neuroticism demonstrate high levels of negative affect causing them to experience strong emotions (i.e., fear, anxiety, sadness, and anger); in response to these emotions, they become distressed, anxious, and uncomfortable so much, so that they—and their caregivers (i.e., via accommodation behaviors)—take actions to suppress, avoid, escape, distract, or otherwise control these unwanted feelings. Because these behaviors quickly reduce discomfort in the short term, the avoidant behaviors increase via negative reinforcement. However, over the long term, using avoidant strategies prevents an individual from learning more helpful behaviors and being positively reinforced for using adaptive and approach-oriented behaviors. For youth, this can have far-reaching implications for the development of one's ability to deal with the situations and experiences that elicit strong emotions [17, 18], as well as the ways

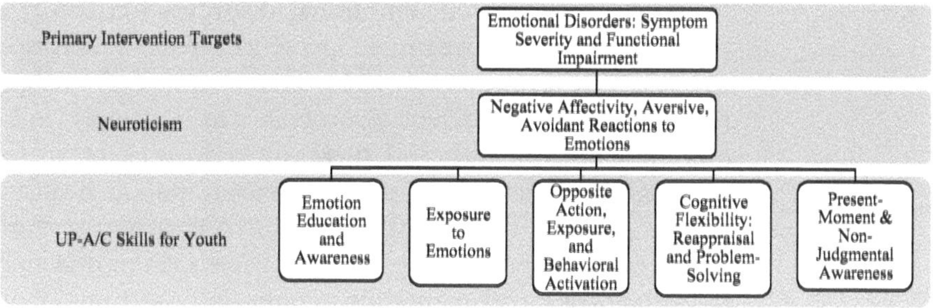

Fig. 1 Facets of neuroticism theorized to be targetted by UP-C/A skills

in which parents/caregivers respond to their children in times of distress [19].

Thus, the UP-C/A's overarching intervention principles include (a) increasing emotional awareness, (b) preventing emotional avoidance by engaging in present-focused awareness during intense emotions, (c) increasing cognitive flexibility, (d) modifying maladaptive action tendencies (i.e., "emotional behaviors") through personalized exposure and activation techniques, and (e) addressing patterns of parental behaviors that maintain youth symptoms. The mechanism of change across these strategies in the UPs has generally been framed as the reduction or extinction of distress and anxiety associated with the experience of strong emotions in circumstances where such experiences are unhelpful to overall functioning. Figure 1 presents the skills included in the UP-C/A and the facets of neuroticism the skills target. By applying these principles in a flexible manner to youth with a range of emotional disorders, youth and their parents/caregivers learn to tolerate their distress in response to strong emotions more effectively and reduce or eliminate negative reinforcement cycles associated with avoidance and other problematic action tendencies (e.g., aggression, etc.).

3 Implications for Dissemination and Implementation

Not only can transdiagnostic approaches achieve multiproblem symptom reduction by targeting features of neuroticism [20], they also offset treatment barriers and burdens for clients and clinicians alike, such as accessibility, training time and costs, treatment length, etc., especially in low- and middle-income populations [21]. Disorder-specific CBT manuals, or single-disorder protocols, effectively treat disorders and are often empirically supported. For example, the Mastery of Anxiety and Panic for Adolescents (MAP-A; [22]) was specifically developed to treat panic disorder, was rigorously tested to prove efficacy, and has been majorly successful in treating individuals with panic disorder.

However, as discussed, emotional disorders frequently co-occur [23]. Moreover, the treatment elements across single-disorder protocols overlap considerably [24]. Thus, "unifying" or shared treatment components across protocols can not only increase the efficiency and breadth and therefore reduce costs, of treatment, but this approach also offers significant public health benefits. Transdiagnostic treatments may help to narrow the "research to practice gap" by increasing the speed and span of dissemination of transdiagnostic treatments into community settings. Community clinics and agencies can expend less resources (e.g., financial, time) by providing a singular training in a transdiagnostic manual in which their clinicians can provide treatment to clients with a variety of presenting concerns as opposed to providing multiple trainings on single-disorder protocols. Transdiagnostic treatments may also help clinicians with heavy caseloads more effectively prepare and deliver such treatments. While more research is needed to evaluate the cost- and time-effectiveness of the UPs specifically, transdiagnostic approaches could be pivotal in extending the reach of evidence-based interventions for clients and clinicians in the most need.

4 Empirical Support

The evidence base for the UPs is growing rapidly and expanding to show positive effects for adults, children, and teens, as well as their participating family members. Researchers around the world study the UPs in individual [20] and group formats [25, 26] and for a variety of disorders [27]. Initial trials supporting the use of the UP-C/A [28–33] have been followed by numerous studies applying the UP-C/A to various problems areas in different treatment settings [27]. Applications across the UPs include treatment for borderline personality disorder in adults [34–36] and borderline features in adolescents [37]; non-suicidal self-injury(NSSI; [38, 39]); eating disorders in adults [40, 41] and youth [42, 43]; early psychosis and bipolar disorder in adolescents [44]; chronic pain in pediatric samples (The Unified Protocol for the Treatment of Emotions in Youth with Pain [UP-YP]; [23]); misophonia in children and adolescents [45, 46]; posttraumatic stress disorder in adolescents (PTSD; [11, 47]); irritable presentations in children with emotional disorders [48, 49]; and for the prevention of emotional disorders in adolescents in school-based settings [50]. The UP-C/A was also adapted for a brief parent-focused intervention during the COVID-19 pandemic (UP-Caregiver; [51]). In addition to being used to treat a variety of clinical presentations, the UP-C/A can be used in a variety of treatment settings and models, such as inpatient settings [39], telehealth delivery, and a stepped-care model [52].

Studies support the notion that the UPs not only decrease symptoms of emotional disorders but that they do so by targeting theoretically linked factors, such as maladaptive emotion regulation strategies [53–56], negative affect, fear of negative emotions, and anxiety sensitivity [57]. Every year, new studies show the versatility and unique treatment effects conferred by the UP-C/A, increasing its use to help children and adolescents, and their parents [58], learn new, more helpful ways to cope with their strong emotions. To note, this chapter refers to "parents" throughout, but this term is intended to be inclusive of all caregivers who play a significant role in a child's treatment and well-being. A fuller review of adaptations for specific clinical presentations and treatment settings for the UP-C and UP-A is available in a clinician-friendly brief companion book [9].

5 The UP-C/A in Practice

5.1 Utilizing a Flexible Structure

The modular structure of the UP-C/A allows for flexible treatment of a variety of disorders, in individual and group formats, with youth and parents. Modules are "containers" of research-validated practice elements, decision-making tools, and delivery techniques [59]. These treatments can improve both externalizing [60–65] and internalizing symptoms [30, 66, 67]. Modular treatments may be better accepted by practitioners than standard manuals [68–70], because they are more flexible, allow for parsimonious implementation of several evidence-based skills, and permit clinician personalization across a wider variety of problems or foci [71]. Modular treatments may also prove more clinically effective than standard manuals [72].

The UP-A consists of eight core modules with options to further expand upon or minimize depending on the needs of the adolescent. Though the UP-A is written and described in the therapist guide and in this chapter as an individual treatment, it is increasingly being used in a group format [44, 50, 73, 74]. The duration of treatment varies by client need; however, it's possible to deliver the entire UP-A in 12–16 sessions. The parent portion of the UP-A manual, Module P, is used as needed and varies greatly depending on the parent's involvement and influence on the adolescent's symptoms and treatment.

The version of the UP for children under 13 (UP-C) is published as a 15-session group treatment, with separate child and parent group curriculum available. Most commonly, clinicians administer the parent and child components in concert with one another. Group sizes typically range from five to seven children with one or both parents participating, though this can vary depending on clinician availability and group members. The UP-C is also frequently modified to be delivered as a more flexibly administered

Table 1
Overview of UP-C/A weekly sessions

1. Client and parent will give their weekly ratings of Top Problems
2. Address any pressing issues from the past week and/or engage in some conversation to build rapport
3. Review the home learning assignment to reinforce the previously learned skill
4. Introduce the new skills of current module/session
5. Practice the new skill first with a neutral example
6. Practice the new skill in the context of the client's personal emotional experience
Assign a home learning assignment

individual treatment. Further guidelines for individual treatment are described in the therapist guide [1].

The UP-C is delivered in the context of an "Emotion Detectives" metaphor. Through the course of the intervention, children in UP-C learn CLUES Skills (*C*onsider How I feel, *L*ook at my Thoughts, *U*se detective thinking and problem-solving, *E*xperience my feelings, and *S*taying healthy and happy) to help "solve the mystery of their strong emotions." When administered similarly to how it has been implemented in clinical trials, both UP-C and UP-A sessions follow the same basic format, outlined in Table 1.

The UP-C/A recognizes that parenting behaviors and practices also (often inadvertently) reinforce youths' emotional disorder symptoms and therefore targets these parenting behaviors throughout treatment. The UP-C/A holds that most parents and caregivers interacting with youth exhibiting strong emotions will normatively act to lessen or minimize the experience of that strong emotion [75]. These actions often arise out of adaptive parenting practices designed to soothe a distressed child. However, in the case of chronically distressed youth, these parenting behaviors may backfire, potentially furthering the notion that situations and experiences associated with strong emotions should be avoided. Therefore, the manuals focus on addressing four "emotional" parenting behaviors: excessive criticism, overcontrol/overprotection, modeling of avoidance, and inconsistent reinforcement and discipline patterns. Parents learn to counter these behaviors through a series of "opposite" parenting strategies. The UP-C features parent-directed content specified for each of its 15 group sessions, while the UP-A offers these materials flexibly, as needed. For example, in Session 1 of the UP-A, the teen and clinician should discuss their parents' involvement, and the most practical and helpful use of Module P with parents. Suggested guidelines for use of Module-P materials in the UP-A manual are outlined in the therapist guide [1].

The UP-C and UP-A share two halves of the same published therapist guide [1], with the UP-C and UP-A each having separate published workbooks [2, 3]. The UP-C workbook contains sections directed at the child client and parent/caregiver separately, whereas the UP-A workbook is directed almost solely at the adolescent client. Each chapter in the UP-A portion of the therapist guide includes a corresponding parent-directed handout summarizing the module content ("Module Summary Forms"). Additional materials that may be helpful in administering the UP-C groups include a roll of butcher paper, markers, small plastic containers to make "Clues Kits," and reinforcement items such as stickers, small prizes, and tokens.

6 UP-C/A: Module by Module

In the section to follow, we describe each of the eight core modules of the UP-A, followed by the corresponding UP-C and parent session content. Table 2 presents a summary of the UP-A modules, including recommendations for potential module length, and the corresponding UP-C sessions and their child and parent-directed content. Importantly, as you review these materials, please keep in mind that the UP-C/A is meant to be used with a good deal of flexibility and attendance to the individual personalized needs of each youth and their family. We present the modules in a recommended order, but there are many instances in which the clinician may wish to move through the modules or sessions with greater (or lesser) rapidity, skip forward to, for example, more saliently exposure-focused content, or eliminate an activity that appears less relevant to a particular client. The authors of the UP-C/A encourage these types of personalization decisions; however, personalization decisions may be easiest to implement following practice with the materials in the order and amounts initially specified in the therapist guide.

6.1 Module 1: Building and Keeping Motivation

The goal of Module 1 is to orient the adolescent to the treatment concepts and structure of each session, including how their parent (s) will be involved. We recommend beginning the session together to introduce both the adolescent and the parent(s) to what the UP-A is, the length of treatment, the use of the workbook, and how to apply what is learned and practiced in therapy in real life. Then the parent(s) can be excused, and the clinician can address the goal of building rapport with their new adolescent client. For a more reluctant adolescent, more time need to be spent on rapport building or discussing their perspective on treatment needs to create a safe space and get to know them. The therapist guide provides several directions for client-centered discussion consistent with principles of motivational enhancement. If appropriate, you can

Table 2
UP-A and UP-C content overview

Module	UP-A title	Session count	Content	Corresponding UP-C session	Parent content for UP-C
1	Building and Keeping Motivation	1 or 2	Rapport Top Problems and goals Motivation to change	Session 1: Intro to UP-C (C)	Introduction to treatment structure and CLUES skills Psychoeducation, including the three-component model of emotions
2	Getting to Know Your Emotions and Behaviors	2 or 3	Psychoeducation about emotions, including the purpose of emotions and the three parts of emotions The cycle of avoidance and other emotional behaviors	Session 2: Getting to Know Your Emotions (C)	Cycle of emotional behaviors Double before, during, and after Introduction to "emotional parenting behaviors" and their "opposite parenting behaviors" Positive reinforcement as the opposite parenting behavior for criticism
3	Introduction to Emotion-Focused Behavioral Experiments	1 or 2	Opposite action and emotion-focused behavioral experiments Tracking emotions and activity levels	Session 3: Using Science Experiments to Change our Emotions and Behavior (C)	Introduction to science experiments Supporting children in completing science experiments, including reinforcement
4	Awareness of Physical Sensations	1 or 2	Connection between physical feelings and strong emotions Awareness of physical feelings Sensational exposures to help tolerate uncomfortable physical feelings	Session 4: Our Body Clues (C)	Introduction to somatization. Body scanning. Sensational exposures group practice. Expressing empathy.
5	Being Flexible in Your Thinking	2 or 3	Thinking flexibly about emotional situations Thinking traps and Detective Thinking	Sessions 5, 6, and 7: Look at my Thoughts (L), Use Detective Thinking (U),	Cognitive flexibility Four common "thinking traps" Inconsistency and its opposite parenting behaviors consistent

			Getting unstuck with problem- solving skills	Problem-Solving and Conflict Management (U)	reinforcement and discipline Detective Thinking. Overcontrol/overprotection and its opposite parenting behavior, healthy independence granting Problem-solving for interpersonal conflicts Reassurance-seeking and accommodation
6	Awareness of Emotional Experiences	1 or 2	Present- moment awareness and nonjudgmental awareness. Generalized emotion exposures using practice awareness skills	Session 8: Awareness of Emotion Exposure (E)	Importance of learning to experience emotions instead of avoiding them Practice present-moment awareness and nonjudgmental awareness Emotional Behavior Form
7	Situational Emotion Exposure	2+	Skills review. Rationale for situational emotion exposures Situational emotion exposures in session and exposures for home learning	Sessions 9–14: Introduction to Emotion Exposure (E), Experience our Emotions Parts 1 and 2 (E)	Introduction to situational emotion exposures Parents' role in practicing exposures at home Modeling of intense emotions and avoidance, and its opposite parenting behavior, healthy emotional modeling Emotional Behaviors Form for upcoming exposures Apply situational emotion exposures to different symptoms Safety behaviors Supporting exposures at home Emotion Ladders

(continued)

Table 2
(continued)

Module	UP-A title	Session count	Content	Corresponding UP-C session	Parent content for UP-C
8	Reviewing Accomplishments and Looking Ahead	1	Skills and goals review Relapse prevention plan	Session 15: Wrap up and Relapse Prevention (S)	Emotion Detective skills review "Opposite parenting behaviors" review Celebrate children's progress Planning for sustaining and furthering progress Distinguish lapses from relapses
P	Parenting the Emotional Adolescent	1–3	Awareness of parental responses to adolescent's distress Emotional parenting behaviors and their opposite parenting behaviors	N/A	N/A

spend less time on this goal and move straight into eliciting the adolescent's experiences with strong emotions and their impression of their main or top problems at present. At minimum, three *Top Problems* should be determined [76, 77]. Youth and their parent (s) will subsequently rate Top Problems weekly to obtain a personalized measure of change. Therefore, Top Problems should be behaviorally focused and appropriate to address within short-term treatment. Some examples of Top Problems consistent with the goals of the UP-C/A include isolating self when upset, engaging in handwashing compulsions, avoiding participating in class, etc. If time permits, a *SMART* goal may be determined for each Top Problem. SMART goals are *s*pecific, *m*easurable, *a*ttainable, *r*elevant, and *t*ime bound. Like Top Problems, SMART goals should be as behaviorally focused as possible to serve as clear indicators of change over time in treatment.

If there is time, or if an adolescent is showing low motivation for change, identify initial steps to achieve the adolescent's SMART goals to help build commitment to treatment and self-efficacy for change. Before the end of the first session, parents/caregivers should be brought in to help finalize and rate Top Problems. Parents' motivation for treatment should also be addressed if this is perceived to be a potential issue for the youth's treatment. By doing so, potential barriers to treatment attendance and engagement can be identified and discussed, as needed. The clinicians may utilize problem-solving techniques from Module 5 to generate solutions to barriers.

Session 1 of the UP-C aims to be fun and enjoyable to reinforce children's attendance and participation. Clinicians should begin with an icebreaker to introduce families to each other. Next, clinicians should establish Top Problems and related SMART goals with each parent and child dyad. After separating into parent and child groups, clinicians should continue to encourage a positive atmosphere by introducing the theme of Emotion Detectives and doing hands-on activities, including some crafting. The child group can establish "Detective Rules," or group rules (e.g., raise your hand to answer questions, listen when group members are speaking), including a developmentally appropriate review of confidentiality. Children make a CLUES Kit using small plastic containers, stickers, letters, and other crafts, in which they will store tokens earned through good behavior and puzzle pieces earned for completing home-learning assignments. Tokens may be exchanged for a small prize of the child's choosing; puzzle pieces are collected in Session 15 to allow children to assemble a puzzle which reveals a group prize for their end of treatment celebration (this is typically a food item like cookies or pizza, as appropriate for each group). For home learning, the clinicians assign children the *My Emotions at Home* worksheet to begin working on identifying emotions [3].

During the first UP-C session, clinicians facilitating the parent group should encourage the parents to introduce themselves and explain why they are seeking treatment for their child. Parents often enjoy sharing and finding commonalities among their parenting experiences; however, it is most beneficial to keep this part of the group brief and structured on a weekly basis to ensure sufficient time to deliver group content. In this first meeting, clinicians inform parents about the structure of group, including recommended attendance, parents' role as "coaches" to practice skills at home, and the emphasis on parenting strategies. The clinicians will also explain the CLUES skills and the three parts of an emotion: thoughts, body feelings, and behaviors. Clinicians introduce parents to the concept of the cycle of avoidance (described in more detail below), including safety behaviors, and how angry behaviors can also be reinforced through negative reinforcement patterns. For home learning, clinicians assign the *C for Parents* worksheet to help parents practice identifying their children's emotional experiences and breaking down experiences into a trigger, thoughts, body clues, and behaviors.

6.2 Module 2: Getting to Know Your Emotions and Behaviors

Module 2 broadly consists of education about emotions and aids the adolescent in better labeling and identifying the components of their emotional experience with this new knowledge. Thus, the goals of Module 2 include teaching the adolescent to begin learning emotion identification skills; increasing the adolescent's insight about the functions of their emotions; explaining how emotions impact behavior; discussing how reinforcement principles maintain certain less adaptive or helpful behaviors; and teaching the adolescent how to break down an emotional experience using a functional assessment tool that references the antecedents, responses, and short- vs. long-term consequences of emotional behaviors (i.e., the before, during, and after form [B/D/A]; [1–3]). Some adolescents may already demonstrate strong emotional identification skills and/or have received psychoeducation about emotions in other therapy experiences. Other adolescents may demonstrate low awareness or insight about their emotional experiences and require all available content and related worksheets in this module. Thus, Module 2 content can be abbreviated, elongated, or repeated as appropriate.

To learn emotion identification skills, the clinician introduces the teen to the concept of an *emotional behavior*, or an emotionally driven behavior, which is any behavior that is motivated by an emotional experience. Some emotional behaviors are positive, like smiling because one is happy; some are necessary and appropriate like reaching out to a teacher because of bullying; and some are maladaptive—the focus of the UP-C/A—like avoiding a party due to fear of interacting with peers. The adolescent also learns that emotions comprise of three parts: the physical feelings in their body, their actions, and their cognitions. Using the *Emotion Twister*, the clinician explains that these three parts of an emotional experience often build up quickly and intensely, whirl around, blur together, and make it difficult to figure out how to deal with an emotional experience. All adolescents are prompted to describe an emotional experience as fully as possible and then collaboratively practice breaking down the experience with their clinician.

Applying the concept of negative reinforcement, the adolescent learns how their emotional behaviors result in short-term relief of their strong emotions but may end up intensifying emotional experiences over the longer term. More specifically, avoiding situations that provoke strong emotions provides immediate relief but lessens the opportunities for positive experiences, adaptive learning, and effective problem-solving in emotional situations. Additionally, engaging in emotional behaviors to avoid the experience of strong emotions can also cause the adolescent to miss out on fun or important opportunities, get them into trouble, bring about unwanted or negative social attention, or cause other functional impairment (e.g., nutritional concerns due to food avoidance). Because the short-term relief is reinforcing and results in the adolescent repeating the emotional behaviors the next time that they experience a trigger, the adolescent gets stuck in the *cycle of avoidance*, which maintains their symptoms. This discussion can be adapted to refer to aggressive, compulsive, or more overtly withdrawal-oriented behaviors, as needed. It is important to consider that some of the adolescent's avoidance or other maladaptive behaviors may be subtle. Sometimes these are safety behaviors like carrying an object with them, like their cell phone, and reassurance seeking when worrying. The clinician should aim to maintain a serious, yet positive and motivating attitude during this discussion. It can be difficult for adolescents to face their emotional behaviors and the realities of their consequences. However, understanding and accurately identifying one's emotional experiences play a pivotal role in choosing to engage in more adaptive behaviors. For home learning, adolescents will track the details of their emotional experiences and cycle of avoidance using the *BDA Form*. From this point on, they will be responsible for tracking the before (trigger), during (thoughts, body clues, behaviors), and after (short- and long-term consequences) of an emotional experience every week for the remainder of treatment.

This is often a good time to bring parents into the discussion of emotional behaviors by introducing the concept of the *Double BDA* [1]. The *Double BDA* is an exercise in functional assessment to examine how parental responses to their adolescent's distress moderate their child's emotional experience. The exercise uses the same monitoring tool their child is being asked to use weekly (the *BDA Form*) but is applied to both their child's emotional behaviors and any emotional behaviors that a parent provides in response to their child. For example, the *Double BDA* form may be used to first identify the trigger for the child's emotional experience (e.g., participating in a soccer game) and analyze a child's thoughts (e.g. "I'm going to mess up and everyone will blame me"), physical sensations (e.g., stomachache), and behaviors (e.g., beg parent to stay home). The second part of the *Double BDA* involves evaluating the parent's experience, for example, the parent's trigger (e.g., my child is crying and upset about going to their soccer game); thoughts (e.g., "Why does this happens every week? I'm too stressed to deal with this today! I just want them to stop crying. It isn't worth them being so upset. I don't want to make it worse"); physical sensations (e.g., tense, chest pressure); and behaviors (e.g., allowing them to skip the soccer game). Finally, the *Double BDA* evaluates the short- and long-term consequences of both the parents' and children's behavior and how the behaviors facilitate the cycle of avoidance.

Clinicians should introduce parents to this concept sensitively. Align with parents over their struggles to parent their adolescent as best they can. Discuss how parents experience their own *Emotion Twister* in response to their children's strong emotions and behaviors. It is often helpful to normalize that many parents engage in emotional behaviors in an effort to reduce their child's distress, which is a normal and natural goal. To illustrate the concept of the bidirectional relationship between a parent's and child's emotional experiences, review an example of a completed *Double BDA* (Fig. 9.1 in the therapist guide) before completing a recent, personal example. Parents can then be introduced to the four emotional parenting behaviors and opposite parenting behaviors, described in greater detail below. However, given time constraints, the clinician is encouraged to consider which of the four parenting behaviors might be most relevant to the adolescent's current functioning and social environment to better focus this discussion during subsequent sessions. During this preliminary discussion and in future discussions about the four emotional parenting behaviors, clinicians should emphasize that these behaviors are common in parents of children with emotional disorders; these behaviors feel natural at times or may be utilized with children and adolescents without emotional disorders without the same implications; and that changing emotional parenting behaviors is challenging and may take time and patience.

In Session 2 of the UP-C, Getting to Know Your Emotions, the clinicians introduce the same concepts as UP-A Module 2 in a developmentally appropriate manner. Children learn to use the "Emotion Thermometer" to take the "temperature" of their emotions in varying hypothetical situations. The clinician uses the activity to facilitate discussion about normative emotional experiences and differences and similarities of the group's emotional reactions. Children then play the "Alarm Game" to illustrate the idea that although emotions are nature's way of encouraging us to act because something dangerous or harmful may be happening, for some kids this "alarm" goes off even when no danger is present, or harm is unlikely. Children learn to better differentiate true and false alarms by identifying the three parts of emotional experiences and their emotional behaviors. The idea of short- and long-term consequences is also explained. To further elucidate this idea, clinicians can assign children the *BDA* worksheet for home learning.

UP-C Session 2 parent group addresses emotional parenting behaviors in a supportive way that normalizes parents' struggles. Parents learn about the *Double BDA* and independently work to expand their *C for Parents* home learning assignment [3]. Clinicians should empathetically point out how certain "emotional parenting" responses may unintentionally reinforce less adaptive child and parent behaviors and encourage observation of this interaction. Clinicians then facilitate a discussion about four emotional parenting behaviors that may reinforce less helpful child responses to emotions (criticism, overprotection/overcontrol, inconsistent reinforcement and discipline, and modeling of emotional behaviors). In this session, clinicians focus initially on criticism specifically, and one of its ***opposite parenting behaviors***: positive reinforcement. In light of anxious children's propensity for heightened sensitivity to criticism, parents learn about the importance of using neutral, descriptive language and praising their children for things they are doing well. Clinicians encourage parents to generate simple reinforcement ideas, such as smiling, nodding, verbal praise, and small rewards, as well as identify behaviors to reinforce. For home learning, clinicians ask parents to complete at least one *Double BDA* for a time their child evidenced an emotional behavior that week. Session ends by rejoining with the children and completing the *Rewards List* [3]. Importantly, some parents may be skeptical of the idea of encouraging new, more adaptive behaviors using positive reinforcers. This could be due to potential expenses or concerns that some behaviors (e.g., talking to peers, doing schoolwork) should be naturally reinforcing. Parents may also have concerns about using extrinsic rewards for behaviors that are developmentally expected. While navigating this discussion, it is crucial to emphasize that rewards in such situations aim to "kick start" behaviors crucial for healthy adaptation, and parents can gradually fade out rewards as these behaviors become more consistent. Subtle modifications—such as referring to praise and rewards as "compliments"—and identifying reinforcers that families feel more comfortable with (e.g., doing an activity together instead of a toy or other reward object) could increase amenability to this strategy.

6.3 Module 3: Introduction to Emotion-Focused Behavioral Experiments

The primary focus of Module 3 is to introduce and reinforce "opposite action" or *acting opposite*, which is doing the opposite of (or something different than) unhelpful emotional behaviors. This is practiced in the context of emotion-focused behavioral experiments to "test out" beliefs about the outcomes of changing one's actions when experiencing a strong emotion. The concept of conducting such experiments and observing their outcome is an essential metaphor throughout the UP-C/A. It can be illustrated in this module through a relatively straightforward application of behavioral activation, although other experiments related to exposure, exposure with response prevention, or opposite actions for anger and aggression should be included.

In terms of behavioral activation, the clinician will initially help identify activities the adolescent enjoys, explain the idea of tracking their emotions and activity levels, and the goal to engage in an "experiment" to see if the adolescent's mood, in fact, changes with increases in personally valued and/or functionally impactful activity. For some adolescents, behavioral activation will be the primary opposite action as you continue moving onward through the modules. Such behavioral experiments for sadness/withdrawal may need to be modified to include more basic vocational behaviors, focus on other emotions, and/or account for barriers in the environment, as time progresses.

For most adolescents, however, behavioral activation should be included as just one type of opposite action. Many clinicians use the "opposite action" and "science experiment" frameworks as a jumping off point for the introduction and implementation of exposures prior to Module 7/Session 9. This can be especially useful for youth necessitating earlier introduction of exposure work, such as children with OCD, challenging exposure content (e.g., disgust, irritability) or those who may need more shaping and time with exposures, or may not benefit as much from pre-exposure skills, such as children with autism spectrum disorder [78]. Opposite actions should be tailored and individualized for the clients' current functioning such that they are encouraged to make any amount of progress toward doing something different than their emotional behaviors. For example, for a child with social anxiety, acting opposite may entail looking up from their desk during class instead of looking down. See Table 3 for examples of opposite actions for a variety of emotional behaviors across problem areas.

Opposite action can also serve as an opportunity to implement skills from other evidence-based manuals, as needed. For example, distress management techniques from Dialectical Behavior Therapy can be introduced as opposite actions to self-harm; content regarding communication strategies, like assertiveness skills, could be introduced as opposite actions to passive or aggressive communication styles; sleep strategies from CBTs for insomnia could be utilized as opposite actions for unhelpful emotional behaviors involving sleep; and so on.

Table 3
Opposite actions for emotional behaviors

Emotion	Emotion behavior	Opposite action
Sad, down, tired	Withdrawal to bedroom, avoid activities/self-care, nap	Engage in enjoyable activities, brush teeth, hang out in a common area
Anger, frustration	Yell, use physical aggression, use mean words or insults, slam door	Speak in a calm voice, walk away, apologize, write down insults
Anxiety	Behavioral avoidance, emotional avoidance, distraction, compulsions/rituals	Approach-oriented behaviors, messing with rituals, present-moment awareness
Guilt	Avoid the person or situation that is causing the guilt	Apologize and admit the wrong (if guilt is justified)
Shame	Hide the thing that is causing shame	Talk about or write about what is causing the shame and try not to hide it
Any emotions	Non-suicidal self-injury	Harm reduction (e.g., snapping rubber band, holding ice cubes)
Emotions associated with enuresis/encopresis	Holding urine/feces, avoiding or ignoring urges to use the bathroom	Scheduled bathroom breaks, listening to urges
Any emotions	Skin-picking, hair pulling	Doing something else with hands, paint fingernails, covering affected areas
Any emotions	Tics	Competing responses
Any emotions	Binging	Throwing out one piece of the meal, eating one bite less
Anxiety, worry	Rumination	Mindfulness, leaves on a stream

In Session 3 of the UP-C, Using Science Experiments to Change our Emotions and Behavior, the clinicians introduce acting opposite and conducting *science experiments* for emotional behaviors. To illustrate the concept, *Nina's Mood and Activity Diary* is discussed, and children set up their own experiments. In the parent group, they are similarly introduced to the concepts and how to support their child in acting opposite. Additionally, ways to reinforce a child during or after a science experiment are offered. Together, the information is reviewed, and home learning is assigned. Through the week, children will complete *My Emotion and Activity Diary*, and parents will complete the *Double BDA* and implement their reinforcement plan [3]. Similar to the UP-A, although much for the content, examples, and worksheets in Session 2 of the UP-C focus on behavioral activation, materials can and should be expanded to include whatever opposite actions are most useful for the clients in the group.

6.4 Module 4: Awareness of Physical Sensations

Module 4 broadly focuses on helping the client become more aware of their physiological sensations and how feelings in their body influence their emotional experiences. The goals of this module are to teach the adolescent about the concept of physiological sensations and their relationship to intense emotions, work with them to identify and reinterpret their feelings during emotional experiences using a body scanning exercise, and finally to use interoceptive exposure to help normalize and increase awareness of these sensations.

By becoming more aware of physical sensations that occur during emotional experiences, teens learn to consider how to best respond to these "body feelings" (as described in Module 2's *BDA*). Because the sensations can feel very intense, make us feel uncomfortable, and try to prompt us to do something to make the sensations go away quickly, it's especially important to understand such sensations, learn to better tolerate them, and manage them more effectively. Body scanning is a key mindful awareness technique that helps the adolescent achieve this tolerance through practice in session and over time. The goal of body scanning is for the adolescent to notice the sensations, say something about them, and to experience them fully [79], allowing themselves to identify any distracting cognitions that occur about sensations in their body or otherwise and return to the present moment by anchoring back in the sensations they are experiencing in the here and now. By continuously practicing attentiveness to their bodily sensations, they can "watch" as the intensity changes and naturally reduces over time. Adolescents should practice rating the intensity of sensations, verbally or mentally describing the sensations, and intentionally staying in the present moment until the feelings are manageable. For some teens, especially those who experience panic attacks, high levels of anxiety sensitivity [80], or unwarranted hyperarousal, it may be helpful to first objectively review the fight-or-flight response and the bodily symptoms that occur or the rationale for exposure (see Module 7) more so in presenting sensational exposure and body scanning skills.

Like individual adolescent clients, in UP-C Session 4, children learn how to recognize "body clues" and how they relate to their emotions. Clinicians introduce children to the idea of *Becoming a Body Detective* and engage in a body drawing and body scanning [3]. The final goal of this session is for children to practice at least three interoceptive exposures (or as they are referred to in the UP-C/A, *sensational)* using examples provided in the therapist guide [1]. By enduring the exposure until stated tolerance of any discomfort, children can see for themselves that the feelings, which are often associated with strong emotions, will lessen, or become more tolerable over time. For children with high anxiety sensitivity or very reactive symptoms to certain body signals (e.g., cases of emetophobia or Avoidant Restrictive Food Intake Disorder, etc.) it is especially important to comprehensively illustrate the relationship between thoughts and body clues and reinforce the use of body scanning and mindful awareness as a means to de-catastrophize any negative interpretations of these experiences.

In this session, clinicians familiarize parents with somatic responses to emotional experiences and help to detect these in their children. Parents also learn the concept of body scanning, how to assist their children in using the skill, and when it might be helpful. To convey the purpose and encourage empathy for uncomfortable feelings, parents will also engage in a sensational exposure. Finally, the importance of expressing empathy is discussed and practiced with a worksheet [3]. At the end of Session 4, children earn their C badge for their CLUES Kit.

6.5 Module 5: Being Flexible in Your Thinking

Module 5 includes several traditional cognitive and behavioral strategies; however, there is stronger emphasis on utilizing cognitive reappraisal in the anticipation of an emotion reaction or difficult situation (*antecedent cognitive reappraisal*), so to preemptively address the difficulty of engaging in this technique during an intensely emotional situation, when it may be less effective or unhelpfully distracting. Additionally, the problem-solving techniques in Module 5 increase flexibility when thinking about solutions in or following emotionally intense situations. Thus, whether referring to reappraisal or problem-solving, the overarching idea of Module 5 is to increase *flexible thinking*.

The clinician discusses with the adolescent that people make automatic, emotionally relevant interpretations of stimuli or situations around or within them. To illustrate this concept, the clinician can first use one of several possible optical illusions. The clinician introduces the adolescent to an optical illusion image and invites them to just tell the clinician what they see first, which may include any number of interpretations. Discuss with the adolescent that there may be several possible interpretations for the pictures, there can be many right answers, and for whatever reason, people will see different things. Typically, the first idea that comes to mind is the one that is most familiar or requires the least mental effort to identify. Whereas additional interpretations may require more time, assistance, or practice to identify. The main idea to convey through this example is that there may also be more than one interpretation of situations causing strong emotions around or within us. By only considering one possible explanation for a situation or for our emotions, one may ignore other possibilities, which may negatively influence an emotional experience.

This experience is normalized by discussing "thinking traps," classically known as cognitive distortions, which may lead youth to routinely identify negative or threat-related cognitions before considering more neutral or positive interpretations. Common thinking traps are introduced: Jumping to Conclusions (Probability

Overestimation), Thinking the Worst (Catastrophizing), and Ignoring the Positive, although others may be discussed. Once the adolescent is more adept at identifying thinking traps, the clinician can explain how the adolescent can use *Detective Thinking*, or cognitive reappraisal, to be more flexible in these interpretations. The steps of Detective Thinking include relatively traditional cognitive restructuring techniques to encourage looking for clues to see if one's thoughts are realistic and what other interpretations may be possible. Steps of Detective Thinking are summarized below:

Identify the Interpretation Use one of the adolescent's own examples of a time their automatic interpretation probably wasn't accurate or realistic and then see if they can identify the trap. For example, "If I don't get straight As on my report card, I'll never get into a good college." In this case, one thinking trap might be Thinking the Worst (i.e., assigning chances of something bad happening are greater than they are).

Evaluate the Evidence This includes using dispute handles (or "detective questioning") to clarify the accuracy and likelihood of one's automatic interpretation, as well as the adolescent's ability to cope with an intense emotional situation. Have the adolescent answer at least two of the questions on the *Being a Detective – Steps for Detective Thinking* worksheet in the UP-A workbook to accomplish this *(2)*. Examples include, but are not limited to: (a) Am I 100% sure that ____ will happen? (b) Can I really tell what someone else is thinking? and (c) How many times has ____ happened in the past?"

After completing Detective Thinking steps, identify the most realistic outcome(s)—even if this includes the "thinking trap" or what may appear to be a negative interpretation—and how they would cope if outcome were to occur. Explain to the adolescent why Detective Thinking is most useful before entering situations and remind them of this as needed. Remind the youth that this skill is about *realistic thinking*, not just thinking more positively, as needed.

The remainder of Module 5 focuses on *Problem-Solving*. This skill can be used to help the adolescent get out of traps where they may feel "stuck" or already enacted an unhelpful emotional behavior. With adolescents, it can be helpful to apply this skill to interpersonal conflicts, but problem-solving can be useful anytime in treatment for determining helpful behavioral solutions, including more crisis-oriented situations.

The UP-C/A therapist guide provides an example of a possible script to practice this skill in a neutral situation and a situation where the adolescent is experiencing a strong emotion [1]. For home learning, the teen is asked to complete the *BDA* and Worksheet 5.4 *Getting Unstuck – Steps for Solving Problem.*

The concepts covered in Module 5 of the UP-A are presented over the course of three sessions in UP-C: Sessions 5 (Look at My Thoughts), Session 6 (Use Detective Thinking), and Session 7 (Problem-Solving and Conflict Management). Like the UP-A, the concept of flexible thinking is first introduced with optical illusions, followed by some nonpersonal examples, and then younger clients can begin to challenge their *snap judgments*, or automatic interpretations. Thinking Traps are then presented as illustrated characters in the workbook: Mind reading or *Psychic Suki*, jumping to conclusions or *Jumping Jack*, thinking the worst or *Disaster Darrell*, and ignoring the positive or *Negative Nina*. To practice recognizing the thinking traps, children can complete the *Match the Thought to the Trap!* Worksheet [3]. In parent group, clinicians familiarize parents with cognitive flexibility, so that they can aid their children in recognizing the Thinking Traps. They also discuss the parenting behavior of inconsistency, along with consistent reinforcement and discipline techniques. Parents learn the basics about positive and negative reinforcement and positive and negative punishment. With this knowledge, the clinician can help parents become more consistent in appropriately praising and punishing their child through consistent behavior management, even in the case of more intense emotional displays. Page 129 of the UP-C Workbook supplies guidelines for such a system (*Guidelines for Creating an Effective Behavior Management System at Home*). Parents' home learning assignment will be to implement this system and complete Session 5 worksheets [3].

Session 6 of the UP-C focuses on Detective Thinking. Children are encouraged to participate in a *Mystery Game* in which they attempt to solve a nonemotional mystery. Throughout the game, children can learn that sometimes they must look for clues to figure out what is going on around them, that there may be more than one right answer, and that they can discover clues to solve such mysteries (alone or in a team) in many ways. Using the Worksheet *U for Children*, the group then learns the Detective Thinking steps to get out of thinking traps—what are referred to as the "Stop, Slow, Go." Then, after going through an example, the group can use their new skill to solve the mysteries in the *Detective Thinking Practice* worksheet. They will practice Detective Thinking for their own experiences at home using the *U for Children at Home* worksheet. In this session, parents will complete Detective Thinking steps using an example from their own lives, followed by a child-focused example. The parenting behavior of overcontrol/overprotection is discussed, along with healthy independence granting and shaping of new behaviors through reinforcement, planned ignoring, and differential reinforcement. By reviewing the origins and consequences of this parenting behavior and how they can engage in opposite parenting behaviors, parents can help their child more independently cope with distress and build self-efficacy.

Session 7 of the UP-C focuses on Problem-Solving, which mirrors the adolescent version of the skill. By the end of Session 7, children will also earn their L and U CLUES Skills badges. For home learning, children complete the *BDA* and the *Problem Solving at Home with Others* worksheets. In the parent portion of the group, clinicians introduce parents to problem-solving as another skill to develop cognitive flexibility. They discuss applications of Problem-Solving in their own lives, specifically to resolve interpersonal conflicts. Finally, the group reviews parents' efforts at healthy independence granting using the *Encouraging Independent Behaviors* worksheet and connects this opposite parenting behavior to the skills learned in L and U. Parents are aided in viewing Detective Thinking and Problem-Solving as alternatives to reassurance-seeking and the provision of accommodation. Parents have learned how providing reassurance, either explicitly or less obviously, is a form of overprotection that while quick and easy may also facilitate children's dependence on their parents to evaluate emotional situations. Because all parents want to protect their children from distress, parents often try to tell their child which solution will work best to help their child avoid choosing the less helpful or less efficient solution. As a result, children don't get as much practice problem-solving, become more inflexible, and may experience feelings of low self-efficacy. Clinicians point out the lay idea that making mistakes often helps children learn best and that such mistakes may not be catastrophic but rather essential for growth and adaptive development. The clinicians introduce the concept of *shaping* and provide examples, suggestions, and prompts for how to shape children's skills instead of providing reassurance (i.e., "What do you think?"). For home learning, parents are assigned *Shaping Detective Thinking and Problem Solving at Home.*

6.6 Module 6: Awareness of Emotional Experiences

Module 6 builds on and enhances the behavioral and cognitive skills learned and practiced thus far by incorporating present-moment awareness and nonjudgmental awareness during emotional experiences. Clinicians should provide adolescents with the rationale for learning these awareness skills. They can explain that in contrast to Module 5's cognitive techniques, practicing present-moment awareness is useful before, during, and after an emotional experience or situation to clarify and deepen the experience itself (as opposed to escape from or avoid it). Clinicians may further introduce the concept by saying:

> Present-moment awareness means we are fully engaged in the "here and now." We are not thinking about the past or the future (although thoughts about these may pass through your mind), but about what is happening in the moment. We are focusing on one thing at a time and letting go of distractions or distracting thoughts. I'd like to suggest that it is okay to just notice and say something to yourself about what you are experiencing and leave it at that, so you can participate in the present moment. In the end,

> this may help you slow down the emotion twisters you experience, which will help you better manage your emotions in the future [1].

Clinicians should describe a situation where this skill would make sense for their client to use, individualizing and relating this concept to the client's specific experience. For some teens, these skills can be thought of as an opposite action to suppression and rumination that may occur during emotionally intense scenarios. Overall, adolescents should understand that practicing present-moment awareness will help them slow down and better utilize their skills to approach problems and manage strong emotions. Similar to the steps of body scanning, adolescents will also use the *Notice it, Say something about it, and Experience it* worksheet as needed for this exercise [79].

Adolescents then put this skill to use in session by practicing at least one present-moment awareness exercise. Some suggestions include General Breathing Awareness, Exploring a Candy Exercise, Mindful Walking, Play-Doh Exercise, and Guessing the Flavor or Scent. See Box 1 for the script included in the therapist guide for the Exploring a Candy Exercise [81].

Box 1 Five Senses Game Option 2: Exploring a Candy Exercise

Allow the children to select one candy (or other small food item, e.g., raisin, M&M, etc.), and ask the children to hold it but not eat it yet. Having a small selection of candies in a dish allows you to complete the second awareness practice referenced in this example as well.

> You are an alien from a planet far, far away. Your first day on Earth, you find some candies in a dish. You have never seen candy before. Pick up one of the candies in the dish. Just notice the candy, looking at it carefully as if you had never seen one before. Feel the candy between your fingers and notice its colors. Use your senses of touch and hearing and sight to notice the candy even better. Be aware of any thoughts you might be having about the candy, even if you don't like this candy. Now, re- focus on the candy and lift it to your nose and smell it for a while. Finally, you are going to use your sense of taste by bringing the candy to your lips, being aware of your arm moving your hand to bring it to your mouth and even your mouth and brain feeling excited about eating it (or disgusted by eating it; there are no right or wrong answers here). Take the candy into your mouth and chew it slowly, experiencing the actual taste of the candy. Hold it in your mouth. Pay close attention as you swallow it and it goes down your throat. When you are ready, pick up a different candy and repeat this process again on your own, as if it is now the first candy you have ever seen.

(continued)

Box 1 (continued)

Adapted from Unified Protocol for Transdiagnostic Treatment of Emotional Disorders in Children and Adolescents: Therapist Guide (p. 317), by J. Ehrenreich-May, S.M. Kennedy, J.A. Sherman, S.M. Bennett, and D.H. Barlow, 2018, New York: Oxford University Press.

This exercise is adapted from Williams et al., 2007.

In this presentation of skills, *nonjudgmental awareness* is framed as a more purposefully compassionate and accepting stance on the notion of present-moment awareness. Instead of evaluating one's experiences or thoughts as good, bad, or wrong as they present in the moment, adolescents are encouraged to approach their emotions with empathy and understanding, as they would with a close friend. Instead of being critical or trying to distract from uncomfortable self-judgment, which can make one avoid situations or feel worse, adolescents can learn to pay attention to their emotions, let go of the judgment, re-engage in the present moment, and acknowledge and accept the experience for what it is. The three steps (Notice it, Say Something About it, Experience it) in present-moment awareness should be applied here as well. For example, when an adolescent is talking to someone, she might notice that her face gets red, and might think, "I'm so stupid! What's wrong with me? This person will never like me if I don't calm down." Using nonjudgmental awareness, this teen might say to herself, "This is how it is right now (notice it), there go my thoughts again (say something about it)," and then gently bring her attention back to the person and conversation (experience it [in the present moment]). Empathize with the adolescent that doing this may be harder than it sounds, but it is not impossible and doesn't need to be done perfectly to be helpful. The more they practice this skill in session and then at home, the more natural and routine it becomes.

Lastly, in preparing for Module 7, teens are introduced to the concept of a *generalized emotion exposure* and practicing nonjudgmental and present-moment awareness in context. Prompt teens to participate in activities that may induce some broad and general emotional experiences, like listening or watching a sad movie or reading or writing a narrative. The goal in engaging in a generalized emotion exposure is to help the adolescent learn that they do not need to avoid experiencing uncomfortable emotions and to help encourage practice of their Module 6 skills. This practice can be introduced using a sample script [1]. Family media rules, rapport, and clinical severity should be considered when coming up with stimuli needed for conducting general emotion exposures. Clinicians may choose to progress from exposures that induce positive, to mixed, to negative emotions, or with significantly depressed youth only induce pleasant emotions, remembering that there is more time to conduct more personalized exposures in Module 7 to come.

In UP-C Session 8, Awareness of Emotion Exposure, introduce children to the concept of being more fully aware of their emotions and teach them the steps of present-moment awareness. The clinician can provide examples of how sitting with an emotion, similar to how they sat with uncomfortable feelings in Session 4, can help them learn that they are not dangerous and help stop the cycle of avoidance. After a neutral exercise, the group plays a "Using My Five Senses" game. The clinician and group can choose between options, one being the "Exploring a Candy Exercise" [3, 81]. Parents discuss the importance of learning to experience emotions instead of avoiding them, practice present-moment and nonjudgmental awareness for situations in parents' lives, and begin to prepare for the upcoming exposure therapy activities [3].

6.7 Module 7: Situational Emotion Exposure

Before moving into exposures, skills from previous modules should be reviewed, along with any remaining emotional behaviors and their triggers, as the clinician works with the adolescent and their parent(s) to create the adolescent's list of exposure situations. Module 7 capitalizes on the principles of exposure therapy in order to help the adolescent engage in increasingly uncomfortable situations they will ideally endure, progressively perceive as harmless, and ultimately alter their cognitions and behaviors to allow for continued engagement without emotional behaviors. *Exposure* and inducing anxiety in treatment, as opposed to relaxation strategies, show positive treatment effects for youth anxiety [82, 83]. Further, much of the improvement in anxiety symptoms in CBT treatment takes place after exposures are initiated [78], supporting the recommendation to initiate exposures sooner in treatment. Additionally, exposures are associated with few side effects [84]; therefore, careful consideration should be made in this flexible, modular application in terms of how soon one can conceivably and comfortably utilize Module 7 with their adolescent client.

Exposure or further opposite action exercises can occupy many sessions and may have already been introduced in treatment at this stage. Some adolescents with depression symptoms may have already been practicing exposure through opposite action. For some irritable or anger-prone adolescents, a combination of irritability/frustration exposures, mindful awareness, and problem-solving steps may be combined to promote more effective engagement with triggering situations and environments. For youth with OCD, combining exposure with response prevention in this module is emphasized, whereas the narrative form in Module 6 may be used as a starting point for exposure to trauma-related cues that may be expanded as needed for trauma-exposed youth. Finally, for those with social skill deficits, instructions are provided in this module for how to combine some basic social skill building with graduated exposure steps. Across triggers and situations, Module 7 is designed to directly reinforce and realistically engage with extinguishing or tolerating distress associated with situations in which emotional behaviors continue to occur. Clinics may vary in their ability to accommodate varying types of exposures or exposure locations other than the office. Regardless, exposures often require creativity in using existing therapy space when limited to such, and creative use of immediately surrounding areas, if possible, such as retail and food stores, schools or classrooms, along with access to confederates, and props. For these reasons, those new to exposure provision are encouraged to seek support and consultation from supervisors and peers to increase their confidence levels in exposure activities. In addition to review and resources about exposure basics in the UP-C/A, other adjunctive materials may be useful.

Some examples of emotional behaviors to target in UP-A exposures include avoidance, escape, withdrawal/isolation, bringing along other people, distraction, emotion suppression, worry, rituals, anger outbursts, and physical aggression. Clinicians and youth should identify and pragmatically list situations and behaviors for exposures on the *Emotional Behavior Form* (Fig. 2). Some clinicians may complete the form hierarchically, with entries listed from lowest degree of emotion elicited to the strongest; however, this not necessary. Most importantly, the clinician should be mindful of where they might start and ask questions that would allow for new safety learning to be easily achieved in exposure (e.g., what would make this harder? easier? etc.).

Exposure may help adolescent learn that nothing truly dangerous or terrible is likely happening when they do not engage in emotional behaviors while in feared situations. There are certainly many theories as to why exposure is effective, including extinction/habituation, learning through increased practice between sessions, disconfirmation of beliefs about potential threats or expectancy violation, and inhibitory learning [85–88]. Inhibitory learning principles suggest that exposure creates new memories to teach youth that their previously feared situations are safe. Repeated practice facing feared situations may also lead to habituation or extinction of distress and anxiety in previously fear-evoking situations. Clinicians can use the *Emotion Curve-Habituation* figure to illustrate habituation-related reductions in distress during exposure to the adolescent, as this rationale may not be readily understood by adolescents. However, clinicians may wish to clarify that it is not always necessary to habituate during an exposure. While traditional

Form 7.1: Emotional Behavior Form

Use this form to identify and describe situations that cause you to feel strong emotions, as well as the emotional behaviors you use in these situations. Using the Emotions Scale thermometer below, rate how much uncomfortable emotion you experience in each situation. When creating this list, think of behaviors like avoidance, escape, or other undesired actions (like aggression) which you may wish to change during treatment. As time goes on in treatment, you can use the last column (Did you work on it?) to see how much progress you've made on these behaviors over time.

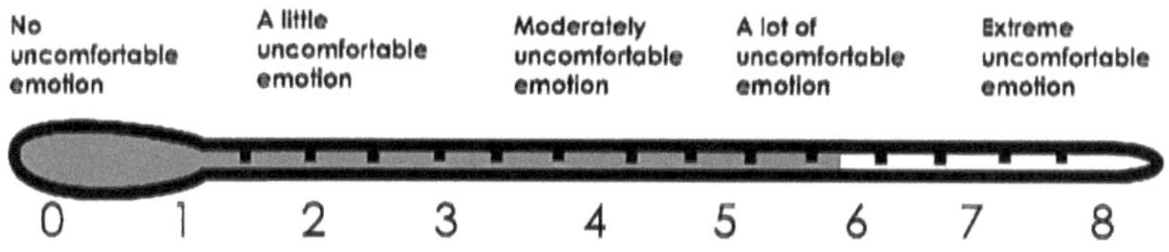

Situation	Emotional Behavior	Emotion (0–8)	Did you work on it? (Y/N)

Jill Ehrenreich-May, Sarah M. Kennedy, Jamie A. Sherman, Shannon M. Bennett, David H. Barlow
Unified Protocol for Transdiagnostic Treatment of Emotional Disorders in Adolescents.

Oxford Clinical Psychology | Oxford University Press

Fig. 2 The Emotional Behavior Form (EBF)

The clinician and adolescent will complete this form to decide which emotional behaviors to concentrate on next. Adapted from Unified Protocol for Transdiagnostic Treatment of Emotional Disorders in Adolescents: Workbook (p. 87), by J. Ehrenreich-May, S.M. Kennedy, J.A. Sherman, S.M. Bennett, and D.H. Barlow, 2018, New York: Oxford University Press. Copyright 2018 by Oxford University Press. Reprinted with permission form models posit that habituation is essential for fear extinction, more recent research shows this may not be valid [89, 90]. An inhibitory learning approach posits that fear tolerance and learning that the situation is safe is central to fear extinction. Habituation is less important in this model, while memory and learning, retrieval, and influences of time and context are key [86, 91]. Yet, if habituation does not occur during an in-session exposure, the clinician should process with parents, as well as the youth, following exposure as to not alarm parents or bring up concerns about teen's ability to handle the exposure activity.

Before exposures commence, the clinician should obtain general consent for planned activity but avoid excess reassurance provision and focus on labeling of any affect and affirmation of the youth's self-efficacy to engage in exposure. The clinician should also pay close attentive to any use of safety behaviors, such as having a phone, another person, and food or drink present during subsequent exposure practices. Dealing with subtle avoidance and/or safety behaviors can be imperative for the efficacy of the exposure practice and long-term exposure efficacy [92, 93]. Adolescents should continue to monitor their progress and outcomes of the exposure through the use of awareness skills throughout each exposure "experiment." The adolescent can rate their distress before, during, and after exposure using a Subjective Units of Distress (SUDS) scale [94]. It is also important to leave enough time to debrief after the exposure activity. Issues to address are as follows:

Box 2 Debriefing After Exposures

- Was there anything that surprised the adolescent? Did what they expect to happen occur?
- Were the adolescent's hypotheses supported?
- Did the adolescent learn anything new about the situation or about their behaviors?
- Did the adolescent's ability to observe the strong emotion experienced improve?
- Was the adolescent's emotion level highest before the exposure started? What level was their emotion at when the exposure was over?

Additionally, after each exposure, you may want to point out any noticeable patterns in how the adolescent's self-ratings changed over the course of the exposure and discuss if the adolescent experienced habituation, referencing the emotion curve to show how in the past their emotion decreased likely at the time they decided to, or took measures to, avoid the feared situation. Discuss how this response has changed since practicing exposure. Encourage the client to verbally express anything new they learned about the feared situation or about their ability to cope with strong emotions if habituation did not occur, understanding that this is not always a necessary precursor to enhanced safety learning.

Home learning should expand upon exposures or other opposition action practice, varying the type and scope. Assign approximately two situational exposure activities or opposite action practices each week between the remaining sessions and instruct the adolescent to monitor reactions during the experiments using the *Tracking the BDA* form. Module P can be important in this stage of treatment in order to plan and conduct exposures. For some adolescents, this may necessitate resources such as materials/ props, transportation, supervised activities, and support. If unavailable or limited, exposure situations should be modified.

In UP-C Session 9, skills are reviewed, opposite action experiments are reviewed, and the concept of a new type of science experiment, "exposure," is introduced. The clinician completes a demonstration of an exposure [3]. Together, families complete the *Emotional Behavior Form* (see Fig. 2). In the parent-only portion of the session, parents are provided with thorough psychoeducation of exposure therapy, as it requires their cooperation to implement exposures, reinforce, and help their child continuously engage in this type of "experiment." When the group returns for Session 10, the concept and rationale for exposures are reviewed, and the group learns about safety behaviors. They practice an exposure, and plans are made for individualized situational emotion exposures. In Sessions 11–14, individualized exposures are conducted. Children then earn their E badge.

6.8 Module 8: Reviewing Accomplishments and Looking Ahead

The end of treatment is a time to celebrate, reinforce the hard work that's been done, ensure the changes endure, and skills continue to be used. The last session of treatment is devoted to such celebration and a review the progress made and skills that have been most useful. Changes in Top Problems are reviewed with parents. Items on the *Emotional Behavior Form* should also be re-rated. The clinician and client should also formulate a plan for the future, which may include ongoing exposures. The difference between lapses and relapses is also instructed, so that clients and their parents can appropriately and adaptively respond to either. In the UP-C, children earn the S badge and use their earned pieces to complete the puzzle. Parents, clinicians, and children then acknowledge youth accomplishments in becoming an "Emotion Detectives" with a party and receipt of a certificate acknowledging this achievement.

7 Conclusion

The UP-C/A offers a unique, flexible approach to the transdiagnostic treatment of emotional disorders in youth. They contain both "common elements" of change for these conditions and present such materials in a singular manner that reflects attendance to potential core dysfunction across these conditions and an overarching theory of change. By presenting common change strategies associated with cognitive behavior therapy and third-wave behavior therapy in a flexible manner that allows clinicians to utilize a range of emotion states (fear, anxiety, anger, and sadness in particular) and progress monitoring tools (Top Problems, Emotional Behaviors Form, continued use of the Before/During/After) to help personalize their treatment approach, we believe the UP-C/A offers a comprehensive and elegant set of tools for treating a range of emotional disorders and related conditions. However, there is certainly more work to be done on the horizon. While initial efficacy trials have shown the adult UP, along with their adolescent and child versions, to be efficacious in research settings, effectiveness and implementation efforts have just begun. It is very likely that some refinement of the protocol to aid clinicians in abbreviating or optimally tailoring the approach to various circumstances and settings will likely result from these efforts, as aspects of the UP-C/A may be considered too lengthy, require too much a priori knowledge of cognitive or behavior therapy principles, or appear cumbersome or nonspecific to some community clinicians initially approaching them at present. More so, there has been substantial interest in adapting the UP-C/A for use in other settings (e.g., pediatric and school settings), and for youth with more severe mental illness, disruptive behavior conditions with strong mood features, borderline disorder features, and eating disorders. Although largely developed in South Florida, with predominantly Hispanic/Latinx samples, there may be needs to also further linguistically or culturally tailor the intervention or its parenting components, in particular, as time goes on. Nevertheless, the UP-C/A shares an important focus on transdiagnostic presentation of skills, a strong parenting component, a fundamental fusion of cognitive-behavioral and third-wave behavioral principles to promote change and a flexibly intended set of treatment elements that, at least in research settings, promote significant improvement in emotional disorders among youth.

Conflicts of Interest

Elizabeth R. Halliday, PhD has no conflicts of interest to declare that are relevant to the content of this chapter.

Jill Ehrenreich-May, Ph.D. receives a royalty from the sales of the Unified Protocols for Transdiagnostic Treatment of Emotional Disorders in Children and Adolescents from Oxford University Press. She also has grant funding related to this intervention from the National Institute of Health, Patient Centered Outcomes Research Institute, Institute of Education Sciences, the Children's Trust, and the Batchelor Foundation. She is also paid for clinical training and implementation support services related to this intervention by a variety of institutions.

References

1. Ehrenreich-May J et al (2017) Unified protocols for transdiagnostic treatment of emotional disorders in children and adolescents: therapist guide. Oxford University Press, New York
2. Ehrenreich-May J et al (2017) Unified protocol for transdiagnostictreatment of emotional disorders in adolescents: workbook. Oxford University Press, New York
3. Ehrenreich-May J et al (2017) Unified protocol for transdiagnostic treatment of emotional disorders in children: workbook. Oxford University Press, New York
4. Barlow DH, Ellard KK, Fairholme CP (2010) Unified protocol for transdiagnostic treatment of emotional disorders: workbook. Oxford University Press
5. Tonarely NA et al (2020) Neuroticism as an underlying construct in youth emotional disorders. Bull Menn Clin 84(3):214–236
6. Barlow DH et al (2014) The origins of neuroticism. Perspect Psychol Sci 9(5):481–496
7. Seager I, Rowley AM, Ehrenreich-May J (2014) Targeting common factors across anxiety and depression using the unified protocol for the treatment of emotional disorders in adolescents. J Ration Emot Cogn Behav Ther 32:67–83
8. Sherman JA, Ehrenreich-May J (2020) Changes in risk factors during the unified protocol for transdiagnostic treatment of emotional disorders in adolescents. Behav Ther 51(6):869–881
9. Ehrenreich-May J, Kennedy SM (2021) Applications of the unified protocols for transdiagnostic treatment of emotional disorders in children and adolescents. Oxford University Press, Oxford
10. Shaw AM, Halliday ER, Ehrenreich-May J (2020) The effect of transdiagnostic emotion-focused treatment on obsessive-compulsive symptoms in children and adolescents. J Obsessive Compuls Relat Disord 26:100552
11. Mohajerin B, Lynn SJ, Cassiello-Robbins C (2023) Unified protocol vs trauma-focused cognitive behavioral therapy among adolescents with PTSD. Behav Ther
12. Melton TH et al (2016) Comorbid anxiety and depressive symptoms in children and adolescents: a systematic review and analysis. J Psychiatr Pract 22(2):84
13. Cummings CM, Caporino NE, Kendall PC (2014) Comorbidity of anxiety and depression in children and adolescents: 20 years after. Psychol Bull 140(3):816–845
14. Kalin NH (2020) The critical relationship between anxiety and depression. Am J Psychiatry 177(5):365–367
15. Marchette LK, Weisz JR (2017) Practitioner review: empirical evolution of youth psychotherapy toward transdiagnostic approaches. J Child Psychol Psychiatry 58(9):970–984
16. Stumpp NE, Southward MW, Sauer-Zavala S (2024) Assessing theories of state and trait change in neuroticism and symptom improvement in the Unified Protocol. Behav Ther 55(1):93–105
17. Chorpita BF, Barlow DH (2018) The development of anxiety: the role of control in the early environment. Neurotic Paradox 2:227–264
18. Kitt ER et al (2022) Family accommodation in pediatric anxiety: relations with avoidance and self-efficacy. Behav Res Ther 154:104107
19. Birk SL et al (2022) Unpacking parental accommodation: relationship to parent distress tolerance and cognitive styles. J Anxiety Disord 92:102639
20. Barlow DH et al (2017) The unified protocol for transdiagnostic treatment of emotional disorders compared with diagnosis-specific protocols for anxiety disorders: a randomized clinical trial. JAMA Psychiatry 74(9):875–884
21. Martin P et al (2018) Transdiagnostic treatment approaches for greater public health impact: Implementing principles of evidence-

based mental health interventions. Clin Psychol Sci Pract 25(4):e12270
22. Pincus DB, Ehrenreich JT, Mattis SG (2008) Mastery of anxiety and panic for adolescents riding the wave, therapist guide. Oxford University Press, Oxford
23. Allen LB et al (2012) A unified, transdiagnostic treatment for adolescents with chronic pain and comorbid anxiety and depression. Cogn Behav Pract 19(1):56–67
24. Schaeuffele C et al (2021) CBT at the crossroads: the rise of transdiagnostic treatments. Int J Cogn Ther 14:86–113
25. Bullis JR et al (2015) The unified protocol for transdiagnostic treatment of emotional disorders: preliminary exploration of effectiveness for group delivery. Behav Modif 39(2): 295–321
26. Osma J et al (2015) The unified protocol for transdiagnostic treatment of emotional disorders in group format in a spanish public mental health setting. Psicol Conduct 23(3):447
27. Carlucci L, Saggino A, Balsamo M (2021) On the efficacy of the unified protocol for transdiagnostic treatment of emotional disorders: a systematic review and meta-analysis. Clin Psychol Rev 87:101999
28. Ehrenreich JT et al (2009) Development of a unified protocol for the treatment of emotional disorders in youth. Child Family Behav Ther 31(1):20–37
29. Trosper SE et al (2009) Emotion regulation in youth with emotional disorders: implications for a unified treatment approach. Clin Child Fam Psychol Rev 12:234–254
30. Ehrenreich-May J et al (2017) An initial waitlist-controlled trial of the unified protocol for the treatment of emotional disorders in adolescents. J Anxiety Disord 46:46–55
31. Queen AH, Barlow DH, Ehrenreich-May J (2014) The trajectories of adolescent anxiety and depressive symptoms over the course of a transdiagnostic treatment. J Anxiety Disord 28(6):511–521
32. Ehrenreich-May J, Bilek EL (2012) The development of a transdiagnostic, cognitive behavioral group intervention for childhood anxiety disorders and co-occurring depression symptoms. Cogn Behav Pract 19(1):41–55
33. Kennedy SM, Bilek EL, Ehrenreich-May J (2019) A randomized controlled pilot trial of the unified protocol for transdiagnostic treatment of emotional disorders in children. Behav Modif 43(3):330–360
34. Lopez ME et al (2015) Examining the efficacy of the unified protocol for transdiagnostic treatment of emotional disorders in the treatment of individuals with borderline personality disorder. Cogn Behav Pract 22(4): 522–533
35. Lopez ME et al (2019) The unified protocol for anxiety and depression with comorbid borderline personality disorder: a single case design clinical series. Cogn Behav Ther 12:e37
36. Sauer-Zavala S, Bentley KH, Wilner JG (2016) Transdiagnostic treatment of borderline personality disorder and comorbid disorders: a clinical replication series. J Personal Disord 30(1):35–51
37. Tonarely NA et al (2021) Targeting elevated borderline features in adolescents using the Unified Protocol for transdiagnostic treatment of emotional disorders in Adolescents (UP-A). Evid Based Pract Child Adolesc Ment Health 6(1):47–64
38. Bentley KH (2017) Applying the unified protocol transdiagnostic treatment to nonsuicidal self-injury and co-occurring emotional disorders: a case illustration. J Clin Psychol 73(5): 547–558
39. Bentley KH et al (2017) Treating suicidal thoughts and behaviors within an emotional disorders framework: acceptability and feasibility of the unified protocol in an inpatient setting. Behav Modif 41(4):529–557
40. Thompson-Brenner H et al (2019) Implementation of transdiagnostic treatment for emotional disorders in residential eating disorder programs: a preliminary pre-post evaluation. Psychother Res 29(8):1045–1061
41. Boisseau C, Boswell J (2018) Unified protocol for eating disorders, in applications of the Unified Protocol for transdiagnostic treatment of emotional disorders. Oxford University Press, Oxford
42. Eckhardt S et al (2019) An ARFID case report combining family-based treatment with the unified protocol for transdiagnostic treatment of emotional disorders in children. J Eat Disord 7(1):1–7
43. Burton C et al (2021) Case presentations combining family-based treatment with the unified protocols for transdiagnostic treatment of emotional disorders in children and adolescents for comorbid avoidant restrictive food intake disorder and autism spectrum disorder. J Can Acad Child Adolesc Psychiatry 30(4): 280
44. Weintraub MJ et al (2020) Applying a transdiagnostic cognitive-behavioral treatment to adolescents at high risk for serious mental illness: rationale and preliminary findings. Cogn Behav Pract 27(2):202–214

45. Lewin AB et al (2021) Transdiagnostic cognitive behavioral therapy for misophonia in youth: methods for a clinical trial and four pilot cases. J Affect Disord 291:400–408
46. Tonarely-Busto NA et al (2023) Applying the unified protocol for transdiagnostic treatment of emotional disorders in children and adolescents to misophonia: a case example. Evid Based Pract Child Adolesc Ment Health 8(1): 133–147
47. Gruen R et al (2023) A trauma-informed approach to the unified protocol for children with exposure to child maltreatment. In: Innovative methods in child maltreatment research and practice: advances in detection, causal estimation, and intervention. Springer
48. Hawks JL et al (2020) Development and application of an innovative transdiagnostic treatment approach for pediatric irritability. Behav Ther 51(2):334–349
49. Grossman RA, Ehrenreich-May J (2020) Using the unified protocol for transdiagnostic treatment of emotional disorders with youth exhibiting anger and irritability. Cogn Behav Pract 27(2):184–201
50. García-Escalera J et al (2020) The unified protocol for transdiagnostic treatment of emotional disorders in adolescents (UP-A) adapted as a school-based anxiety and depression prevention program: an initial cluster randomized wait-list-controlled trial. Behav Ther 51(3):461–473
51. Halliday ER et al (2024) Initial effects of a brief transdiagnostic intervention on parent emotion management during COVID-19. Child Psychiatry Hum Dev 55(2):372–383
52. Tonarely NA et al (2022) Tailoring the unified protocol for adolescents for a stepped-care approach: case exemplars. J Cogn Psychother 35(4):290–307
53. Farchione TJ et al (2012) Unified protocol for transdiagnostic treatment of emotional disorders: a randomized controlled trial. Behav Ther 43(3):666–678
54. Sakiris N, Berle D (2019) A systematic review and meta-analysis of the Unified Protocol as a transdiagnostic emotion regulation based intervention. Clin Psychol Rev 72:101751
55. Conklin LR et al (2015) Relationships among adaptive and maladaptive emotion regulation strategies and psychopathology during the treatment of comorbid anxiety and alcohol use disorders. Behav Res Ther 73:124–130
56. Arunya B, Sudhir PM, Arumugham SS (2023) Effect of Unified Protocol on neuroticism, emotion regulation skills, and symptom severity – a case series. J Affect Disord Rep 14:100619
57. Sauer-Zavala S et al (2012) The role of negative affectivity and negative reactivity to emotions in predicting outcomes in the unified protocol for the transdiagnostic treatment of emotional disorders. Behav Res Ther 50(9):551–557
58. Tonarely NA et al (2021) Impact of youth transdiagnostic treatment on parents' own emotional responding and socialization behaviors. J Child Fam Stud 30:1141–1155
59. Weisz JR, Chorpita BF, Kendall P (2011) "Mod squad" for youth psychotherapy. In: Child and adolescent therapy: cognitive-behavioral procedures, pp 379–397
60. Kolko DJ et al (2007) Internalizing symptoms and affective reactivity in relation to the severity of aggression in clinically referred, behavior-disordered children. J Child Fam Stud 16: 745–759
61. Kolko DJ et al (2009) Community vs. clinic-based modular treatment of children with early-onset ODD or CD: a clinical trial with 3-year follow-up. J Abnorm Child Psychol 37(5):591–609
62. Kolko DJ, Pardini DA (2010) ODD dimensions, ADHD, and callous–unemotional traits as predictors of treatment response in children with disruptive behavior disorders. J Abnorm Psychol 119(4):713
63. Shaffer A et al (2013) Bidirectional relations between parenting practices and child externalizing behavior: a cross-lagged panel analysis in the context of a psychosocial treatment and 3-year follow-up. J Abnorm Child Psychol 41: 199–210
64. Evans SC et al (2020) Effects of standard and modular psychotherapies in the treatment of youth with severe irritability. J Consult Clin Psychol 88(3):255
65. Evans SC et al (2021) Modular psychotherapy outcomes for youth with different latent profiles of irritability and emotion dysregulation. Front Psych 12:618455
66. Chorpita BF (2007) Modular cognitive-behavioral therapy for childhood anxiety disorders. Guilford Press
67. Harmon SL et al (2021) Evaluating a modular approach to therapy for children with anxiety, depression, trauma, or conduct problems (MATCH) in school-based mental health care: Study protocol for a randomized controlled trial. Front Psychol 12:639493
68. Bartuska AD et al (2024) Provider adherence to modular cognitive behavioral therapy for children and adolescents. Admin Pol Ment Health:1–13
69. Borntrager CF et al (2009) Provider attitudes toward evidence-based practices: are the

concerns with the evidence or with the manuals? Psychiatr Serv 60(5):677–681
70. Corteselli KA et al (2020) School counselor perspectives on implementing a modular treatment for youth. Evid Based Pract Child Adolesc Ment Health 5(3):271–287
71. Weisz JR et al (2023) Research review: the internalizing paradox–youth anxiety and depression symptoms, psychotherapy outcomes, and implications for research and practice. J Child Psychol Psychiatry 64(12): 1720–1734
72. Weisz JR et al (2012) Testing standard and modular designs for psychotherapy treating depression, anxiety, and conduct problems in youth: a randomized effectiveness trial. Arch Gen Psychiatry 69(3):274–282
73. Fonseca-Pedrero E et al (2023) PSICE project protocol: evaluation of the unified protocol for transdiagnostic treatment for adolescents with emotional symptoms in school settings. Clínica y Salud 34(1)
74. Zemestani M et al (2024) A culturally adapted unified protocol for transdiagnostic treatment of anxiety disorders in adolescents (UP-A): a randomized waitlist-controlled trial. Psychol Med 54(2):385–398
75. Hajal NJ, Paley B (2020) Parental emotion and emotion regulation: a critical target of study for research and intervention to promote child emotion socialization. Dev Psychol 56(3):403
76. Frederick J et al (2023) Do specific modules of cognitive behavioral therapy for depression have measurable effects on youth internalizing symptoms? An idiographic analysis. Psychother Res 33(3):265–281
77. Weisz JR et al (2011) Youth top problems: using idiographic, consumer-guided assessment to identify treatment needs and to track change during psychotherapy. J Consult Clin Psychol 79(3):369
78. Guzick AG et al (2022) Change during cognitive and exposure phases of cognitive–behavioral therapy for autistic youth with anxiety disorders. J Consult Clin Psychol 90(9):709
79. Linehan MM et al (2015) Dialectical behavior therapy for high suicide risk in individuals with borderline personality disorder: a randomized clinical trial and component analysis. JAMA Psychiatry 72(5):475–482
80. Qi J et al (2021) Differential associations of distress tolerance and anxiety sensitivity with adolescent internalizing psychopathology. J Clin Child Adolesc Psychol
81. Williams M et al (2012) The mindful way through depression: freeing yourself from chronic unhappiness. Guilford Press
82. Whiteside SP et al (2020) A meta-analysis to guide the enhancement of CBT for childhood anxiety: exposure over anxiety management. Clin Child Fam Psychol Rev 23:102–121
83. Bilek E et al (2022) Exposure-focused CBT outperforms relaxation-based control in an RCT of treatment for child and adolescent anxiety. J Clin Child Adolesc Psychol 51(4): 410–418
84. Walter HJ et al (2020) Clinical practice guideline for the assessment and treatment of children and adolescents with anxiety disorders. J Am Acad Child Adolesc Psychiatry 59(10): 1107–1124
85. Kennedy SM, Hawks JL (2021) Lessons learned from a randomized controlled pilot trial testing an inhibitory learning approach to exposure in anxious youth. Cogn Behav Pract 28(2):241–258
86. Craske MG et al (2014) Maximizing exposure therapy: an inhibitory learning approach. Behav Res Ther 58:10–23
87. de Jong R et al (2024) Effectiveness of exposure-based treatment for childhood anxiety disorders: an open clinical trial to test its relation with indices of emotional processing and inhibitory learning. J Behav Ther Exp Psychiatry 83:101942
88. Guzick AG et al (2020) That was easy! Expectancy violations during exposure and response prevention for childhood obsessive-compulsive disorder. Behav Modif 44(3):319–342
89. Baker A et al (2010) Does habituation matter? Emotional processing theory and exposure therapy for acrophobia. Behav Res Ther 48(11):1139–1143
90. Jacoby RJ, Abramowitz JS (2016) Inhibitory learning approaches to exposure therapy: a critical review and translation to obsessive-compulsive disorder. Clin Psychol Rev 49:28–40
91. Abramowitz JS, Arch JJ (2014) Strategies for improving long-term outcomes in cognitive behavioral therapy for obsessive-compulsive disorder: insights from learning theory. Cogn Behav Pract 21(1):20–31
92. Blakey SM et al (2019) A randomized controlled trial of the judicious use of safety behaviors during exposure therapy. Behav Res Ther 112:28–35
93. Hedtke KA, Kendall PC, Tiwari S (2009) Safety-seeking and coping behavior during exposure tasks with anxious youth. J Clin Child Adolesc Psychol 38(1):1–15
94. Wolpe J (1969) Subjective units of distress scale. J EMDR Pract Res

Chapter 16

CBT with Youth: Catchers in the Rye

Micaela A. Thordarson and Robert D. Friedberg

Abstract

CBT is a long-standing frontrunner in treatment of youth with psychological disorders. This text enumerates all the ways in which that is still true, as well as all the growth that is happening. Any science unwilling to undergo regular and repeated evaluation is destined to be left in the past. Contrary to resting on their laurels, CBT clinical researchers continue to investigate relevance and identify novel applications to extant treatment protocols. This chapter illustrates the numerous contributions that the text makes to today's landscape of CBT for young patients. Consistent across chapters are authors' insistence that case conceptualization is imperative, developmental application is essential, and, above all, before assuming treatment is failing, double check whether therapy is faithfully attending to the requisite targets. Each chapter includes a range of creative activities, innovative exercises, astute scripts, and troubleshooting tips. Readers of this text are sure to keep their copy on a shelf very close to their treatment space, for easy reference and apt guidance!

Key words Innovation, Creativity, Evolution, Case conceptualization, Developmentally tailored, CBT, Children, Adolescents, Third wave

1 Introduction

There is nothing so stable as change
Bob Dylan

Film and TV histories are scarred with failed sequels. Reboots of old favorites are commonly disappointing. Second editions are kind of sequels and similarly often fail to live up to the first edition's legacy. Perhaps, the letdowns are due to nothing being really new is in the sequel or maybe it's just regression to the mean. When we undertook this project, we were acutely aware of this problem. Fortunately, our worries were unfounded. CBT with youth: Tradition and innovation are a fresh take on old favorites. In this final chapter, we review the book's landscape and offer some concluding thoughts.

Robert D. Friedberg and Micaela A. Thordarson (eds.), *Cognitive Behavioral Therapy with Youth: Tradition and Innovation*, Neuromethods, vol. 231, https://doi.org/10.1007/978-1-0716-5134-6_16,

2 Reviewing the Landscape

This book opens with a bang with Violante and colleague's (Chap. 1) powerful overview of the empirical support for CBT. Setting the stage for subsequent chapters, Violante et al. highlight the transdiagnostic effectiveness of CBTs for youth. The first section of this chapter brilliantly details the science and theory upon which CBT is designed. As noted, no further chapter may be adequately used if readers fail to metabolize the mechanisms of action of this potent treatment. The second section outlines the various techniques used to help youth accomplish the behavioral, cognitive, and emotional outcomes they seek from therapy. Strategies that were popularized by third-wave behavioral treatments are incorporated into this chapter (and throughout the book), illustrating conceptually faithful development within the CBT umbrella. Formats for service delivery are also reviewed, giving brief guidance for when CBT may be executed in individual, group, family, and virtual methods. Finally, the authors illustrate vital process elements for successful CBT with youth, emphasizing the need to conceptualize care and tailor therapy to meet the unique needs of each youth. This introductory chapter is an excellent gateway into the CBT universe! Violante et al. depict the ongoing relevance of CBT and demonstrate that recent innovations to CBT platforms are consistent and cohesive with the original model.

Korte and Friedberg then take readers into a deep dive of case conceptualization in Chap. 2. While clinicians of all experience levels often speak of conceptualization, anecdotally many use the term interchangeably with diagnostic impressions. Understanding a case solely in terms of diagnosis is a recipe for sterile, mechanical treatment. It is imperative that a clinician first understands children's inner and external worlds before a targeted, cogent plan of attack can be devised: How did the pathology develop, and what are the maintenance factors that are keeping the youth stuck? Once the how and why are elucidated, the next steps are relatively straightforward. Korte and Friedberg use a clinical case example to illustrate the ways that pathology can be translated into mechanisms of action, showing readers the ways to build a recipe from identified ingredients. The chapter represents a compelling bridge between bench science and bedside practice. As every subsequent chapter will emphasize, conceptualization is vital to effective care. This second chapter is the handbook to be able to do just that.

One of the most voiced complaints about CBT is the neglect toward the therapeutic relationship. Bose and Pettit (Chap. 3) firmly put these criticisms to rest in their thorough review of the crucial role that alliance plays in the treatment outcomes for youth. A strong therapeutic alliance improves the patient's belief that treatment will work, increases motivation for change, keeps patients coming back, and promotes commitment to doing the work. Ensuring there is a connection creates a better foundation for the challenging clinical work to be done. Bose and Pettit break down therapeutic alliance into three domains: relationship and emotional bond, collaborative development of goals, and unified agreement on therapeutic tasks. Moving beyond platitudes, the authors deliver a succinct guide with concrete steps for the establishment and maintenance of a healthy therapeutic alliance.

Recent studies investigate moderators of therapeutic alliance and the effect on treatment outcomes. Studies suggest that a strong relationship promotes improved emotion regulation, which then in turn yields better progress on therapy tasks. Bose and Pettit further review the importance of considering caregiver alliance as well. While youth are typically considered the primary consumer of therapy services, caregivers must also be engaged to ensure session attendance (at a minimum) and support for therapeutic work outside of session. Strategies for building a trusting relationship with parents are outlined, with great attention to the delicate balance of navigating the differences between youth and caregiver needs. Bose and Pettit put to rest the myths about alliance in CBT and perfectly display that, far from being an afterthought, the therapeutic relationship is an essential element of good cognitive behavioral practice.

Riise, Haugland, and Wergeland (Chap. 4) move the text into specific clinical domains of CBT treatment with a thorough review of the treatment of depression. Unsurprisingly, Riise and colleagues emphasize the value of the therapeutic alliance, noting how hard it can be to engage kids with depression in treatment. The chapter divides CBT for depression into three phases, loosely described as (a) beginning, understanding, and building; (b) learning, practicing, and doing; and (c) repeating, relapse prevention, and saying goodbye. Considerations for the outset of care include thorough assessment, case conceptualization, relationship building, establishing clear goals, and delivering clear education about depression and CBT. Importantly, the authors recommend soliciting myths about depression, so that any stigma, self-blame, or other unhelpful beliefs are promptly addressed. For the central phase of care, Riise et al. include creative examples for a variety of interventions, including sample handouts that may be used. As interventions are described, positive reinforcement is highlighted as particularly salient for depressed youth. Given the ways depression affects young patients, clinicians must maintain consistent attention to reinforcement opportunities to keep youth engaged and working. As treatment comes to a close, Riise et al. illustrate the potent ways in which graduation-style celebrations can further dispel any lingering hints of depression by focusing on the mastery, self-efficacy, and myriad accomplishments made in care. Authors beautifully describe the components of CBT for depression while weaving in the conceptual and process principles throughout, giving readers a clear guide for how to deliver treatment.

Suicidal ideation is often considered an element of severe depression; however, studies show that suicidal thoughts may affect youth across a spectrum of diagnoses. Given the grave nature of suicide risk, Padron, Balser, and Pettit (Chap. 5) deliver a specifically tailored CBT approach to address suicidal thoughts and behaviors in youth. The chapter opens with a fantastic review of the prevailing theoretical models of suicide, so that readers may conceptualize suicide with a nuanced understanding of the drivers of risk. Clear ties are drawn between the theory and the science underpinning CBT, illustrating the salience of CBT for suicidal youth. Padron et al. synthesize the extant literature on the outcomes of treatment and cogently summarize what the field does—and does not—know about effective care. Of particular note, this chapter gives readers a detailed review of screening tools and assessment measures that may be employed for suicidal patients. Such instruments allow for close monitoring of risk as well as clear treatment target identification. Authors illustrate the ways that well-known cognitive and behavioral strategies may be specifically applied to suicide risk. For example, hopelessness (a known driver of suicidal ideation) may be targeted by cognitive interventions related to distorted thinking or behavioral experiments to combat pessimistic expectancies. Relapse prevention strategies conclude this chapter and emphasize the importance of normalizing hard days as expected, not a sign of regression. This chapter demystifies the treatment of suicide risk in such a way that readers will feel empowered to address this often-feared experience.

Next, Chu et al. (Chap. 6) dive into fears and give a fulsome update to the CBT for anxiety chapter. CBT has long been the gold standard for anxiety disorders in youth, so how could there be meaningful updates? Chu and colleagues begin by acknowledging that today's youth are facing real dangers in novel ways. Climate change, racism, and other forms of identity-related aggression, social media, and a seemingly unending series of world events provide cues to youth that would reasonably activate anxiety. Ceding functioning and living in heightened distress, however, remain maladaptive. Thus, the chapter illustrates approaches to deconstruct worry thoughts and decrease avoidance behaviors, amplified by interventions to validate the reasonable fears. CBT for anxious youth is not just to stop feeling anxious about maladaptive things; it is also about how to have an adaptive approach to coping with very real, very scary things. The authors also include a section detailing innovative new exercises for anxious youth such as values-based assessment, "life buckets," and "celebration toasts." These activities are meant to complement the vital modules of cognitive interventions and exposures. A case example used throughout the chapter emphatically combats the stereotype that CBT is robotic, bringing to life the ways in which CBT may be flexibly and creatively applied with youth. One might think that CBT for anxious youth clearly maxed out its potential. After going through this chapter, readers will be amazed at the continued evolution of this gold (maybe we call it platinum now?) standard approach.

Obsessive compulsive disorder is another diagnosis where CBT boasts long-standing recognition as the clear forerunner of care for youth. Abramowitz and Juel (Chap. 7) demonstrate exactly what makes CBT so valuable for youth with OCD. Two particular barriers important to discern in the treatment of young patients are the limited insight into the irrationality of the symptoms and the impact of accommodating caregivers. Magical thinking is developmentally appropriate for younger children, making therapy further distressing as treatment moves into exposures and changes to rituals. Parents are biologically wired to want to prevent distress in their children...and yet therapy will counterintuitively require them to withhold the "protection" from distress they reflexively practice. The updated chapter gives readers a detailed conceptual guide to be able to share with parents, so that they understand the importance and efficacy of the CBT approach. Abramowitz and Juel share several scripts that illustrate ways to introduce OCD and therapy—specifically the mechanisms of action for ideal engagement—and emphasize the central significance of psychoeducation at the outset of care. The highlight of this chapter is the series of case examples that depict common ways that CBT for OCD can seem stalled or totally derailed...and the solutions for how to effectively proceed. These common pitfalls are likely experiences that cause many clinicians to give up or to believe themselves or the treatment to be ineffective; thus, this chapter serves as an inoculation against defeat! With ongoing troubleshooting guides, the CBT literature indicates that when treatment seems to fail, minor adjustments help to regain impact. If we all could adapt like this, the dinosaurs might still be among us!

In Chap. 8, Marquez, Cruthirds, Sirois-Geddie, and Deblinger walk readers through the substantially supported trauma-focused CBT model. TF-CBT was originally designed as a CBT application for childhood sexual abuse and is now empirically supported for a broad range of traumatic experiences. TF-CBT research demonstrates internationally validation, for a range of identities, indicating robust effectiveness for youth with a history of trauma. Because of the rigorous and numerous studies that exist, TF-CBT is known to be effectively delivered in clinics, schools, homes, and residential treatment programs. Therapy can also be done in individual, group, in person, and telehealth formats with comparable results. TF-CBT uses the PRACTICE acronym to outline treatment: psychoeducation, parenting skills, relaxation strategies, affective modulation skills, cognitive coping, trauma narrative and processing, in vivo mastery, conjoint sessions, and enhancing safety.

The authors repeatedly emphasize the integral role that non-offending caregivers have in treatment. A particular strength of TF-CBT is the identification that any important adult in a child's life may serve as the participating caregiver. With this feature, every youth's specific needs and family structure may be honored as there can be numerous barriers to incorporating parents into treatment. Furthermore, TF-CBT heavily focuses on affirming the family's values, broadening the focus of trauma care to include strengths and successes. Throughout the chapter, authors call attention to the need to tailor all interventions to the developmental level of the youth, including the incorporation of creative activities and play. This format of CBT is notable for the fact that as studies generalize to new settings, clinical presentations, and populations, efficacy stands true. Considering the numerous challenges our youth face today (see Chap. 6), this treatment will be needed for years to come.

Banneyer and Fein (Chap. 9) write an astute summary of the ways in which CBT must be adapted to meet the needs of youth on the autism spectrum who experience anxiety and OCD. They highlight the importance of diagnostic considerations for anxiety, as many symptoms may be misunderstood as Autism related, leaving youth to remain untreated and in persistent distress. The chapter provides a compelling description of how CBT is a good fit for ASD in combination with anxiety/OCD due to the empirical approach and behaviorally specific interventions. Research indicates that tailoring specific interventions to recognize the unique needs yields a more powerful treatment than the traditional model. Authors indicate it is essential to assess language abilities and cognitive functioning at the outset of treatment to best understand how to measure anxiety symptoms. As a primarily internal experience, evaluation of anxiety heavily relies on linguistic communication of distress. Youth with lower language functioning possess attenuated abilities to describe their distress, not necessarily less anxiety! Thus, clinicians must turn to creative alternative metrics and communication strategies to accurately conceptualize a youth's anxiety. Banneyer and Fein further share the ways in which characteristics of autism (e.g., restricted and repetitive interests) may be mischaracterized as OCD (e.g., compulsions). The authors hammer home the essential function of behavior science principles to delineate—restricted and repetitive interests are typically positively reinforcing, while compulsive rituals are negatively reinforcing. When conducting interventions for anxiety, authors recommend the use of visual aids to support teaching and communication, to include concrete step-by-step instructions, and to avoid the use of metaphors. As Autism and other neurodivergent conditions are diagnosed in higher rates, the need to tailor interventions properly will be as relevant as adapting adult approaches for youth—different brains need different packaging! Chapter 9 paves the way for this work.

Prolific researchers in CBT for angry and aggressive youth, Lochman, Kassing, Romero, and Stromeyer (Chap. 10), deliver a top-notch clinical chapter. The authors open with a thorough review of the contextual social cognitive model, so that readers can fully understand the etiology of these behaviors. Additionally, the chapter describes the relevant risk factors and reiterates that it is truly a combination of genetic and environmental elements that produces aggression in youth. Armed with this information, clinicians will be more effective at conceptualizing the patients they serve. The ability to distill a youth's problematic behaviors into this framework further helps parents and youth understand they are not "bad" kids; they simply developed a pattern of behavior that is getting them in trouble and can be changed. The intervention options for patients with anger include options that target youth alone, caregivers alone, and both. Among those, the Coping Power, the clear winner, is a model all clinicians who work with kids need.

Coping Power employs cognitive restructuring strategies to shift hostile attribution of neutral/ambiguous stimuli, improve perspective taking, develop coping thoughts, and enhance problem-solving skills. This armory of techniques means youth now can choose from a plethora of options rather than resort to aggression to get their needs met. Parent components include the typical behavior management strategies used for this population. However, Coping Power is a comprehensive treatment program aimed at both decreasing unwanted behaviors and increasing resilience and connection. Other caregiver elements emphasize the need for parents to learn to better regulate their own emotions, address family stressors (e.g., homework), improve communication, and build family cohesion. This approach yields a wealth of positive outcomes for both kids and caregivers that are maintained over time. The Coping Power model has been rigorously tested across a variety of cultural applications such as with Deaf children, rural populations, and validated for Dutch, Pakistani, Swedish, and Italian youth. Recent studies noted that the addition of mindfulness skills further improves outcomes. Ongoing research continues to prove the continued prowess of this CBT approach for angry youngsters. Proper intervention for aggression is a life-altering treatment, truly transforming the trajectory of a youth's life; dissemination efforts are a top priority for ensuring more clinicians are prepared to deliver such feasible services.

Diverging slightly from aggressive youth, those diagnosed with Disruptive Mood Dysregulation Disorder suffer from chronic irritability and broad-spectrum emotion regulation challenges. In Chap. 11, German et al. introduce a new treatment approach for youth with DMDD. The CBT model applies exposure-based sessions to generate improved tolerance of anger and learn more effective ways of responding to intense emotions. Creative tools such as the "motivation ruler" keep a clinician's awareness on the child's engagement in therapy. Psychoeducation about anger and self-monitoring of the signs of anger are also covered using fun worksheets. During exposures, the primary goal is to normalize the experience of anger and shape the child's response to more adaptive ways of frustration expression. These sessions are combined with parent management training sessions for caregivers, so that they simultaneously learn more effective ways of responding to their child's behavior. This novel model is currently generating data, and authors report an extremely low rate of study participant dropout! This data point indicates the high acceptability of this model and is totally counter to typical clinician beliefs about exposures. The expansion of CBT into this youth population further demonstrates the transdiagnostic applicability of CBT for youth.

Essayli welcomes readers into the world of eating disorder treatment in Chap. 12. The majority of research in eating disorders is done with adult populations, despite many ED conditions beginning in adolescence. Authors beautifully illustrate eating disorder conceptualization from a cognitive-behavioral theoretical framework to identify the salient targets for care. While family-based treatment is often considered the frontline approach, CBT for eating disorders boasts several differences that make it a uniquely beneficial approach for some youth. True to CBT's roots, this model is highly collaborative and consistently emphasizes the youth's autonomy, a departure from typical ED care. CBT is considered a much more flexible model for EDs compared to FBT and scaffolds the inclusion of caregivers. Additionally, CBT targets the cognitive distortions associated with EDs and focuses on the perceived consequences of weight gain rather than the probability of weight gain. The chapter provides a number of examples of cognitive interventions. Authors also review in detail the ways in which exposures are employed in a CBT model for ED, including open weights—an intervention that is in direct contradiction to the typical approach to ED care. By targeting key CBT mechanisms of action—maladaptive thoughts and avoidance behaviors—EDs can be effectively treated. Many clinicians consider eating disorders a specialty population and decline to work with these patients, which lead to frequent access problems for youth seeking care. Chapter 12 demonstrates the ready and effective application of CBT.

Semple and Madni (Chap. 13) give readers a primer on the use of mindfulness-based CBT for children, a group treatment model. At this point, mindfulness pervades many CBT models, and for good reason. The outcomes associated with MCBT-C for youth experiencing stress and anxiety are profound! The highlights of this chapter are the numerous ways in which Semple and Madni teach readers to truly adapt mindfulness activities for young patients. Activities must be brief; language must be more concrete; and attention targets must vary. Youth are taught to be curious and simply notice without changing...their bodies, then their thoughts, then their emotions. Exercises may switch from imagination, to movement and to a sensory focus. Authors go into detail about the nuanced application of developmental adaptations for younger elementary age kids and older adolescents. The chapter also completes an extraordinary review around the fundamental differences in mechanisms of action for MCBT-C compared with traditional CBT. This is a fabulous opportunity to depict the similarities, differences, and ultimate compatibility between CBT and third-wave applications.

As our world becomes increasingly diverse, our treatments must be adequate to serve our youth. And let's be real—most CBT models were initially designed and validated with extremely limited diversity. Park, Velez, and Kim (Chap. 14) discuss effective ways to attend to the needs of different youth and their families. Despite the flaws of early research conducted primarily with white, middle-class youth, CBT maintains effectiveness when tested with youth with a variety of intersectional identities. Thus, the most relevant considerations in treatment of diverse patients are whether and when to adapt. Most importantly, adaptations must not create fundamental departures from cognitive and behavioral science in order to preserve value. This chapter introduces a sequential model for approaching treatment, beginning with "CBT as usual." In other words, clinicians are best served by starting care exactly as it was validated. Authors instruct clinicians to use an iterative assessment to determine if adaptation is then needed, make the shift. Once therapy feels back on track, therapists are to return to an unadapted approach. Situations such as when a child is specifically experiencing a stressor related to a marginalized identity or when treatment fails to progress as would be expected may indicate a need for adaptation. Authors caution that youth are highly unlikely to disclose identity-based stressors at the outset of care, and clinicians must be attuned to progress and attendance to determine if there may be hidden factors driving distress. As thoroughly demonstrated throughout this text, CBT in its ideal format is already flexible and designed to meet the needs of specific youth. It is heartening to learn that even when considering complex intersectional identities, CBT requires minimal and time-limited adjustment.

Halliday and Ehrenreich-May (Chap. 15) round out the text with the Unified Protocol, a transdiagnostic treatment validated for children and adolescents. One of the earliest formats of transdiagnostic modular CBT for youth, the UP is rife with creative activities and teaching specifically designed for youth. This update incorporates third-wave techniques such as nonjudgmental awareness and present-moment attention. UP can be delivered either individually or in a group. When working with children (under age 13), caregivers are enrolled in a parallel group. The chapter provides immense detail in the differences needed across the developmental trajectory as young children, tweens, and teens all require pertinent content. The UP emphasizes emotion awareness, understanding, and tolerance to reduce maladaptive coping strategies. A comprehensive review of the mechanisms of action of treatment makes it easy for readers to grasp the ways one would apply care across a combination of disorders. Authors share the exciting news that further implementation and dissemination trials are underway, including opportunities to expand into new settings (e.g., schools) and populations (e.g., international samples). Additional data to support the relevance of UP to include more young patients would indicate the substantial value of a larger transdiagnostic and developmental tailored protocol.

As far as paying homage to tradition and celebrating innovation—this text nails it! It is incredibly rewarding for some of the greatest minds in CBT for youth to contribute individual chapters that come together so seamlessly. Every author emphasizes the integral function of case conceptualization. Each chapter specifies the need for attunement to unique needs. All sections celebrate the new ways that CBT models continue to evolve to remain the frontline approach to the treatment of myriad psychopathologies in children and adolescents.

3 Last Stop: Implications for Would-Be Catchers in the Rye

> What I have to do, I have to catch everybody if they start to go over the cliff-I mean if they're running and they don't look where they are going I have to come out of somewhere and catch them. That's all I do all day. I'd just be the catcher in the rye and all. I know it's crazy, but that's the only thing I'd really like to be (p. 172).
>
> J.D. Salinger
>
> The Catcher in the Rye (1945)

The Catcher in the Rye [1] was the first book I (RDF) read as a freshman in high school that transformed me, and it remains my favorite book of all time. As the passage above implies, the book's protagonist, Holden Caulfield, was passionate about the well-being of children. So are most, if not all, CBT clinicians who work with youth. Aren't we all then catchers in the rye to a certain degree?

The chapters in this second edition reflect both tradition and innovation in CBT. They were written by experts in their respective areas who embrace the faithful and flexible application of CBT. Their work is grounded in sound theory and solid empirical science yet remains pliant enough to fully embrace contextual vicissitudes. Practicing with fidelity and flexibility is necessary when trying to catch diverse youth coming through the rye [2].

Good training is essential for potential catchers in the rye. They need to be in the right place at the right time with precise skills. Training in case conceptualization helps clinicians know what, when, and how to mitigate young patients' distress [3]. Skill building via supervised practice facilitates proper interventions. The chapters in this second edition seamlessly integrate case formulation and clinical procedures.

The issue of type III error in clinical research and practice is too frequently neglected [4]. This type of error refers to drawing the inaccurate conclusion that the treatment is ineffective when in fact the treatment is poorly or incompletely delivered. Often, in medical research, type III error occurs if a pill compound is flawed or a syringe is marred, preventing adequate delivery. This type of error is readily recognized when training and supervising behavioral health clinicians. A novice therapist may attempt a CBT intervention, implement it improperly, deliver it at too low a dose, and/or target the improper problem. Subsequently, the therapist may conclude CBT does not work with this patient. Alternatively, in some group comparisons evaluating CBT outcomes, studies may yield less than satisfying outcomes. But the question arises, how well-trained and competent were the practitioners?

When providing psychotherapy, the vessel that provides the conduit between the intervention and the patient is human. Human vessels vary tremendously. Therefore, training in a faithful and flexible approach as espoused by the expert authors of the preceding chapters is absolutely essential. AI-based interventions ostensibly seek to attenuate the variation in clinical service delivery. While these technological innovations may potentially hold promise for standardized implementation, they can only approximate the flexibility born from a human's critical reasoning and creative capacities. The chapters in this second edition ideally provide a framework for training in genuine CBT spectrum approaches with children.

Honoring tradition yet seeking innovation invites a necessary dialectical tension. Traditional theories and procedures must evolve to survive in changing contexts. Innovations must be firmly tethered to a coherent, robust, and established conceptual paradigm; otherwise, they are just rogue options. Viable innovations need to be infused with empirically established theories. If they are not, creative interventions risk becoming fads rather than fashions. Each chapter represents *de rigueur* thinking and practice. Tradition and innovation are two poles on the same continuum, like darkness and light. There cannot be light without darkness. Similarly, there cannot be innovation without a foothold in tradition and tradition lives on via innovation. Like Dylan said in the beginning of this chapter, there is nothing so stable as change.

Chapter 14 cogently and explicit addresses CBT with diverse with youth and various chapters within the second edition also attend to this important topic. It is important to note that this edition takes place after the October 7, 2023, pogrom perpetrated by Hamas in Israel and the exploding rate of global antisemitism. Michelli and colleagues [5] emphasized the disproportionate hateful targeting of Jews explaining that while Jewish Americans represent a total of a little over 2% of the entire US population, antisemitic incidents account for 60% of all religious hate crimes, and 10% of total hate crimes [5, 6]. Antisemitism is not new. Rather, it is an ancient form of Jew Hatred [7]. However, the recent escalation is particularly alarming, pervasive, and significantly neglected. Recently scholars have noted that antisemitism has bled into academic and clinical settings [8, 9]. More specifically, they found that nearly 90% of Jewish healthcare workers reported experiencing antisemitic incidents. Sadly, these experiences have been regularly discounted and invalidated [10].

Accordingly, the added allostatic load that heightened antisemitism places on Jewish youth needs wide recognition. Consequently, there is an urgent need to ensure that meritorious social justice initiatives identify Jewish children, adolescents, and their caregivers as a vulnerable and marginalized group [11]. In this way, the aspirational goals of diversity, equity, and inclusiveness can be fully realized rather than restricted to the injustices of some vulnerable groups. We urge all cognitive behaviorally minded professionals as well as other allies to advocate for greater genuine inclusiveness in their work. Indeed, we have to catchers in the rye for ALL children and adolescents.

CBT offers a sturdy and inclusive theoretical and clinical model. The theory is sturdy yet supple enough to welcome diverse procedures and populations. Accordingly, this characteristic allows for necessary evolution [12]. Paradigms that systematically shift with new empirical science and dynamic circumstance endure, while iron-bound approaches stubbornly stand pat and risk antiquity.

Psychotherapy theory, research, and practice are iterative processes. New developments stand on the shoulders of past paragons. CBT has come a long way [13]. We hope this second edition of *CBT with youth: Tradition and innovation* advances this ever-widening path, serves future progress, and catalyzes better clinical care while remaining faithful to established maxims. May all CBT clinicians continue to be reliable and accountable catchers in the rye.

Conflicts of Interest

Micaela A. Thordarson receives royalties from Guilford Publications.

Robert D. Friedberg has received speaker and consultant honoraria from the Beck Institute for Cognitive and Behavioral Therapy and

Research, Kinark Family and Youth Services, and the Rockland County Department of Mental Health. He receives royalties from Springer, Guilford, John Wiley, Routledge, and Professional Resource Press.

References

1. Salinger JD (1945) The catcher in the rye. Little, Brown, Boston
2. Kendall PC, Frank HE (2018) Implementing evidence-based treatment protocols: flexibility within fidelity. Clin Psychol Sci Pract 25(4):40
3. Friedberg RD (2015) Where's the beef? Concrete elements when supervising cognitive-behavioral therapy with youth. J Am Acad Child Adolesc Psychiatry 54(7):527–531
4. VandenBos GR (2007) APA dictionary of psychology. American Psychological Association
5. Michelli NM, Campo S, Jacobowitz TJ, Jahnsen D (2023) Discrimination and repression: slavery, segregation, racism, antisemitism, and their relation to democracy. In: Education for social justice. Routledge, pp 41–70
6. LaFreniere Tamez HD, Anastasio N, Perliger A (2023) Explaining the rise of antisemitism in the United States. Stud Conflict Terrorism 20: 1–22. https://doi.org/10.1080/1057610X.2023.2297317
7. Kaufman CC, Paladino AJ, Porter DV, Thurston IB (2020) Psychological research examining antisemitism in the United States: a literature review. Antisemitism Stud 4(2): 237–269
8. Gottlieb JD, Steigerwald A (2024) Social work's role in ending antisemitism: if not us, who? Int J Soc Work Values Ethics 21(1): 14–27
9. Schwartz DM, Leiba R, Feldman CL, Spence NZ, Oratz R, Wald HS, Roth S (2025) Social media, survey, and medical literature data reveal escalating antisemitism within the United States healthcare community. J Relig Health 64(1):206–223
10. Bar-Halpern M, Wolfman J (2025) Traumatic invalidation in the Jewish community after October 7. J Hum Behav Soc Environ 15:1–28. https://doi.org/10.1080/10911359.20252503441
11. Getzoff Testa E, Becker JA, Brosbe MS, Fiat AE, Kaufman CC, Berger C, Fishbein D, Gross Y (2024) Widening the tent: Jewish inclusivity, antisemitism, and recommendations for action in pediatric psychology. Clin Pract Pediatr Psychol 12(4):237
12. Rozmid EV, Friedberg RD (2022) Incorporating humor, superheroes, and improvisational theatre exercises into cbt with youth: "just because something works, it doesn't mean it can't be improved". In: Creative CBT with youth: clinical applications using humor, play, superheroes, and improvisation. Springer, Cham, pp 1–7
13. Friedberg RD, Hoyman LC, Behar S, Tabbarah S, Pacholec NM, Keller M, Thordarson MA (2014) We've come a long way, baby!: Evolution and revolution in CBT with youth. J Ration Emot Cogn Behav Ther 32(1):4–14

Index

Robert D. Friedberg and Micaela A. Thordarson (eds.), *Cognitive Behavioral Therapy with Youth: Tradition and Innovation*, Neuromethods, vol. 231, https://doi.org/10.1007/978-1-0716-5134-6,

Index

G

I

M

N

O

P

S

T

U

Y

MIX
Papier aus verantwortungsvollen Quellen
Paper from responsible sources
FSC® C105338

If you have any concerns about our products,
you can contact us on
ProductSafety@springernature.com

In case Publisher is established outside the EU,
the EU authorized representative is:
Springer Nature Customer Service Center GmbH
Europaplatz 3, 69115 Heidelberg, Germany

Printed by Libri Plureos GmbH
in Hamburg, Germany